# SELECTED TABLES IN MATHEMATICAL STATISTICS

## Volume VI

### THE DISTRIBUTION OF THE SIZE OF THE MAXIMUM CLUSTER
### OF POINTS ON A LINE

*by*

**NORMAN D. NEFF and JOSEPH I. NAUS**

**Edited by the Institute of Mathematical Statistics**

*Coeditors*

**W. J. Kennedy**

*Iowa State University*

*and*

**R. E. Odeh**

*University of Victoria*

*Managing Editor*

**J. M. Davenport**

*Texas Tech University*

**AMERICAN MATHEMATICAL SOCIETY**

**PROVIDENCE, RHODE ISLAND**

**This volume was prepared with the aid of:**

D. R. Cook, University of Minnesota

J. E. Gentle, International Mathematical & Statistical Libraries, Inc.

W. J. Hemmerle, University of Rhode Island

K. Hinklemann, Virginia Polytechnic Institute and State University

R. L. Iman, Sandia Laboratories

A. Londhe, Iowa State University

D. Meeter, Florida State University

D. B. Owen, Southern Methodist University

S. Pearson, Bell Laboratories

N. S. Urquhart, New Mexico State University

R. H. Wampler, National Bureau of Standards

1980 *Mathematics Subject Classification*
Primary 62Q05; Secondary 62E15.

International Standard Serial Number 0094-8837
International Standard Book Number 0-8218-1906-2
Library of Congress Card Number 74-6283

# PREFACE

This volume of mathematical tables has been prepared under the aegis of the Institute of Mathematical Statistics. The Institute of Mathematical Statistics is a professional society for mathematically oriented statisticians. The purpose of the Institute is to encourage the development, dissemination, and application of mathematical statistics. The Committee on Mathematical Tables of the Institute of Mathematical Statistics is responsible for preparing and editing this series of tables. The Institute of Mathematical Statistics has entered into an agreement with the American Mathematical Society to jointly publish this series of volumes. At the time of this writing, submissions for future volumes are being solicited. No set number of volumes has been established for this series. The editors will consider publishing as many volumes as are necessary to disseminate meritorious material.

Potential authors should consider the following rules when submitting material.

1. The manuscript must be prepared by the author in a form acceptable for photo-offset. This includes both the tables and introductory material. The author should assume that nothing will be set in type although the editors reserve the right to make editorial changes.

2. While there are no fixed upper and lower limits on the length of tables, authors should be aware that the purpose of this series is to provide an outlet for tables of high quality and utility which are too long to be accepted by a technical journal but too short for separate publication in book form.

3. The author must, wherever applicable, include in his introduction the following:

(a) He should give the formula used in the calculation, and the computational procedure (or algorithm) used to generate his tables. Generally speaking, FORTRAN or ALGOL programs will not be included but the description of the algorithm used should be complete enough that such programs can be easily prepared.

(b) A recommendation for interpolation in the tables should be given. The author should give the number of figures of accuracy which can be obtained with linear (and higher degree) interpolation.

(c) Adequate references must be given.

(d) The author should give the accuracy of the table and his method of rounding.

(e) In considering possible formats for his tables, the author should attempt to give as much information as possible in as little space as possible. Generally speaking, critical values of a distribution convey more information than the distribution itself, but each case must be judged on its own merits. The text portion of the tables (including column headings, titles, etc.) must be proportional to the size 5--1/4″ by 8--1/4″. Tables may be printed proportional to the size 8−1/4″ by 5−1/4″ (i.e., turned sideways on the page) when absolutely necessary; but this should be avoided and every attempt made to orient the tables in a vertical manner.

(f) The table should adequately cover the entire function. Asymptotic results should be given and tabulated if informative.

(g) An example or examples of the use of the tables should be included.

4. The author should submit as accurate a tabulation as he can. The table will be checked before publication, and any excess of errors will be considered grounds for rejection. The manuscript introduction will be subjected to refereeing and an inadequate introduction may also lead to rejection.

5. Authors having tables they wish to submit should send two copies to:

> Dr. Robert E. Odeh, Coeditor
> Department of Mathematics
> University of Victoria
> Victoria, B. C., Canada   V8W 2Y2

At the same time, a third copy should be sent to:

> Dr. William J. Kennedy, Coeditor
> 117 Snedecor Hall
> Statistical Laboratory
> Iowa State University
> Ames, Iowa 50011

Additional copies may be required, as needed for the editorial process. After the editorial process is complete, a camera-ready copy must be prepared for the publisher.

Authors should check several current issues of *The Institute of Mathematical Statistics Bulletin* and *The AMSTAT News* for any up-to-date announcements about submissions to this series.

## ACKNOWLEDGMENTS

The tables included in the present volume were checked at the University of Victoria. Dr. R. E. Odeh arranged for, and directed this checking with the assistance of Mr. Bruce Wilson. The editors and the Institute of Mathematical Statistics wish to express their great appreciation for this invaluable assistance. So many other people have contributed to the instigation and preparation of this volume that it would be impossible to record their names here. To all these people, who will remain anonymous, the editors and the Institute also wish to express their thanks.

*To*: **PATRICIA and SARAH**

# Contents of VOLUMES I, II, III, IV and V of this Series

# TABLE OF CONTENTS

# THE DISTRIBUTION OF THE SIZE OF THE MAXIMUM CLUSTER
## OF POINTS ON A LINE

Norman D. Neff  Trenton State College

Joseph I. Naus  Rutgers University

## ABSTRACT

Given N points randomly drawn from the unit line, let $\tilde{n}_p$ be the size of the largest number of points clustered within an interval of length p. Let $\tilde{p}_n$ be the size of the smallest interval that contains n out of the N points. The distributions of $\tilde{n}_p$ and $\tilde{p}_n$ are related: $\Pr(\tilde{p}_n \leq p) = \Pr(\tilde{n}_p \geq n)$. We denote the common probability $P(n;N,p)$. Tables are given for $P(n;N,p)$ and for the expectation over N of $P(n;N,p)$ where N is a Poisson random variate.

## 1. INTRODUCTION

Researchers in many fields deal with the clustering of events in time and space. A quality control expert investigates clusters of defectives. A communications engineer designs system capacity to accommodate clusters. An educational psychologist sets success quotas to gauge learning. Experts in accident prevention, reliability, traffic control, ecology, epidemiology and many other fields focus on unusually large clusters. The probabilities of large clusters under various models are tools of the natural, physical, and social sciences.

The present tables provide probabilities for the size of the largest cluster of random points on the line. Given N points independently drawn from the uniform distribution on (0,1), let $\tilde{n}_p$ be the largest number of points to be found in any subinterval of (0,1) of length p. Let $P(n;N,p)$ denote the probability $\Pr(\tilde{n}_p \geq n)$.

Received by the editors December 1978 and in revised form June 1979.
AMS (MOS) Subject Classifications (1970): Primary 62Q05; Secondary 62E15.

1

Certain applications deal with points generated from a Poisson process, where the total number of points in the unit interval is a Poisson random variate with expectation $\lambda$. Let $P'(n;\lambda,p)$ denote the expectation taken over N of $P(n;N,p)$, where N is a Poisson random variate with expectation $\lambda$. Tables are given for $P(n;N,p)$ and $P'(n;\lambda,p)$. Section two describes applications of these probabilities to a wide variety of scientific and statistical areas. Section three summarizes known results on the probabilities and details approaches that were used to compute the tables.

## 2.    APPLICATIONS

This section gives a variety of applications for $P(n;N,p)$ and for $P'(n;\lambda,p)$.

### 2.1  Applications of $P(n;N,p)$

(a) Developability of silver specks. The photographic signal recorded on a film depends not only on the total number of photons during exposure, but also on their time sequence. Hamilton, Lawton, and Trabka (1972, p. 855) note that this dependency is called reciprocity failure, a key aspect of photographic science. They note that $P(n;N,p)$ was studied early in photographic history by Silberstein (1939) and others and plays a role in low intensity reciprocity failure (a decrease in sensitivity at less than optimum intensities). A classical model for the developability of silver specks is that a speck will develop if k or more photons are absorbed within the decay time, $t_o$, of the nucleus.

Silberstein (1945), Berg (1945), and Mack (1948, 1950) find the expected number of n-aggregates, or clusters, where n points form an aggregate if "they are all contained within a subinterval, p, of (0,1), no matter how placed in (0,1)."[1] Berg (1945, p. 340) and Mack (1948, p. 784) indicate their interest in $P(n;N,p)$ and derive an approximate formula for it in the case of sufficiently rare aggregates. Silberstein (1945, p. 319) states that

> The rigorous determination of the probability P of any given number of aggregates, n-ets, among N points is exceedingly complicated and becomes to all purposes impracticable when N exceeds a few units.

---
[1] Silberstein (1945, p. 319).

Silberstein makes two exceptions, namely, the probability of no aggregate, $1-P(2;N,p)$, and the probability of the largest possible aggregate, $P(N;N,p)$.

(b) <u>Clustering of leukemia cases</u>. Childhood leukemia is a relatively rare disease. Scientists seeking clues as to common causative agents investigate unusual clusters of cases. They ask whether clusters of a given size within a given time period are likely to occur by chance. Ederer, Myers, and Mantel (1962, p. 9), who develop approaches to this problem, state:

> In considering temporal clusters we chose the calendar year as the unit of time, and in this way obtained 5 non-overlapping years in a 5-year period. The reader will recognize that a 5-year period in fact contains a continuum of overlapping periods one year in length. However, the distribution of the maximum number of cases in a year under the null hypothesis [of randomness] cannot be readily determined unless the number of periods is restricted in some way.

It is natural to look at clusters where they fall, be it in a calendar year, or in a year period that overlaps two calendar years. The above authors note that Pinkel and Nefzeger (1959) did look at the unrestricted continuum. However, if one does this, then one must use the appropriate null distribution in checking for significance. For the above example, $P(n;N,0.2)$ is the distribution of the maximum number of cases in a year, under the null hypothesis that the N cases are distributed at random over the five-year period, for the unrestricted continuum of 1-year periods. For example, given N = 15 cases over the five year period, the chance that there is any calendar year that contains as many as seven cases is 0.07 (from Appendix - Table 1 in Ederer, Myers, and Mantel). This is much less likely than the probability that there is a one-year period (calendar or not) that contains at least 7 cases. From Table 1a of our appendix, $P(7;15,0.2) = 0.30$.

(c) <u>Dialing calls</u>. Dialing is started for fifteen phone calls at times that are distributed at random over a one-minute period. The dialing time for a call is ten seconds. Find the probability that eight or more phone calls are being dialed at the same time.

Feller (1958, p. 397) notes that compound events such as "seven calls within a minute on a certain day" involve complicated sample spaces and goes on to state,

We cannot deal here with such complicated sample spaces and must defer the study of the more delicate aspects of the theory.

Given 15 phone calls initiated during a minute, the event "eight or more calls initiated during a ten-second interval" is equivalent to the event that for some i=1,2,...,8, dialing for the (7+i)th phone call started less than ten seconds after dialing started for the ith phone call. Interpolating in Table 1 we find the desired probability P(8;15,1/6) = 0.037.

## 2.2 Applications of $P'(n;\lambda,p)$

(d) <u>Visual perception</u>. Photons arrive at a receptor in the eye according to a Poisson process. Under conditions of low illumination it is observed that the retinal neurones do not discharge for each photon, but rather it is conjectured that there is a triggering effect from several photons. One of the classical theories of perception is that if n or more photons arrive within an integration time t, then this triggers an impulse and the neurone discharges. What is frequently of interest is the distribution of waiting times till discharge. Leslie (1969, p. 379) states that the distribution of the waiting time till discharge under the preceding model is intractable and gives an alternative model. Van de Grind et al. (1971) note that an analytic solution is lacking and use a Monte Carlo approach to estimate the distribution under the classical model. The probability that the waiting time till first discharge is less than or equal to T is $P'(n;\lambda,t/T)$, where

$$P'(n;\lambda,t/T) = \Sigma_{N=n}^{\infty}\, e^{-\lambda T}(\lambda T)^{N} P(n;N,t/T)/N!. \tag{2.1}$$

Tables 2 and 2a provide values for $P'(n;\lambda,p)$. Outside the range of these tables we can use approximations of the form

$$P'(n;\lambda,1/2L) \doteq 1 - [1 - P'(n;2\lambda/L,\tfrac{1}{4})]^{L-1}[1 - P'(n;\lambda/L,\tfrac{1}{2})]^{-L+2}.$$

The reasoning behind this approximation and its generalization for other p values is detailed in the appendix under the discussion of the use of Tables 2 and 2a.

Van de Grind et al. (1971) use a Monte Carlo approach to chart the values of $\lambda$ such that $P'(k;\lambda,t/T) = 0.60$. Based on a simulation of $10^{4}$ flashes they find $P'(4;\lambda,1/200) = 0.60$ for $\lambda$ somewhere between 90 and 100 (given the roughness of the plot).  From the above approximation and using values from Table 2 we find

$$P'(4;100,1/200) \doteq 1 - [1 - P'(4;2,\tfrac{1}{4})]^{99}[1 - P'(4;1,\tfrac{1}{2})]^{-98}$$

$$= 1 - [1 - 0.018456295]^{99}[1 - 0.007452004]^{-98} \doteq 0.67 \ .$$

Similarly, $P'(4;90,1/200) \doteq 0.54$.

(e)  Pedestrians on a street.  Furth (1920) did an empirical study of the number of people on a certain segment of a street.  Feller (1958, pp. 370-372) discusses this study and shows how it is related to a moving average process, an example of a nonmarkovian process.  Feller points out how this process is related to our problem.

If the pedestrians are walking at a fixed rate, then the time each pedestrian spends on the segment is a constant, C.  If the arrival times are exponential, then the probability that the street has had at any time before T as many as n pedestrians on it is $P'(n;\lambda,C/T)$.

(f)  A counter problem.  A Poisson process with mean rate $\lambda$ generates impulses that are received by a counter.  The counter registers whenever n impulses have occurred together in an interval of length less than t. $P'(n;\lambda,t/T)$ gives the c.d.f. of the waiting time until the first registration of the counter.  Janossy (1944) and Schroedinger (1944) consider the rate of n-fold accidental coincidences in counters, and find asymptotic expressions for the probability.  Domb (1950) studies various counter problems and derives an implicit formula for the probability that k "n-clusters" occur in (0,T), where he defines an n-cluster as a "recorded event whose dead period contains n-1 other nonrecorded events."  Domb's solution for this type of cluster is closely related to Mack's (1948, p. 784) solution for a different type of cluster (see application (a)).  Both solutions assume that overlapping effects are negligible.  Domb's cluster differs in the way an earlier cluster can prevent a later cluster.

(g)  A queueing problem.  Customers arrive at an n-server system according to a Poisson process with mean $\lambda$.  Customers are served by any free server on a first come first served basis.  Service time is a constant, $t_o$.  If k or more

customers arrive within a time $t_0$, then service will stagnate (with the result
that some customers will have to wait, or some impatient customers will leave).
Solov'ev (1966) deals with this problem and gives some results for the expected
waiting time till stagnation and some approximate results for $P'(n;\lambda,p)$, the
c.d.f. of the waiting time. Newell (1963) also mentions the queueing applications of $P'(n;\lambda,p)$ and various generalizations and derives asymptotic results.
Our experience with various examples suggests that the approximation of example
(d) together with values from Tables 2 and 2a give more accurate approximations
than these asymptotic results. For example, from Table 2 find $P'(4;10,0.1) =$
0.3741. This is the exact result to the number of places we are able to compute it by averaging $P(4;N,0.1)$ for N up to 19, and then treating the remaining
probabilities as 1. The number of significant places results from averaging
over rapidly shrinking Poisson terms. Applying the approximation of example
(d) together with the values from Table 2 gives

$$P'(4;10,0.1) = 1 - [1 - P'(4;4,\tfrac{1}{4})]^4[1 - P'(4;2,\tfrac{1}{2})]^{-3}$$
$$= 1 - [1 - 0.156709254]^4[1 - 0.068630099]^{-3} = 0.3740456.$$

The best previous approximation, that of Newell (1963) and Ikeda (1965) is
$P'(n;\lambda,p) \doteq 1 - \exp(-\lambda^n p^{n-1}/(n-1)!)$, which for this example gives
$P'(4;10,0.1) \doteq 0.811$.

(h) <u>Breaking strength of materials</u>. Several models postulate the breaking
of an object whenever there are several flaws within a certain distance: the
breakaway due to several close dislocations in crystals; a multistrand thread
breaking when several strands have weaknesses within a common length; earth-
quakes triggered by shift flaws within a certain distance; nervous breakdowns
resulting from several severe personal problems occurring within a short period
of time.

2.3 <u>$P(n;N,p)$ and $P'(n;\lambda,p)$ as approximations to the probability of a generalized run.</u>

Let N one's and M-N zeros be randomly arranged in a row. Let the random
variable $\tilde{k}_m$ be the maximum number of one's within any m consecutive positions
in the arrangement. An n:m quota is defined as at least n one's within some m

consecutive trials.  $Pr(\tilde{k}_m \geq n)$ is the probability of a quota, and we denote
this probability as P(n in m|N in M).  Saperstein (1969, 1972, 1973), Naus
(1974), Greenberg (1970), Huntington (1974, 1976) and others derive various
results for P(n in m|N in M).

Saperstein (1972) gives bounds on the upper 5% points for $\tilde{k}_m$, for
N = 5(1)15, m/M = 0.2(0.1)0.7, and certain values of M.  The bounds are based
on the formula for the case n > N/2.  Huntington (1974, pp. 204-285) gives ta-
bles for the distribution of P(n in m|N in M) for two classes of cases.

Case 1:  m/M = 1/L, L = 2(1)5, n > [N/2] + 1.

Case 2:  m/M = 0.35(.05)0.50, and full range of n.

Naus (1974) gives a simple expression for the case M/m = L, L an integer, where
n > N/2:

$$Pr(\tilde{k}_m \geq n) = 2\sum_{s=n}^{N}\binom{m}{s}\binom{M-m}{N-s}/\binom{M}{N} + (Ln-N-1)\binom{m}{n}\binom{M-m}{N-n}/\binom{M}{N}. \tag{2.2}$$

The formula for the other case n ≤ N/2 is not computationally simple and is in
terms of sums of LxL determinants.  For this other case, and the cases where
m/M ≠ 1/L, L an integer, one requires either a high speed computer and substan-
tial programming effort, or an approximation.  For the case where P(n in m|N in
M) is small (less than 0.1), equation (2.2) provides a good approximation for
the case n ≤ N/2.  (It is an approximation for this case because equation (2.2)
fails to count the possibility that there might be several non-overlapping sets
of n successes each within m trials.  Of course, for n > N/2, this cannot hap-
pen.  It is a good approximation for n ≤ N/2 and small P(n in m|N in M), be-
cause the chance of two non-overlapping quotas is of smaller order of magnitude.
Equation (2.2) does take account of the nonnegligible chance that there might
be overlapping sets of n successes each within m trials.)

Another approximation is available when m and M are large relative to n
and N.  The limit of P(n in m|N in M) as m → ∞, M → ∞, such that m/M = p, is
P(n;N,p).  We can use P(n;N,p) to approximate P(n in m|N in M) in much the same
way that the normal distribution is used to approximate the binomial.  Table 2
in Saperstein (1972) is based on this approximation.  For example, in a manu-
facturing process there were eleven defectives in 500 items coming off an

assembly line. After reviewing the defectives it is found that six of the de-
fectives occurred within 100 consecutive trials. We approximate P(6 in 100|11
in 500) by P(6;11, 100/500) which from Table 1 equals 0.198. Equation (2.2)
gives the exact probability P(6 in 100|11 in 500) = 0.185. In a similar way
$P'(n;\lambda,p)$ approximates the expectation over N of P(n in m|N in M), where N is a
binomial random variable. We now give applications of P(n in m|N in M).

(i) A learning criterion. Psychologists studying transfer and learning
sometimes use a generalized run criterion to decide when to terminate a particular treatment. At each trial, the psychologist counts the number of successes
in the last m trials. If at any trial this number exceeds a critical number n
(the criterion), this signals a change in the underlying process. For an experiment with M(> m) trials, the critical number n is chosen to give a speci-
fied experiment wide level of significance. To set the experiment wide criterion, we require the probability that within M trials there exists a subse-
quence of m consecutive trials with at least n successes. This is the proba-
bility P(n in m|N in M), or the related probability where N is a binomial ran-
dom variate.

The special case n = m deals with the event of a run of m successes.
Bogartz (1965) derives probabilities for the special case of m-2 and m-1 cor-
rect responses out of m trials. Bogartz notes that an approach based on
Markov chains is practical only for very simple cases, and gives an approximation. Runnels, Thompson, and Runnels (1968) reexamine Bogartz's approaches
and give additional calculations for the case m-1 out of m trials.

(j) Target detection systems. Various target detection systems react to
a "quota" of responses where a quota is a set of m consecutive trials in which
there are at least k successes. Goldman and Bender (1962) consider events
where a run immediately follows a quota, and derive the distribution of waiting
times till the event. The derivation is done for the case of where the run
size is at least equal to the number of successes in the quota, thereby allow-
ing treatment of the event as a recurrent event.

For the case of run size less than the number of successes in the quota, Brookner (1966) describes a general procedure. The procedure is to transform the original sequence of overlapping states to an m-state first-order Markov chain. The method is fairly general in that the original trials need not be independent. However, the method becomes unwieldy for m and n of moderate size because of the resulting large number of states in the transformed matrix. The method is feasible for quotas of a few successes in a few trials.

(k) <u>Quality control</u>. In quality control, a point outside the control chart limits for the mean signals that the process may be out-of-control. In addition, runs of observations above or below the mean, even if within the control chart limits, are additional warning signs. Roberts (1958) develops a series of zone tests that combine these two ideas. Roberts sets up zones within the usual control limits such that if too many observations within a consecutive group of observations fall within the zone, the process is called out-of-control. (For example, "too many" might be 4 out of 7 consecutive observations falling between 1.28 and 3 sigma limits above the mean). To study the operating characteristics of such plans, one needs the probabilities of quotas. Saperstein (1976) gives a variety of other zone tests.

(l) <u>Acceptance sampling</u>. Troxell (1972) develops acceptance sampling plans based on a quota of batches being unacceptable. The special cases considered are 2 out of m, and 3 out of m batches. Given present results for $P(n \text{ in } m | N \text{ in } M)$, the acceptance sampling plans can be generalized to the case of n out of m batches.

(m) <u>Faults in a sequence of trials</u>. Leslie (1967) describes some ingenious uses of a type of generalized run different than but related to quotas. The occurrence of many leaky pipe joints within a given distance may justify replacement. Faulty sleepers in railway tracks can cause problems if too many occur within a given spacing. Several genetic mutations within a given distance on a chromosome may lead to defects.

(n) <u>Other applications</u>. Ecologists study clustering of diseased plants in transects through a field. Meteorologists investigate the alternating of dry and wet days during the rainy season.  Economists predict clustering of stores of a particular type.  Sportscasters comment on winning or hitting streaks (quotas), such as hitting successfully in 18 out of the last twenty times at bat.

## 3.  COMPUTING THE CLUSTER PROBABILITY $P(n;N,p)$

### 3.1  Existing Formulas and Approaches

In this section we discuss the available formulas and major approaches to computing exact values of $P(n;N,p)$.

Let $x_1 \leq x_2 \leq \ldots \leq x_N$ be the ordered values of N independent random variables from the uniform distribution on $(0,1)$.  For any $2 \leq n \leq N$ let $\tilde{p}_n$ denote $\min_{1 \leq i \leq N-n+1} \{x_{n+i-1} - x_i\}$.  The statistic $\tilde{p}_n$ measures the length of the smallest interval that contains n points.  Let $\tilde{n}_p$ be the largest number of points clustered within an interval of length p.  $P(n;N,p)$ denotes the equivalent probabilities $Pr(\tilde{p}_n \leq p)$ and $Pr(\tilde{n}_p \geq n)$.

The probability $1-P(n;N,p)$ can be written as the iterated integral of $N!$ with respect to $x_1,\ldots,x_N$ over the region:

$$0 \leq x_1 \leq x_2 \leq \ldots \leq x_N \leq 1$$

and

$$x_{n+i-1} - x_i > p \text{ for } i=1,\ldots,N-n+1.$$

Naus (1963) demonstrated that the above integral could, in theory, be evaluated by a finite algorithm and that the result would be piecewise polynomial in p with finitely many pieces, each piece having as domain either $[0,1/L]$ or $[1/(L+1),1/L]$, L an integer.  Unfortunately, direct evaluation of the integral is hopelessly complicated, except for the very simplest cases:  the distribution of the sample range, $\tilde{p}_N$, and of the smallest gap between any of the points, $\tilde{p}_2$, which were known for a long time.  Parzen (1960) illustrates the integration approach for deriving

$$P(2;N,p) = 1-(1-(N-1)p)^N \text{ for } 0 < p < 1/(N-1). \tag{3.1}$$

Naus (1963) uses an integration approach to find a formula for $n=N-1$.  The most ambitious effort in this direction was carried out by Van Elteren and Gerritis

(1961) who evaluate the integral for n=3, N ≤ 8.

Naus (1963, 1965) developed an alternative combinatorial approach. The combinatorial approach divided (0,1) into subintervals and conditioned on the number of points within each subinterval. The cluster event $\tilde{n}_p \geq n$ was related to certain relative orders of points within subintervals; random walk techniques were used to count relevant relative orders. This approach solved the case n > (N+1)/2. This case has the two p domains, 0 < p ≤ 1/2 and 1/2 < p ≤ 1. The polynomials are:

$$P(n;N,p) = C(n;N,p) - R(n;N,p), \quad p \geq 1/2, \ n > (N+1)/2 \qquad (3.2)$$
$$= C(n;N,p) \qquad\qquad , \quad p \leq 1/2, \ n > N/2$$

where

$$C(n;N,p) = 2G_b(n;N,p) + (np^{-1} - N - 1)b(n;N,p)$$
$$R(n;N,p) = \Sigma_{y=n}^{N} b(y;N,p)F_b(N-n;y,q/p) + H(n;N,p)b(n;N,p)$$

where

$$q = 1 - p$$
$$b(n;N,p) = \binom{N}{n}p^n(1-p)^{N-n}$$
$$F_b(n;N,p) = \Sigma_{i=0}^{n} b(i;N,p)$$
$$G_b(n;N,p) = \Sigma_{i=n}^{N} b(i;N,p)$$
$$H(n;N,p) = (nq/p)F_b(N-n;n-1,q/p) - (N-n+1)F_b(N-n+1;n,q/p).$$

We will refer to a result for P(n;N,p) as a "tail" result if it is a simple polynomial that agrees with P(n;N,p) for all p sufficiently small. Formula (3.2) for P(n;N,p) for n > N/2 in the range p ≤ 1/2 is an example of a tail result. Each subsequent generalization of the Naus results was either a tail result or a "p-range" result, a polynomial that works only within some range of p values bounded away from zero.

For example, Naus (1966) obtained the following formula for P(n;N,1/L), where L is a given positive integer:

$$1 - P(n;N,1/L) = N!L^{-N}\Sigma \det|1/c_{ij}!|, \qquad (3.3)$$

where

$$c_{ij} = (j-i)n - \Sigma_{r=i}^{j-1} n_r + n_i \qquad i < j$$
$$= (j-i)n + \Sigma_{r=j}^{i} n_r \qquad i \geq j$$

and where the sum is over the set of all partitions of N into L positive integers $n_i$ < n, i=1,...,L. Since the orders of the determinants as well as the range of summation depend on L, this formula is not a simple polynomial. It is not a tail result, but a countably infinite set of p-range results, where each p-range contains just one p value. This formula was evaluated by computer in simple cases; its chief drawbacks were excessive computer time for large N or small p and the restriction to p values of the form 1/L.

The method of Wallenstein and Naus (1974) was a (restricted) tail result. Even though it was limited to p=1/L, the polynomial form of the formula did not change for small p. Because of the complexity of the method, it was possible to program the formula only for n in the range N/3 < n ≤ N/2. However, within this restriction, the Wallenstein and Naus method was quite effective:  cases with N values as large as 100 were readily computable. Wallenstein (1971) tabulates P(n;N,1/L) for N = 6(1)20, L = 3(1)10(5)50 and for N = 20(1)100, L = 3(1)10 where N/2 ≥ n > N/3.

In deriving formulas for cases with p = 1/L, the approach was to partition the unit interval into L equal subintervals and then condition on the occupancy numbers of the subintervals. By using a different method of partitioning, Huntington and Naus (1975) extended the Naus (1966) result to a general formula for P(n;N,p). Let L = 1/p, and let Q be the set of partitions of N into 2L+1 nonnegative integers $n_i$ satisfying $n_i + n_{i+1}$ < n, i=1,2,...,2L. Denote 1 - P(n;N,p) by $f_L$(n;N,p). Huntington and Naus (1975) find

$$f_L(n;N,p) = \Sigma_Q R \det\left|1/h_{ij}!\right|\det\left|1/\ell_{ij}!\right| \qquad (3.4)$$

where

$$R = N!(1-pL)^F(-1+p(L+1))^E$$

$$F = \Sigma_{i=0}^{L} n_{2i+1} \quad ; \qquad E = N - F$$

and where the determinants are of order L+1 and L with

$$h_{ij} = \Sigma_{k=2j-1}^{2i-1} n_k - (i-j)n \qquad 1 \le j \le i \le L+1 \qquad (3.5)$$

$$= -\Sigma_{k=2i}^{2j-2} n_k + (j-i)n \qquad 1 \le i < j \le L+1$$

and

$$\ell_{ij} = \Sigma_{k=2j}^{2i}\, n_k - (i-j)n \qquad 1 \le j \le i \le L$$

$$= -\Sigma_{k=2i+1}^{2j-1}\, n_k + (j-i)n \qquad 1 \le i < j \le L.$$

Like equation (3.3), formula (3.4) is a countable set of p-range results. Within each p-range $[1/(L+1),1/L]$, it is a polynomial in p. Huntington (1974) tabulated $P(n;N,p)$ for $N = 3(1)30$, $n = 2(1)N$, $p = 0.26(0.01)0.50$. Large N values or small p values greatly increase the difficulty of computation. Certain values computed by Huntington (in the range $1/4 < p < 1/3$) were in error and are corrected by the present tables.

Hwang (1977) generalized the Karlin-McGregor Theorem and obtained another expression for $P(n;N,p)$:

$$1 - P(n;N,p) = N!\Sigma_F \det\left|g_{ij}/(\beta_j- \alpha_i)!\right| \qquad (3.6)$$

where

$$\alpha_i = \Sigma_{j=1}^{i-1} n_j - (i-1)n$$

$$\beta_i = \alpha_i + n_i$$

$$g_{ij} = \Sigma_{x=\max(0,\beta_{L+1}+1-\alpha_i)}^{\beta_j-\alpha_i} \binom{\beta_j-\alpha_i}{x}(p')^x(p-p')^{\beta_j-\alpha_i-x}, \qquad *$$

$$= (p')^{\beta_j-\alpha_i} \qquad j = 1,2,\ldots,L,\ \beta_j-\alpha_i \ge 0$$

$$\qquad\qquad j = L+1,\ \beta_j-\alpha_i \ge 0$$

$$= 0 \qquad\qquad \beta_j-\alpha_i < 0$$

$L = [1/p]$, $p' = 1 - pL$,

where the summation is over the set F of all partitions of N into L+1 integers $n_1,\ldots,n_{L+1}$ such that

$$0 \le n_i \le n-1 \qquad i = 1,\ldots,L+1$$

and where $g_{ij}/(\beta_j - \alpha_i)!$ is defined zero for $\beta_j - \alpha_i < 0$. The set of partitions here defined is much smaller than in formula (3.4). Unfortunately, many of the computational refinements of subsection 3.2 cannot be applied to Hwang's formula. Note that the Hwang formula is also a countable set of p-range results.

---

*In the Hwang article, the lower limit of summation was erroneously given as $\beta_{L+1}+ 1$.

## 3.2  An alternative approach to computing $P(n;N,p)$

The available general formulas (3.4) and (3.6) are difficult or impossible
to compute for small p.  Both of the general formulas are sets of p-range results; as $p \to 0$ we pass through infinitely many p domains $1/(L+1) < p < 1/L$,
and complexity of computation increases exponentially with $L = [1/p]$.  Neff
(1978) shows that $P(n;N,p)$ is a polynomial on the domain $0 \le p \le T(n,N)$, where
$T(n,N) = [N/n] + [(N-1)/n]$, the brackets denoting the greatest integer function.  This implies that only finitely many p domains need be computed.

For example, consider computing $P(3;9,0.01)$ by the Huntington-Naus formula
(3.4).  The formula involves a huge number of determinants of order L and L+1,
where $L = [1/p] = 100$.  Neff's result implies that $P(3;9,p)$ is a polynomial in
the domain $0 \le p \le 1/5$.  To compute $P(3;9,p)$ in this domain, it suffices to
find a polynomial $\phi(p)$ such that $\phi(p) = P(3;9,p)$ for infinitely many p values
within $0 \le p \le 1/5$.  Set L=5 in the Huntington-Naus formula (3.4).  The result
is a polynomial $1 - f_5(3;9,p)$ that agrees with $P(3;9,p)$ over the infinite set
$1/6 \le p < 1/5$.  Thus $1 - f_5(3;9,p)$ may serve as $\phi(p)$, and $P(3;9,0.01) = 1 - f_5$
$(3;9,0.01)$.  In summary, we have the following simplification:  In both formulas (3.4) and (3.6), the definition of L may be changed from $L = [1/p]$ to
$L = \min([1/p], ([N/n] + [(N-1)/n]))$.

Even with the above simplification, computations of $P(n;N,p)$ can be extremely expensive.  Rather than going through the entire computation for each p
value, we first compute and store intermediate quantities $d_{n,N,F,L}$, essentially
equivalent to the coefficients of $P(n;N,p)$ within each p domain.  To extract
the quantities $d_{n,N,F,L}$, Neff (1978) rearranges the Huntington-Naus formula
(3.4).

Let $Q(F)$ denote the set of all partitions of N into 2L+1 nonnegative integers $n_i$, $i=1,2,\ldots,2L+1$, that satisfy both

$$n_i + n_{i+1} < n, \text{ for } i = 1,2,\ldots,2L$$

and

$$n_1 + n_3 + \ldots + n_{2L+1} = F.$$

In equation (3.4), write $Q = \bigcup_{F=0}^{N} Q(F)$.  Q is a disjoint union of sets $Q(F)$.

Replace the summation $\Sigma_Q$ by the double summation $\Sigma_{F=0}^N \Sigma_{Q(F)}$.  Factor out $(1-pL)^F(-1+p(L+1))^{N-F}$ to get another form:

$$f_L(n,N,p) = \Sigma_{F=0}^N (1-pL)^F(-1+p(L+1))^{N-F}\, d_{n,N,F,L} \tag{3.7}$$

where

$$d_{n,N,F,L} = \Sigma_{Q(F)}\, N!\, \det|1/h_{ij}!|\det|1/\ell_{ij}!| \tag{3.8}$$

with $1/h_{ij}!$ and $1/\ell_{ij}!$ defined as in equation (3.5).

This new form for $f_L$ has the advantage of dividing the computation into two very distinct phases.  The computation of the $d_{n,N,F,L}$ by equation (3.8) is extremely long and complex.  Once that computation is complete, however, probabilities may be computed with relative ease for many different values of p.  In particular, the set of $d_{n,N,F,L}$ for a given n, N, and L, and for $F = O(1)N$ allow rapid computation of $P(n;N,p)$ for all p in the range $1/(L+1) \le p \le 1/L$.  Observe that the computation of the set of $d_{n,N,F,L}$ for $F = O(1)N$ is no more difficult than the computation of a single value of $P(n;N,p)$ by formula (3.4).

Expansion of formula (3.7) leads to an expression for $f_L(n;N,p)$ as a polynomial in p.

$$f_L(n;N,p) = \Sigma_{t=0}^N b_{t,n,N,L}\, p^t \tag{3.9}$$

where

$$b_{t,n,N,L} = \Sigma_{F=0}^N r_{t,N,F,L}\, d_{n,N,F,L} \tag{3.10}$$

$$r_{t,N,F,L} = \Sigma_{i+j=t} \binom{F}{i}\binom{N-F}{j}(-1)^{N-F+i-j} L^i (L+1)^j.$$

This result is used to generate the polynomials in p of Table 3 of the appendix.

Computation of formulas (3.7), (3.8), (3.9), and (3.10) demands high precision, sometimes exceeding the 32 significant digit limit of the PL/1 Optimizer Compiler.  Neff shows that the fundamental quantities $d_{n,N,F,L}$ are always integers.  Our strategy is first to compute and store the exact values of the d's.  All tables are then generated directly from the d's.

We would have liked to use quadruple precision (32 significant digits) throughout.  However, this would have been too inefficient for the expensive computation of the d's by formula (3.8).  The program used represents a

compromise in which the bulk of the operations are performed in double precision, while quadruple precision is used to make corrections at crucial points
in the calculation.  This program runs in about one-third the time of a pure
quadruple precision program, and produces quadruple precision accuracy for N
and L within the range of the tables.  Beyond a certain point accuracy deterio-
rates, but the program has built in checks to detect this point and issue a
warning.

We now describe the basic algorithm used to compute $d_{n,N,F,L}$ by formula
(3.8).  Neff (1978) contains further details and a program listing.

Enumerating the partitions of N:  Since $Q(F)$ is required for all F, the algorithm enumerates all partitions in Q and adds the result of each determinant's
calculation to a register containing a running total of the appropriate
$d_{n,N,F,L}$.  Recall that Q is defined to be the set of all partitions of N into
2L+1 integers $n_i$ satisfying $n_i + n_{i+1} < n$, i=1,...,2L.

We make the transformation
$$s_j = \Sigma_{i=1}^{j} n_i, \quad j=1,2,\ldots,2L+1.$$
The condition that the $n_i$ be nonnegative integers becomes the condition that
the $s_j$ be nonnegative integers with
$$s_j \geq s_{j-1}, \quad j=2,3,\ldots,2L+1 \tag{3.11}$$
The condition that $n_i + n_{i+1} < n$ becomes
$$s_j \leq s_{j-2} + n - 1, \quad j=1,2,\ldots,2L+1 \tag{3.12}$$
if we define $s_{-1} = s_0 = 0$.  The condition that the $n_i$ sum to N becomes
$$s_{2L+1} = N. \tag{3.13}$$

In the computer program, we delimit the $s_j$ values using DO loops.  $s_j$ is assigned a value and then all sequence completions $s_{j+1},\ldots,s_{2L+1}$ satisfying Q
are enumerated.  $s_j$ should never be assigned a value that allows no sequence
completion satisfying Q.  Thus we have the additional conditions:
$$N \geq s_{2L} \geq N-(n-1)$$
$$N \geq s_{2L-1} \geq N-(n-1)$$
$$N \geq s_{2L-2} \geq N-2(n-1)$$

.................

$$N \geq s_{2L+1-r} \geq N - [(r+1)/2](n-1), \tag{3.14}$$

the brackets denoting the greatest integer function.

Conditions (3.11) through (3.14) express Q in terms of the $s_j$. In practice, it is more efficient to consider a __range__ of N values, $N = N_1, N_1+1, \ldots, N_2$. Let $Q_{N_1,N_2} = \bigcup_{N=N_1}^{N_2} Q_N$, where $Q_N$ is the set of partitions of N satisfying

$n_i + n_{i+1} < n$. To express $Q_{N_1,N_2}$ in terms of the $s_j$, replace equation (3.13) by

$$N_2 \geq s_{2L+1} \geq N_1 \tag{3.15}$$

and replace equation (3.14) by

$$N_2 \geq s_{2L+1-r} \geq N_1 - [(r+1)/2](n-1). \tag{3.16}$$

Combine equations (3.15) and (3.16) to find

$$N_1 - [(2L+2-j)/2](n-1) \leq s_j \leq N_2. \tag{3.17}$$

The conditions (3.11) and (3.12) remain unchanged.

THEOREM 1:   Let $Q_{N_1,N_2}$ denote the set of nonnegative integer sequences

$\{n_i\}$, $i=1,2,\ldots,2L+1$, satisfying $n_i + n_{i+1} < n$, $i=1,\ldots,2L$, and $N_1 \leq n_1 + n_2 + \ldots + n_{2L+1} \leq N_2$. Then $Q_{N_1,N_2}$ is expressed in terms of the $s_j$ by the inequalities

$$\max(s_{j-1}, \ell_j) \leq s_j \leq \min(s_{j-2} + n-1, N_2)$$

where $s_{-1} = s_0 = 0$ and

$$\ell_j = N_1 - [(2L+2-j)/2](n-1),$$

the brackets denoting the greatest integer function.

Efficient calculation of determinants.

For each partition in $Q_{N_1,N_2}$ we must compute

$$N! \det|1/h_{ij}!| \det|1/\ell_{ij}!|,$$

where H and L are matrices of order L+1 and L, respectively. The determinants are calculated by Gaussian elimination, which requires about $L^3/3$ arithmetic operations for an L by L matrix. Our discussion will concentrate on H, but similar considerations apply to the matrix L.

The key idea of our method is the avoidance of unnecessary repetition of operations. Suppose the program enumerates $Q_{N_1,N_2}$ using DO loops and Theorem

1. At some point in the calculation, suppose $s_1$ equals 2. Then $n_1 = 2$, and

equation (3.5) implies $h_{11} = 2$. Hence $1/h_{11}! = 0.5$. At the end of the computa-

tion, the matrix $(1/h_{ij}!)$ will be row-reduced, and the determinant will be the

product of the diagonal elements. But we already know, at the stage where $s_1$

is set equal to 2, that the first factor of the product equals 0.5. Therefore

this 0.5 value should be saved, and for all $s_2, s_3, \ldots, s_{2L+1}$ in the enumeration

we need only concern ourselves with computing the remaining diagonal elements

of the row-reduced matrix. The algorithm should perform this simplification

not only for $s_1$, but also at all stages. When $s_j$ <u>is assigned, all possible</u>

<u>steps of the Gaussian elimination are performed immediately</u>.

Suppose that $s_1, s_2, \ldots, s_{j-1}$ have been previously assigned and that all

possible computations have been completed. We now detail the next steps of the

computation that can be completed.

The equations (3.5) for $h_{ij}$ and $\ell_{ij}$ may be written in terms of the $s_j$ as

follows:

$$h_{ij} = s_{2i-1} - s_{2j-2} - (i-j)n \tag{3.18}$$

$$\ell_{ij} = s_{2i} - s_{2j-1} - (i-j)n \tag{3.19}$$

where $s_0$ is 0 by definition.

When $s_1$ is known, $h_{11}$ may be computed. When $s_1$ and $s_2$ are known, $h_{12}$ and

$\ell_{11}$ may also be computed. When $s_3$ also becomes known, $h_{21}$, $h_{22}$ and $\ell_{12}$ may

also be computed. The general pattern is shown in Figure 1.

Figure 1

| | H | | | | | | L | | | |
|-----|---|---|---|---|---|-----|---|---|---|---|
| i/j | 1 | 2 | 3 | 4 | 5 | i/j | 1 | 2 | 3 | 4 |
| 1 | 1 | 2 | 4 | 6 | 8 | 1 | 2 | 3 | 5 | 7 |
| 2 | 3 | 3 | 4 | 6 | 8 | 2 | 4 | 4 | 5 | 7 |
| 3 | 5 | 5 | 5 | 6 | 8 | 3 | 6 | 6 | 6 | 7 |
| 4 | 7 | 7 | ⑦ | 7 | 8 | 4 | 8 | 8 | 8 | 8 |
| 5 | 9 | 9 | 9 | 9 | 9 | | | | | |

The circled entry in Figure 1 is interpreted "$h_{43}$ first becomes computable when

$s_7$ is defined." In fact, when $s_7$ is assigned, $h_{13}$, $h_{23}$, $h_{33}$, and $h_{43}$ are all

computed.

To carry out Gaussian elimination in accordance with the pattern in Figure 1, we use a variant of Doolittle's method known as Crout's method. Crout's method (Stoutmeyer, 1971) starts with a matrix A and recursively generates a matrix B. If the elements above the main diagonal of B are set equal to zero, the resulting matrix is the column reduced version of matrix A. The elements of B above the main diagonal are the column multipliers used during the process. The formulas used to generate B are:

$$b_{i1} = a_{i1}$$

$$b_{ij} = a_{ij} - \Sigma_{k=1}^{j-1} b_{ik}b_{kj} \qquad\qquad i \geq j \geq 2 \qquad\qquad (3.20)$$

$$= (a_{ij} - \Sigma_{k=1}^{j-1} b_{ik}b_{kj})/b_{ii} \qquad\qquad i < j$$

In most versions of Crout's method, the $b_{ij}$ are computed in the following order: first column, remainder of first row, remainder of second column, remainder of second row. However, the only requirement of the method is that elements above and to the left of the desired $b_{ij}$ be available, so we use the order in Figure 2, which is consistent with Figure 1.

Figure 2

| i/j | 1 | 2 | 3 | 4 |  | l/j | 1 | 2 | 3 |
|---|---|---|---|---|---|---|---|---|---|
| 1 | 1.0 | 2.0 | 4.1 | 6.1 |  | 1 | 2.0 | 3.0 | 5.1 |
| 2 | 3.1 | 3.2 | 4.2 | 6.2 |  | 2 | 4.1 | 4.2 | 5.2 |
| 3 | 5.1 | 5.2 | 5.3 | 6.3 |  | 3 | 6.1 | 6.2 | 6.3 |
| 4 | 7.1 | 7.2 | 7.3 | 7.4 |  |  |  |  |  |

H (left), L (right)

## Refinement: zero suppression

The first refinement rests on the fact that the condition $n_i + n_{i+1} < n$ implies that the columns of H (and L) must be strictly decreasing and that the rows must be strictly increasing. Sinc $1/\nu!$ is zero when $\nu < 0$, this means that once zero occurs at a position in the matrix $(1/h_{ij}!)$ or $(1/\ell_{ij}!)$, all entries below and to the left of the position must also equal zero.

In terms of equation (3.20), suppose $1/h_{ij}! = a_{ij} = 0$, with $i > j$. Then $a_{i1} = a_{i2} = \ldots = a_{ij} = 0$. An induction argument shows that $b_{i1} = b_{i2} = \ldots = b_{ij} = 0$ and that $b_{i,j+1} = a_{i,j+1}$. Furthermore, in calculating the rest of row i, the summations need not begin at k = 1 but at k = j+1. In other words, the computation proceeds as if row i started in column j+1 rather than column 1. This is true not only for row i, but also for all subsequent rows, so that the simplification will affect all partitions at higher levels of recursion.

Let $c(i) \leq i$ be the column number of the first non-zero element of row i of A. The formula for the revised version of Crout's method is

$$
\begin{aligned}
b_{ij} &= \text{(Not computed)} & j &< c(i) \\
&= a_{i,c(i)} & j &= c(i) \\
&= a_{ij} - \Sigma_{k=c(i)}^{j-1} b_{ik} b_{kj} & c(i) &< j \leq i \\
&= (a_{ij} - \Sigma_{k=c(i)}^{j-1} b_{ij} b_{kj})/b_{ii} & i &< j.
\end{aligned} \tag{3.21}
$$

## Refinement: Wolf's observation

$c(i)$ can never exceed i because the diagonal elements of H (and L) are reciprocal factorials of the occupany numbers $n_{2i+1}$ (or $n_{2i}$), which are non-negative. However, $c(i)$ may equal i, and when this happens there is an opportunity for a second refinement.

For example, suppose that in the matrix $A = (1/h_{ij}!)$, $c(3) = 3$. This implies $a_{31} = a_{32} = 0$. Also zero are the entries lower in the first two columns: $a_{i1} = a_{i2} = 0$, $i=3,4,\ldots,L+1$. It may be shown that $\det|A|$ is the product of the determinants of two submatrices of A. One submatrix is the 2×2 matrix in the upper left corner of A; the other submatrix is the (L-1)×(L-1) matrix in the lower right corner of A. Wolf (1968) found a similar factoring pattern in the earlier formula for $P(n;N,1/L)$.

To implement Wolf's observation, we store for each column j of the matrix $(1/h_{ij}!)$, the starting row $r(j)$. $r(j)$ is initially set equal to 1 and is reset to i when $c(i) = i$. The equation for this final version of the algorithm is:

$$
\begin{aligned}
b_{ij} &= \text{(Not computed)} & j &< c(i) \text{ or } i < r(j) \\
&= a_{i,c(i)} & j &= c(i) \\
&= a_{ij} - \Sigma_{k=c(i)}^{j-1} b_{ik} b_{kj} & c(i) &< j \leq i \\
&= (a_{ij} - \Sigma_{k=c(i)}^{j-1} b_{ik} b_{kj})/b_{ii} & r(j) &\leq i < j
\end{aligned} \tag{3.23}
$$

where $c(i)$ is the column number of the first nonzero entry in row i of A,

$$r(1) = 1$$
$$r(j) = j \text{ if } c(j) = j$$
$$= r(j-1) \text{ otherwise.}$$

There is no limitation on the number of resettings of $r(j)$; in some cases a determinant may split into more than two factors.

## Refinement:  Precision improvement

The precision improvement rests on the fact that $E!$ $\det|1/\ell_{ij}!|$ and $F!$ $\det|1/h_{ij}!|$ are integers.  These quantities are computed in double precision arithmetic, using the refined algorithm (3.23).  The program then shifts to quadruple precision and corrects these quantities by rounding to the nearest integer.  Remaining in quadruple precision, $N!$ $\det|1/h_{ij}!||\det|1/\ell_{ij}!|$ is computed and added to the running total of $d_{n,N,F,L}$.

These steps will give quadruple precision accuracy unless accuracy has already deteriorated at the time of rounding.  The program detects this by performing a significance test at the 0.125 level along with each rounding operation.  If the number to be rounded is not within ±0.0625 of the nearest integer, a warning is printed.

## REFERENCES

Berg, W. (1945).  Aggregates in one- and two-dimensional random distributions. London, Edinburgh and Dublin, Philosophical Magazine and Journal of Science 36, 319-336.

Bogartz, R. S. (1965).  The criterion method:Some analyses and remarks.  Psych. Bull. 64, 1-14.

Brookner, E. (1966).  Recurrent events in a Markov chain.  Information and Control 9, 215-229.

Domb, C. (1950).  Some probability distributions connected with recording apparatus II.  Proc. Cambridge Philosophical Soc. 46, 429-435.

Ederer, F., Myers, E. and Mantel, N. (1964).  A statistical problem in space and time: Do leukemia cases come in clusters.  Biometrics 20, 626-638.

Feller, W. (1958).  An introduction to probability theory and its applications, Vol. I, Second Edition, New York:  John Wiley and Sons.

Furth, R. (1920).  Schwankungserscheinungen un der Physik.  Braunschweig: Sammlung Vieweg, 1-17.

Goldman, A. J. and Bender, B. K. (1962).  The first run preceded by a quota. J. Res. Nat. Bur. Standards Sect. B, 77-89.

Greenberg, I. (1970).  The first occurrence of N successes in M trials. Technometrics 12, 627-634.

Grind, W. A. Van De. Schalm, T, Van, and Bowman, M. A. (1968).  A coincidence model of the processing of quantum signals by the human retina. Kybernetik 4, 141-146.

Hamilton, J. F., Lawton, W. H. and Trabka, E. A. (1972).  Some spatial and temporal point processes in photographic science.  In Lewis, P. A. W. Ed. Stochastic Point Processes: Statistical Analysis, Theory and Applications. Wiley Interscience: New York, 1972, 817-867.

Huntington, R. J. (1974).  Distributions and expectations for clusters in continuous and discrete cases, with applications.  Ph.D. Thesis, Rutgers University.

Huntington, R. J. and Naus, J. I. (1975).  A simpler expression for kth nearest neighbor coincidence probabilities.  Ann. Probability 3, 894-896.

Huntington, R. J. (1976).  Mean recurrence times for k successes within M trials.  J. Appl. Probability 13, 604-607.

Hwang, F. K. (1977).  A generalization of the Karlin-McGregor Theorem on coincidence Probabilities and an application to clustering.  Ann. Probability 5, 814-817.

Ikeda, S. (1965).  On Bouman-Velden-Yamamoto's asymptotic evaluation formula for the probability of visual response in a certain experimental research in quantum biophysics of vision.  Ann. Inst. Statist. Math. 17, 295-310.

Janossy, L. (1944).  Rate of N-fold accidental coincidences.  Nature 153, 165.

Leslie, R. T. (1967).  Recurrent composite events.  J. Appl. Probability 4, 34-61.

Leslie, R. T. (1969).  Recurrence times of clusters of Poisson points.  J. Appl. Probability 6, 372-383.

Mack, C. (1948).  An exact formula for Q*K(N), the probable number of K-aggregates in a random distribution of N points.  London, Edinburgh, Dublin Philos. J. Science 39, 778-790.

Mack, C. (1950).  The expected number of aggregates in a random distribution of N points.  Proc. Cambridge Philosophical Soc. 46, 285-292.

Naus, J. (1963).  Clustering of random points in line and plane.  Ph.D. Thesis, Department of Statistics, Harvard University.

Naus, J. (1965).  The distribution of the size of the maximum cluster of points on a line.  J. Amer. Statist. Assoc. 60, 532-538.

Naus, J. (1966).  Some probabilities, expectations, and variances for the size of largest clusters and smallest intervals.  J. Amer. Statist. Assoc. 61, 1191-1199.

Naus, J. (1966).  A power comparison of two tests of nonrandom clustering.  Technometrics 8, 493-517.

Naus, J. (1968).  An extension of the birthday problem.  Amer. Statist. 22(1), 27-29.

Naus, J. (1974).  Probabilities for a generalized birthday problem.  J. Amer. Statist. Assoc. 69, 810-815.

Neff, N. (1978).  Piecewise polynomials for the probability of clustering on the unit interval.  Ph.D. Thesis, Department of Statistics, Rutgers University.

Newell, G. F. (1958).  Some statistical problems encountered in a theory of pinning and break away of dislocations.  Quart. Appl. Math. 16, 155-168.

Newell, G. F. (1963).  Distribution for the smallest distance between any pair of kth nearest-neighbor random points on a line.  Proc. of symposium on time series, Brown Univ., 89-103.  John Wiley and Sons:  New York.

Parzen, E. (1960).  Modern probability theory and its applications.  John Wiley and Sons: New York.

Pinkel, D. and Nefzger, D. (1959).  Some epidemiological features of childhood Leukemia.  Cancer, 12, 351-358.

Roberts, S. W. (1958).  Properties of control chart zone tests.  Bell System Tech. J. 37, 83-114.

Roberts, S. W. (1958).  Duration of an M-player game of chance that ends when a player achieves k successes within N consecutave trials.  Unpublished memo, Bell Labs., MM58-5214-1.

Roberts, S. W. (1958).  On the first occurrence of any of a selected set of outcome patterns in a sequence of repeated trials.  Unpublished manuscript.

Runnels, L. K., Thompson, R. and Runnels, P. (1968).  Near perfect runs as a learning criterion.  J. Mathematical Psychology 5, 362-368.

Saperstein, B. (1969).  Some generalizations of the birthday problem and related problems with applications.  Ph.D. Thesis, New York University.

Saperstein, B. (1972).  The generalized birthday problem.  J. Amer. Statist.
    Assoc. 67, 425-428.

Saperstein, B. (1973).  On the occurrence of n successes within N Bernoulli
    trials.  Technometrics 15, 809-818.

Saperstein, B. (1975).  Note on a clustering problem.  J. Appl. Probability 12,
    629-632.

Saperstein, B. (1976).  The analysis of attribute moving averages: MIL-STD-105D
    reduced inspection plans.  Paper given at Sixth Conference on Stochastic
    Processes and Applications:  Tel-Aviv.

Schroedinger, E. (1944).  Rate of N-fold accidental coincidence.  Nature 153,
    592-593.

Silberstein, L. (1945).  The probable number of aggregates in random distri-
    butions of points.  London, Edinburgh, Dublin Philos. Magazine J. Sci. 36,
    319-336.

Solov'ev, A. D. (1966).  A combinatorial identity and its application to the
    problem concerning the first occurrence of a rare event.  Theor. Proba-
    bility Appl. 11, 276-282.

Stoutmeyer, D. R. (1971).  PL/1 programming for engineering and science.
    Prentice-Hall, Englewood Cliffs: New Jersey.

Troxell, J. R. (1972).  An investigation of suspension systems for small scale
    inspections.  Ph.D. Thesis, Department of Statistics, Rutgers University.

Van Elteren, P. H. and Gerritis, H. J. M. (1961).  Een Wachtprobleem Voorko-
    mende Bij Drempelwaardemetingen Aan Het Oog.  Statistica Neerlandica 15,
    385-401.  (English Summary).

Wallenstein, S. R. (1971).  Coincidence probabilities used in nearest neighbor
    problems on the line and circle.  Ph.D. Thesis, Rutgers University.

Wallenstein, S. R. and Naus, J. I. (1973).  Probabilities for a kth nearest
    neighbor problem on the line.  Ann. Probability 1, 188-190.

Wallenstein, S. R. and Naus, J. I. (1974).  Probabilities for the size of larg-
    est clusters and smallest intervals.  J. Amer. Statist. Assoc. 69, 690-697.

Wolf, E. H. (1968).  Test for randomness on the line and a related k-sample
    test for homogeneity.  Ph.D. Thesis, Rutgers University.

APPENDIX

We summarize the contents of the tables. Interpolation procedures and extrapolation approximations are discussed.

Tables 1 and 1a.  P(n;N,p):

P(n;N,p) is the probability that there exists a subinterval of (0,1) of length p that contains at least n out of a fixed number N of randomly distributed points. Table 1 gives probabilities for p changing by steps of 0.01, while in Table 1a p changes by steps of 0.001. The finer subdivision is necessary because for some (n,N) the interesting values of p are over a narrow range. For example, from Table 1a, P(3;10,0.100) = 0.91, P(3;10,0.001) = 0.00036.

Table 1a presents fine subdivisions over the range of the distribution for the difficult to compute case n ≤ N/2, p < 0.5. For the case n > N/2, equation (3.2) allows ready computation with binomial tables and a calculator. Equation (3.2) also handles the case p > 0.5; this case being of less practical interest, tables are omitted. Equation (3.1) gives an easily computable form for the case n = 2. Together with equations (3.1) and (3.2), Tables 1 and 1a give the distribution for all (n,N) combinations for N < 19.

Within the table, linear interpolation is reasonable for P(n;N,p) between 0.1 and 0.9. For example, interpolating between P(3;6,0.200) = 0.80768, and P(3;6,0.190) = 0.77480, gives and interpolated figure of 0.79124 for P(3;6,0.195) = 0.79164. Interpolating for small values of P(n;N,p) can lead to relatively large errors. For example, interpolating between P(3;6,0.001) and P(3;6,0.011) gives an interpolated figure of 0.0035 for P(3;6,0.006) = 0.0021. For such cases or where one needs greater precision, the polynomials given in Table 3 allow computation to any number of significant places.

For N ≥ 19 one can refer to the tables of Wallenstein (1971) for P(n;N,1/L) for N = 20(1)100, L = 3(1)10 for N/3 < n ≤ N/2. Selected values from this table appear in Wallenstein and Naus (1974). Outside of the range of these tables one can use equation (3.2) to approximate P(n;N,p) for n ≤ N/2, p < 1/2. This approximation is reasonable when P(n;N,p) is less than 0.1.

Tables 2 and 2a.   $P'(n;\lambda,p)$:

$$P'(n;\lambda,p) = \Sigma_{N-n}^{\infty} e^{-\lambda}\lambda^N P(n;N,p)/N!$$

Table 2 gives values of $P'(n;\lambda,p)$ for $\lambda = 1(1)12$, $n = 3(1)9$, $p = 0.01(0.01)0.50$.
The tabled probabilities are accurate within one unit in the least significant
digit printed.  The number of significant places varies with p because some of
the $P(n;N,p)$ values are unknown and must be bounded by $P(n;N_o,p) \le P(n;N,p) \le 1$,
where $N_o < N$ is such that $P(n;N_o,p)$ is known.  If the unknown $P(n;N,p)$ values
are replaced by 1, we get an upper bound for $P'(n;\lambda,p)$; if the values are replaced by $P(n;N_o,p)$ we get a lower bound.  The difference between the bounds is
given by $(1 - P(n;N_o,p)) \Sigma_{N=N_o+1}^{\infty} e^{-\lambda}\lambda^N/N!$, which is quite small in many cases.

For values of $P'(n;\lambda,p)$ beyond the range of Table 2, we recommend the
approximation given in application (d), or more generally, formula (A.1).  If p
is of the form 1/L, L an integer, the approximation requires values from Table
2a.  Table 2a gives exact values of $P'(n;\lambda,p)$ for $\lambda = 0.1(0.1)10$, $n = 3(1)9$,
$p = 1/2$, $1/3$, $1/4$.  The general approximation is as follows:

Let $p = (1-b)/2L$, where L is an integer and b is a constant $0 \le b \le p$.

$$P'(n;\lambda,p) \doteq 1-[1-P'(n;(2p+b)\lambda,p/(2p+b)]H(L) \qquad (A.1)$$

where

$$H(L) = [1-P'(n;4p\lambda,\tfrac{1}{4})]^{L-1}/[1-P'(n;2p\lambda,\tfrac{1}{2})]^{L-1}.$$

Two special cases of approximation (A.1) are of particular interest:  For b=0,
$p = 1/2L$; for b=p, $p = 1/(2L+1)$.  For these cases the approximation (A.1) only
requires $P'(n;\lambda,p)$ for $p = 1/2$, $1/3$, $1/4$.

We illustrate the reasoning behind the approximation (A.1) for the special
case $p = 1/2L$.  Divide the unit interval into 2L equal cells, and define the
events:

$A_i$:   cell i contains less than r points, i=1,2,...,2L.

$B_i$:   there is no interval of length 1/2L overlapping cells i and
         i+1 that contains as many as r points, i=1,2,...,2L-1.

The approximation views the 2L cells as grouped into L pairs of adjacent cells
and approximates $1-P'(n;\lambda,1/2L)$ by

$$[\Pi_{i=1}^{L} \; Pr(A_{2i-1} \cap A_{2i} \cap B_{2i-1})][\Pi_{i=1}^{L-1} \; Pr(B_{2i} | \cap_{i=1}^{2L} A_i \cap_{\substack{j=1 \\ j \neq 2i}}^{2L-1} B_j)]. \tag{A.2}$$

A further approximation is made by assuming that the conditional probability in expression (A.2) can be approximated by $Pr(B_{2i} | \cap_{j=2i-1}^{2i+2} A_j \cap B_{2i-1} \cap B_{2i+1})$. Given the symmetry in the problem, expression (A.2) then reduces to $C^L D^{L-1}$ where

$$C = Pr(A_1 \cap A_2 \cap B_1) = 1 - P'(n; \lambda/L, \tfrac{1}{2}) \tag{A.3}$$

and

$$D = Pr(B_2 \cap A_1 \cap A_2 \cap A_3 \cap A_4 \cap B_1 \cap B_3)/Pr(A_1 \cap A_2 \cap A_3 \cap A_4 \cap B_1 \cap B_3) \tag{A.4}$$

The final step is to write

$$Pr(A_1 \cap A_2 \cap A_3 \cap A_4 \cap B_1 \cap B_3) = Pr(A_1 \cap A_2 \cap B_1)Pr(A_3 \cap A_4 \cap B_3). \tag{A.5}$$

With these approximations $1 - P'(n; \lambda, 1/2L)$ is estimated by

$$[1 - P'(n; \lambda/L, \tfrac{1}{2})]^L [1 - P'(n; 2\lambda/L, \tfrac{1}{4})]^{L-1}/[1 - P'(n; \lambda/L, \tfrac{1}{2})]^{2L-2}.$$

Cancelling common terms in numerator and denominator leads to the special case of approximation (A.1) as detailed in application (d). Application (g) illustrates the remarkable accuracy of this approximation.

## Table 3: The Piecewise Polynomials for P(n;N,p)

Table 3 gives the piecewise polynomials for $P(n;N,p)$ for various ranges of p. The example n = 3, N = 4 illustrates the information in the table. In the table headed n = 3 there are two entries for the case N = 4. The first entry is

| N | p range | F | coeff. $p^{2F}$ | coeff. $p^{2F+1}$ |
|---|---------|---|-----------------|-------------------|
| 4 | $(0,\tfrac{1}{2})$ | 0 | 0 | 0 |
|   |         | 1 | 12 | -24 |
|   |         | 2 | 14 |    |

this entry states that

$$P(3;4,p) = 12p^2 - 24p^3 + 14p^4 \qquad \text{for } 0 \le p \le \tfrac{1}{2}.$$

The second entry is

n = 3

| N | p range | F | coeff. $p^{2F}$ | coeff. $p^{2F+1}$ |
|---|---------|---|-----------------|-------------------|
| 4 | $(\tfrac{1}{2},1)$ | 0 | -1 | 8 |
|   |         | 1 | -12 | 8 |
|   |         | 2 | -2 |    |

this entry states that

$$P(3;4,p) = -1 + 8p - 12p^2 + 8p^3 - 2p^4 \qquad \text{for } \tfrac{1}{2} \le p \le 1.$$

The two entries together give the two pieces of the polynomial for $P(3;4,p)$.
As N becomes large the coefficients become big making the chosen format convenient. For example, for $n = 3$, $N = 10$ there are three different polynomials
corresponding to the three different ranges for p: $(0,1/6)$, $(1/6,1/5)$, $(1/5,$
$1/4)$. The polynomial for the range $1/5 \le p \le 1/4$ is

$$P(3;10,p) = - 41 + 1680p - 30240p^2 + 322560p^3 - 2257920p^4$$
$$+ 10838016p^5 - 36126720p^6 + 82575360p^7$$
$$- 123863040p^8 + 110100480p^9 - 44040192p^{10}.$$

The polynomials of Table 3 can be used to compute individual probabilities
$P(n;N,p)$ to any required number of significant places. The coefficients of the
polynomial are given alternately in two columns. This is convenient for the
non-tail cases where the signs of the coefficients usually alternate. For example, to compute $P(3;4,0.6)$ one would evaluate $(-1) + (-12)(0.6)^2 + (-2)(0.6)^4$
$= -5.5792$. We would then evaluate $8(0.6) + 8(0.6)^3 = + 6.528$. $P(3;4,0.6) =$
$6.528 - 5.5792 = 0.9488$. The polynomials should be used in applications where
greater precision is required than is available by linear interpolation in
Tables 1 and 1a.

$P(n;N,p)$ is the cumulative distribution function of $\widetilde{p}_n$ the size of the
smallest interval that contains n out of the N points. The polynomial expressions for $Pr(\widetilde{p}_n \le p)$ given in Table 3 allow evaluation of the moments of the
distribution of $\widetilde{p}_n$ and allow one to derive the distribution and moments of
functions of $\widetilde{p}_n$. The polynomials can also be used to solve for critical values
$p(\alpha)$ of the smallest interval that satisfy $Pr(\widetilde{p}_n \le p(\alpha)) = \alpha$.

Table 4:   Mean and Variance of Shortest Interval:

Table 4 gives the expectation and variance of $\widetilde{p}_n$, the length of the
smallest subinterval of $(0,1)$ that contains at least n out of a fixed number N
of points distributed at random over $(0,1)$.

TABLE 1:　CLUSTERING PROBABILITY　P(n;N,p)

$n = 3$

| P | N = 3 | N = 4 | N = 5 | N = 6 |
|---|---|---|---|---|
| 0.50 | 0.500000000 | 0.875000000 | 1 | 1 |
| 0.49 | 0.485002000 | 0.864696140 | 0.999980005 | 1.000000000 |
| 0.48 | 0.470016000 | 0.853770240 | 0.999840154 | 0.999999980 |
| 0.47 | 0.455054000 | 0.842203340 | 0.999461166 | 0.999999767 |
| 0.46 | 0.440128000 | 0.829979840 | 0.998724915 | 0.999998689 |
| 0.45 | 0.425250000 | 0.817087500 | 0.997515000 | 0.999995000 |
| 0.44 | 0.410432000 | 0.803517440 | 0.995717325 | 0.999985070 |
| 0.43 | 0.395686000 | 0.789264140 | 0.993220674 | 0.999962352 |
| 0.42 | 0.381024000 | 0.774325440 | 0.989917286 | 0.999916114 |
| 0.41 | 0.366458000 | 0.758702540 | 0.985703435 | 0.999829939 |
| 0.40 | 0.352000000 | 0.742400000 | 0.980480000 | 0.999680000 |
| 0.39 | 0.337662000 | 0.725425740 | 0.974153045 | 0.999433100 |
| 0.38 | 0.323456000 | 0.707791040 | 0.966634394 | 0.999044485 |
| 0.37 | 0.309394000 | 0.689510540 | 0.957842206 | 0.998455421 |
| 0.36 | 0.295488000 | 0.670602240 | 0.947701555 | 0.997590548 |
| 0.35 | 0.281750000 | 0.651087500 | 0.936145000 | 0.996355000 |
| 0.34 | 0.268192000 | 0.630991040 | 0.923113165 | 0.994631291 |
| 0.33 | 0.254826000 | 0.610340940 | 0.908555314 | 0.992275978 |
| 0.32 | 0.241664000 | 0.589168640 | 0.892429926 | 0.989116498 |
| 0.31 | 0.228718000 | 0.567508940 | 0.874705275 | 0.984952041 |
| 0.30 | 0.216000000 | 0.545400000 | 0.855360000 | 0.979560000 |
| 0.29 | 0.203522000 | 0.522883340 | 0.834383685 | 0.972703358 |
| 0.28 | 0.191296000 | 0.500003840 | 0.811777434 | 0.964137861 |
| 0.27 | 0.179334000 | 0.476809740 | 0.787554446 | 0.953618955 |
| 0.26 | 0.167648000 | 0.453352640 | 0.761740595 | 0.940908500 |
| 0.25 | 0.156250000 | 0.429687500 | 0.734375000 | 0.925781250 |
| 0.24 | 0.145152000 | 0.405872640 | 0.705510605 | 0.908031099 |
| 0.23 | 0.134366000 | 0.381969740 | 0.675214754 | 0.887477104 |
| 0.22 | 0.123904000 | 0.358043840 | 0.643569766 | 0.863969272 |
| 0.21 | 0.113778000 | 0.334163340 | 0.610673515 | 0.837394120 |
| 0.20 | 0.104000000 | 0.310400000 | 0.576640000 | 0.807680000 |
| 0.19 | 0.094582000 | 0.286828940 | 0.541599925 | 0.774802199 |
| 0.18 | 0.085536000 | 0.263528640 | 0.505701274 | 0.738787807 |
| 0.17 | 0.076874000 | 0.240580940 | 0.469109886 | 0.699720350 |
| 0.16 | 0.068608000 | 0.218071040 | 0.432010035 | 0.657744200 |
| 0.15 | 0.060750000 | 0.196087500 | 0.394605000 | 0.613068750 |
| 0.14 | 0.053312000 | 0.174722240 | 0.357117645 | 0.565972360 |
| 0.13 | 0.046306000 | 0.154070540 | 0.319790994 | 0.516806070 |
| 0.12 | 0.039744000 | 0.134231040 | 0.282888806 | 0.465997087 |
| 0.11 | 0.033638000 | 0.115305740 | 0.246696155 | 0.414052039 |
| 0.10 | 0.028000000 | 0.097400000 | 0.211520000 | 0.361560000 |
| 0.09 | 0.022842000 | 0.080622540 | 0.177689765 | 0.309195280 |
| 0.08 | 0.018176000 | 0.065085440 | 0.145557914 | 0.257719992 |
| 0.07 | 0.014014000 | 0.050904140 | 0.115500526 | 0.207986384 |
| 0.06 | 0.010368000 | 0.038197440 | 0.087917875 | 0.160938939 |
| 0.05 | 0.007250000 | 0.027087500 | 0.063235000 | 0.117616250 |
| 0.04 | 0.004672000 | 0.017699840 | 0.041902285 | 0.079152660 |
| 0.03 | 0.002646000 | 0.010163340 | 0.024396034 | 0.046779675 |
| 0.02 | 0.001184000 | 0.004610240 | 0.011219046 | 0.021827141 |
| 0.01 | 0.000298000 | 0.001176140 | 0.002901195 | 0.005724198 |

TABLE 1: CLUSTERING PROBABILITY $P(n;N,p)$

$n = 4$

| P | N = 4 | N = 5 | N = 6 | N = 7 |
|---|---|---|---|---|
| 0.50 | 0.312500000 | 0.687500000 | 0.921875000 | 1 |
| 0.49 | 0.297651970 | 0.668505148 | 0.912019166 | 0.999965017 |
| 0.48 | 0.283115520 | 0.649042330 | 0.901156700 | 0.999720537 |
| 0.47 | 0.268901570 | 0.629146615 | 0.889227902 | 0.999059077 |
| 0.46 | 0.255020320 | 0.608856147 | 0.876183631 | 0.997777164 |
| 0.45 | 0.241481250 | 0.588211875 | 0.861985547 | 0.995677312 |
| 0.44 | 0.228293120 | 0.567257293 | 0.846606295 | 0.992569965 |
| 0.43 | 0.215463970 | 0.546038175 | 0.830029625 | 0.988275381 |
| 0.42 | 0.203001120 | 0.524602310 | 0.812250452 | 0.982625469 |
| 0.41 | 0.190911170 | 0.502999242 | 0.793274852 | 0.975465544 |
| 0.40 | 0.179200000 | 0.481280000 | 0.773120000 | 0.966656000 |
| 0.39 | 0.167872770 | 0.459496838 | 0.751814048 | 0.956073888 |
| 0.38 | 0.156933920 | 0.437702970 | 0.729395937 | 0.943614381 |
| 0.37 | 0.146387170 | 0.415952305 | 0.705915151 | 0.929192126 |
| 0.36 | 0.136235520 | 0.394299187 | 0.681431409 | 0.912742451 |
| 0.35 | 0.126481250 | 0.372798125 | 0.656014297 | 0.894222437 |
| 0.34 | 0.117125920 | 0.351503533 | 0.629742837 | 0.873611832 |
| 0.33 | 0.108170370 | 0.330469465 | 0.602704997 | 0.850913785 |
| 0.32 | 0.099614720 | 0.309749350 | 0.574997135 | 0.826155414 |
| 0.31 | 0.091458370 | 0.289395732 | 0.546723389 | 0.799388171 |
| 0.30 | 0.083700000 | 0.269460000 | 0.517995000 | 0.770688000 |
| 0.29 | 0.076337570 | 0.249992128 | 0.488929576 | 0.740155284 |
| 0.28 | 0.069368320 | 0.231040410 | 0.459650294 | 0.707914559 |
| 0.27 | 0.062788770 | 0.212651195 | 0.430285044 | 0.674113981 |
| 0.26 | 0.056594720 | 0.194868627 | 0.400965506 | 0.638924552 |
| 0.25 | 0.050781250 | 0.177734375 | 0.371826172 | 0.602539063 |
| 0.24 | 0.045342720 | 0.161287373 | 0.343003300 | 0.565170772 |
| 0.23 | 0.040272770 | 0.145563555 | 0.314633814 | 0.527051790 |
| 0.22 | 0.035564320 | 0.130595590 | 0.286854139 | 0.488431154 |
| 0.21 | 0.031209570 | 0.116412622 | 0.259798972 | 0.449572600 |
| 0.20 | 0.027200000 | 0.103040000 | 0.233600000 | 0.410752000 |
| 0.19 | 0.023526370 | 0.090499018 | 0.208384548 | 0.372254463 |
| 0.18 | 0.020178720 | 0.078806650 | 0.184274171 | 0.334371086 |
| 0.17 | 0.017146370 | 0.067975285 | 0.161383182 | 0.297395339 |
| 0.16 | 0.014417920 | 0.058012467 | 0.139817124 | 0.261619083 |
| 0.15 | 0.011981250 | 0.048920625 | 0.119671172 | 0.227328188 |
| 0.14 | 0.009823520 | 0.040696813 | 0.101028482 | 0.194797763 |
| 0.13 | 0.007931170 | 0.033332445 | 0.083958476 | 0.164286972 |
| 0.12 | 0.006289920 | 0.026813030 | 0.068515062 | 0.136033424 |
| 0.11 | 0.004884770 | 0.021117912 | 0.054734800 | 0.110247127 |
| 0.10 | 0.003700000 | 0.016220000 | 0.042635000 | 0.087104000 |
| 0.09 | 0.002719170 | 0.012085508 | 0.032211765 | 0.066738919 |
| 0.08 | 0.001925120 | 0.008673690 | 0.023437967 | 0.049238298 |
| 0.07 | 0.001299970 | 0.005936575 | 0.016261165 | 0.034632176 |
| 0.06 | 0.000825120 | 0.003818707 | 0.010601461 | 0.022885820 |
| 0.05 | 0.000481250 | 0.002256875 | 0.006349297 | 0.013890813 |
| 0.04 | 0.000248320 | 0.001179853 | 0.003363185 | 0.007455619 |
| 0.03 | 0.000105570 | 0.000508135 | 0.001467383 | 0.003295628 |
| 0.02 | 0.000031520 | 0.000153670 | 0.000449505 | 0.001022641 |
| 0.01 | 0.000003970 | 0.000019602 | 0.000058072 | 0.000133808 |

TABLE 1:   CLUSTERING PROBABILITY   P(n;N,p)

n = 4

| P | N = 8 | N = 9 | N = 10 | N = 11 |
|------|-------------|-------------|-------------|-------------|
| 0.50 | 1 | 1 | | |
| 0.49 | 0.999999998 | 1.000000000 | | |
| 0.48 | 0.999999864 | 1.000000000 | | |
| 0.47 | 0.999998486 | 1.000000000 | | |
| 0.46 | 0.999991718 | 0.999999994 | | |
| 0.45 | 0.999969270 | 0.999999958 | | |
| 0.44 | 0.999910841 | 0.999999783 | | |
| 0.43 | 0.999781773 | 0.999999132 | | |
| 0.42 | 0.999528547 | 0.999997114 | | |
| 0.41 | 0.999074419 | 0.999991669 | | |
| 0.40 | 0.998315520 | 0.999978496 | | |
| 0.39 | 0.997117747 | 0.999949295 | | |
| 0.38 | 0.995314767 | 0.999889044 | | |
| 0.37 | 0.992707487 | 0.999771961 | | |
| 0.36 | 0.989065331 | 0.999555705 | | |
| 0.35 | 0.984129670 | 0.999173314 | | |
| 0.34 | 0.977619784 | 0.998522256 | | |
| 0.33 | 0.969241708 | 0.997449887 | 0.999999985 | 1.000000000 |
| 0.32 | 0.958700293 | 0.995735540 | 0.999995583 | 0.999999999 |
| 0.31 | 0.945713747 | 0.993079578 | 0.999953228 | 0.999999951 |
| 0.30 | 0.930027690 | 0.989108632 | 0.999782580 | 0.999999168 |
| 0.29 | 0.911427019 | 0.983389233 | 0.999314455 | 0.999993353 |
| 0.28 | 0.889745562 | 0.975446478 | 0.998281566 | 0.999965920 |
| 0.27 | 0.864873681 | 0.964786498 | 0.996304941 | 0.999869653 |
| 0.26 | 0.836763972 | 0.950921414 | 0.992889103 | 0.999595690 |
| 0.25 | 0.805435181 | 0.933395386 | 0.987428665 | 0.998929977 |
| 0.24 | 0.770974471 | 0.911810410 | 0.979227986 | 0.997503308 |
| 0.23 | 0.733538126 | 0.885850614 | 0.967533866 | 0.994749646 |
| 0.22 | 0.693350798 | 0.855303930 | 0.951579240 | 0.989890471 |
| 0.21 | 0.650703358 | 0.820080274 | 0.930634318 | 0.981952620 |
| 0.20 | 0.605949440 | 0.780225536 | 0.904061235 | 0.969818317 |
| 0.19 | 0.559500699 | 0.735930887 | 0.871368309 | 0.952302740 |
| 0.18 | 0.511820843 | 0.687537086 | 0.832260220 | 0.928252352 |
| 0.17 | 0.463418456 | 0.635533588 | 0.786680774 | 0.896655561 |
| 0.16 | 0.414838617 | 0.580552401 | 0.734845376 | 0.856756295 |
| 0.15 | 0.366653321 | 0.523356707 | 0.677260860 | 0.808160654 |
| 0.14 | 0.319450681 | 0.464824372 | 0.614730897 | 0.750927025 |
| 0.13 | 0.273822888 | 0.405926466 | 0.548345798 | 0.685630833 |
| 0.12 | 0.230352869 | 0.347700993 | 0.479456093 | 0.613396414 |
| 0.11 | 0.189599617 | 0.291222000 | 0.409629817 | 0.535890203 |
| 0.10 | 0.152082090 | 0.237564208 | 0.340593900 | 0.455271502 |
| 0.09 | 0.118261624 | 0.187763297 | 0.274160403 | 0.374099242 |
| 0.08 | 0.088522744 | 0.142771861 | 0.212138610 | 0.295195372 |
| 0.07 | 0.063152279 | 0.103410995 | 0.156234049 | 0.221467429 |
| 0.06 | 0.042316647 | 0.070317337 | 0.107935400 | 0.155694353 |
| 0.05 | 0.026037177 | 0.043885252 | 0.068389974 | 0.100280416 |
| 0.04 | 0.014163324 | 0.024203696 | 0.038267812 | 0.056981889 |
| 0.03 | 0.006343608 | 0.010987084 | 0.017613692 | 0.026609572 |
| 0.02 | 0.001994099 | 0.003499298 | 0.005685135 | 0.008707076 |
| 0.01 | 0.000264270 | 0.000469733 | 0.000773075 | 0.001199574 |

NEFF and NAUS


                    TABLE 1:   CLUSTERING PROBABILITY   P(n;N,p)


                                                                         n = 5

| P | N = 5 | N = 6 | N = 7 | N = 8 |
|------|-------------|-------------|-------------|-------------|
| 0.50 | 0.187500000 | 0.500000000 | 0.781250000 | 0.945312500 |
| 0.49 | 0.175249950 | 0.477509996 | 0.761042395 | 0.935926300 |
| 0.48 | 0.163499213 | 0.455079887 | 0.739818225 | 0.925179482 |
| 0.47 | 0.152246047 | 0.432769149 | 0.717621796 | 0.912955111 |
| 0.46 | 0.141487610 | 0.410636445 | 0.694508588 | 0.899157656 |
| 0.45 | 0.131220000 | 0.388739250 | 0.670544452 | 0.883713904 |
| 0.44 | 0.121438310 | 0.367133499 | 0.645804737 | 0.866573612 |
| 0.43 | 0.112136673 | 0.345873260 | 0.620373373 | 0.847709908 |
| 0.42 | 0.103308307 | 0.325010424 | 0.594341902 | 0.827119436 |
| 0.41 | 0.094945570 | 0.304594430 | 0.567808472 | 0.804822248 |
| 0.40 | 0.087040000 | 0.284672000 | 0.540876800 | 0.780861440 |
| 0.39 | 0.079582370 | 0.265286906 | 0.513655110 | 0.755302534 |
| 0.38 | 0.072562733 | 0.246479756 | 0.486255050 | 0.728232627 |
| 0.37 | 0.065970467 | 0.228287803 | 0.458790602 | 0.699759292 |
| 0.36 | 0.059794330 | 0.210744779 | 0.431376987 | 0.670009260 |
| 0.35 | 0.054022500 | 0.193880750 | 0.404129567 | 0.639126879 |
| 0.34 | 0.048642630 | 0.177721997 | 0.377162769 | 0.607272374 |
| 0.33 | 0.043641893 | 0.162290917 | 0.350589010 | 0.574619913 |
| 0.32 | 0.039007027 | 0.147605946 | 0.324517663 | 0.541355502 |
| 0.31 | 0.034724390 | 0.133681512 | 0.299054035 | 0.507674724 |
| 0.30 | 0.030780000 | 0.120528000 | 0.274298400 | 0.473780340 |
| 0.29 | 0.027159590 | 0.108151752 | 0.250345066 | 0.439879778 |
| 0.28 | 0.023848653 | 0.096555082 | 0.227281497 | 0.406182529 |
| 0.27 | 0.020832487 | 0.085736314 | 0.205187489 | 0.372897471 |
| 0.26 | 0.018096250 | 0.075689849 | 0.184134421 | 0.340230165 |
| 0.25 | 0.015625000 | 0.066406250 | 0.164184570 | 0.308380127 |
| 0.24 | 0.013403750 | 0.057872351 | 0.145390507 | 0.277538122 |
| 0.23 | 0.011417513 | 0.050071390 | 0.127794581 | 0.247883510 |
| 0.22 | 0.009651347 | 0.042983165 | 0.111428495 | 0.219581677 |
| 0.21 | 0.008090410 | 0.036584210 | 0.096312985 | 0.192781577 |
| 0.20 | 0.006720000 | 0.030848000 | 0.082457600 | 0.167613440 |
| 0.19 | 0.005525610 | 0.025745174 | 0.069860602 | 0.144186666 |
| 0.18 | 0.004492973 | 0.021243783 | 0.058508987 | 0.122587962 |
| 0.17 | 0.003608107 | 0.017309560 | 0.048378626 | 0.102879752 |
| 0.16 | 0.002857370 | 0.013906215 | 0.039434553 | 0.085098912 |
| 0.15 | 0.002227500 | 0.010995750 | 0.031631386 | 0.069255867 |
| 0.14 | 0.001705670 | 0.008538801 | 0.024913901 | 0.055334113 |
| 0.13 | 0.001279533 | 0.006495000 | 0.019217758 | 0.043290196 |
| 0.12 | 0.000937267 | 0.004823359 | 0.014470397 | 0.033054208 |
| 0.11 | 0.000667630 | 0.003482684 | 0.010592093 | 0.024530853 |
| 0.10 | 0.000460000 | 0.002432000 | 0.007497200 | 0.017601140 |
| 0.09 | 0.000304430 | 0.001631012 | 0.005095568 | 0.012124746 |
| 0.08 | 0.000191693 | 0.001040581 | 0.003294157 | 0.007943125 |
| 0.07 | 0.000113327 | 0.000623223 | 0.001998846 | 0.004883415 |
| 0.06 | 0.000061690 | 0.000343637 | 0.001116449 | 0.002763196 |
| 0.05 | 0.000030000 | 0.000169250 | 0.000556939 | 0.001396179 |
| 0.04 | 0.000012390 | 0.000070787 | 0.000235891 | 0.000598879 |
| 0.03 | 0.000003953 | 0.000022865 | 0.000077153 | 0.000198339 |
| 0.02 | 0.000000787 | 0.000004610 | 0.000015748 | 0.000040988 |
| 0.01 | 0.000000050 | 0.000000294 | 0.000001017 | 0.000002679 |

TABLE 1:   CLUSTERING PROBABILITY   P(n;N,p)

n = 5

| P | N = 9 | N = 10 | N = 11 | N = 12 |
|------|------|------|------|------|
| 0.50 | 1 | 1 | 1 | 1 |
| 0.49 | 0.999947538 | 0.999999992 | 1.000000000 | 1.000000000 |
| 0.48 | 0.999581208 | 0.999999513 | 1.000000000 | 1.000000000 |
| 0.47 | 0.998591662 | 0.999994731 | 0.999999995 | 1.000000000 |
| 0.46 | 0.996678531 | 0.999971929 | 0.999999935 | 1.000000000 |
| 0.45 | 0.993554783 | 0.999898626 | 0.999999535 | 1.000000000 |
| 0.44 | 0.988950921 | 0.999713932 | 0.999997712 | 0.999999996 |
| 0.43 | 0.982618956 | 0.999319517 | 0.999991282 | 0.999999974 |
| 0.42 | 0.974336131 | 0.998572403 | 0.999972462 | 0.999999870 |
| 0.41 | 0.963908330 | 0.997280532 | 0.999924672 | 0.999999466 |
| 0.40 | 0.951173120 | 0.995201843 | 0.999816172 | 0.999998108 |
| 0.39 | 0.936002407 | 0.992047334 | 0.999591251 | 0.999994061 |
| 0.38 | 0.918304645 | 0.987488300 | 0.999158960 | 0.999983128 |
| 0.37 | 0.898026568 | 0.981167656 | 0.998379876 | 0.999955912 |
| 0.36 | 0.875154422 | 0.972714915 | 0.997052013 | 0.999892715 |
| 0.35 | 0.849714650 | 0.961764078 | 0.994897871 | 0.999754474 |
| 0.34 | 0.821774036 | 0.947973300 | 0.991555666 | 0.999467350 |
| 0.33 | 0.791439255 | 0.931044830 | 0.986579079 | 0.998897475 |
| 0.32 | 0.758855856 | 0.910743282 | 0.979451214 | 0.997812778 |
| 0.31 | 0.724206638 | 0.886910119 | 0.969615580 | 0.995847792 |
| 0.30 | 0.687709440 | 0.859473212 | 0.956516058 | 0.992495077 |
| 0.29 | 0.649614355 | 0.828451796 | 0.939637378 | 0.987118417 |
| 0.28 | 0.610200366 | 0.793957545 | 0.918542552 | 0.978984196 |
| 0.27 | 0.569771442 | 0.756192411 | 0.892904984 | 0.967308265 |
| 0.26 | 0.528652120 | 0.715443748 | 0.862533682 | 0.951313723 |
| 0.25 | 0.487182617 | 0.672077179 | 0.827390671 | 0.930293798 |
| 0.24 | 0.445713507 | 0.626527589 | 0.787600274 | 0.903673648 |
| 0.23 | 0.404600039 | 0.579288566 | 0.743450422 | 0.871065351 |
| 0.22 | 0.364196153 | 0.530900573 | 0.695386474 | 0.832311492 |
| 0.21 | 0.324848272 | 0.481938087 | 0.643998319 | 0.787514166 |
| 0.20 | 0.286888960 | 0.432995942 | 0.590001644 | 0.737047667 |
| 0.19 | 0.250630552 | 0.384675077 | 0.534214369 | 0.681554430 |
| 0.18 | 0.216358852 | 0.337567907 | 0.477529252 | 0.621924898 |
| 0.17 | 0.184327034 | 0.292243541 | 0.420883664 | 0.559262880 |
| 0.16 | 0.154749878 | 0.249233083 | 0.365227521 | 0.494838624 |
| 0.15 | 0.127798484 | 0.209015289 | 0.311490301 | 0.430032260 |
| 0.14 | 0.103595632 | 0.172002878 | 0.260548055 | 0.366270526 |
| 0.13 | 0.082211950 | 0.138529867 | 0.213191316 | 0.304959788 |
| 0.12 | 0.063663092 | 0.108840309 | 0.170094828 | 0.247418320 |
| 0.11 | 0.047908115 | 0.083078915 | 0.131790034 | 0.194810773 |
| 0.10 | 0.034849280 | 0.061284088 | 0.098641420 | 0.148087659 |
| 0.09 | 0.024333506 | 0.043383958 | 0.070827875 | 0.107932667 |
| 0.08 | 0.016155720 | 0.029196123 | 0.048330470 | 0.074720737 |
| 0.07 | 0.010064371 | 0.018431838 | 0.030928267 | 0.048490027 |
| 0.06 | 0.005769383 | 0.010705521 | 0.018204056 | 0.028931423 |
| 0.05 | 0.002952851 | 0.005550519 | 0.009562260 | 0.015399860 |
| 0.04 | 0.001282777 | 0.002442181 | 0.004261612 | 0.006952717 |
| 0.03 | 0.000430193 | 0.000829371 | 0.001465633 | 0.002421685 |
| 0.02 | 0.000090010 | 0.000175695 | 0.000314365 | 0.000525945 |
| 0.01 | 0.000005955 | 0.000011767 | 0.000021314 | 0.000036100 |

TABLE 1:   CLUSTERING PROBABILITY   $P(n;N,p)$

n = 6

| P | N = 6 | N = 7 | N = 8 | N = 9 |
|---|---|---|---|---|
| 0.50 | 0.109375000 | 0.343750000 | 0.617187500 | 0.835937500 |
| 0.49 | 0.100278713 | 0.322146073 | 0.590692750 | 0.815463596 |
| 0.48 | 0.091729428 | 0.301117137 | 0.563781298 | 0.793429909 |
| 0.47 | 0.083710928 | 0.280709115 | 0.536565717 | 0.769874945 |
| 0.46 | 0.076206301 | 0.260963329 | 0.509161897 | 0.744861471 |
| 0.45 | 0.069198047 | 0.241916372 | 0.481687604 | 0.718475017 |
| 0.44 | 0.062668165 | 0.223600013 | 0.454261079 | 0.690822104 |
| 0.43 | 0.056598251 | 0.206041153 | 0.426999694 | 0.662028256 |
| 0.42 | 0.050969580 | 0.189261815 | 0.400018677 | 0.632235810 |
| 0.41 | 0.045763199 | 0.173279168 | 0.373429908 | 0.601601560 |
| 0.40 | 0.040960000 | 0.158105600 | 0.347340800 | 0.570294272 |
| 0.39 | 0.036540801 | 0.143748803 | 0.321853266 | 0.538492093 |
| 0.38 | 0.032486419 | 0.130211906 | 0.297062787 | 0.506379900 |
| 0.37 | 0.028777742 | 0.117493627 | 0.273057572 | 0.474146615 |
| 0.36 | 0.025395794 | 0.105588455 | 0.249917831 | 0.441982523 |
| 0.35 | 0.022321797 | 0.094486853 | 0.227715154 | 0.410076619 |
| 0.34 | 0.019537232 | 0.084175484 | 0.206511999 | 0.378614034 |
| 0.33 | 0.017023896 | 0.074637456 | 0.186361293 | 0.347773550 |
| 0.32 | 0.014763950 | 0.065852586 | 0.167306156 | 0.317725257 |
| 0.31 | 0.012739972 | 0.057797675 | 0.149379724 | 0.288628356 |
| 0.30 | 0.010935000 | 0.050446800 | 0.132605100 | 0.260629164 |
| 0.29 | 0.009332573 | 0.043771612 | 0.116995409 | 0.233859325 |
| 0.28 | 0.007916769 | 0.037741649 | 0.102553967 | 0.208434257 |
| 0.27 | 0.006672242 | 0.032324644 | 0.089274552 | 0.184451860 |
| 0.26 | 0.005584247 | 0.027486850 | 0.077141782 | 0.161991499 |
| 0.25 | 0.004638672 | 0.023193359 | 0.066131592 | 0.141113281 |
| 0.24 | 0.003822060 | 0.019408418 | 0.056211794 | 0.121857629 |
| 0.23 | 0.003121626 | 0.016095749 | 0.047342733 | 0.104245174 |
| 0.22 | 0.002525280 | 0.013218860 | 0.039478017 | 0.088276962 |
| 0.21 | 0.002021630 | 0.010741349 | 0.032565310 | 0.073934977 |
| 0.20 | 0.001600000 | 0.008627200 | 0.026547200 | 0.061182976 |
| 0.19 | 0.001250430 | 0.006841065 | 0.021362101 | 0.049967634 |
| 0.18 | 0.000963680 | 0.005348536 | 0.016945207 | 0.040219975 |
| 0.17 | 0.000731226 | 0.004116393 | 0.013229471 | 0.031857081 |
| 0.16 | 0.000545260 | 0.003112845 | 0.010146592 | 0.024784047 |
| 0.15 | 0.000398672 | 0.002307741 | 0.007628017 | 0.018896158 |
| 0.14 | 0.000285047 | 0.001672762 | 0.005605920 | 0.014081257 |
| 0.13 | 0.000198642 | 0.001181588 | 0.004014161 | 0.010222241 |
| 0.12 | 0.000134369 | 0.000810038 | 0.002789196 | 0.007199672 |
| 0.11 | 0.000087773 | 0.000536177 | 0.001870931 | 0.004894417 |
| 0.10 | 0.000055000 | 0.000340400 | 0.001203500 | 0.003190268 |
| 0.09 | 0.000032772 | 0.000205469 | 0.000735943 | 0.001976481 |
| 0.08 | 0.000018350 | 0.000116528 | 0.000422769 | 0.001150141 |
| 0.07 | 0.000009496 | 0.000061069 | 0.000224391 | 0.000618279 |
| 0.06 | 0.000004432 | 0.000028863 | 0.000107393 | 0.000299654 |
| 0.05 | 0.000001797 | 0.000011847 | 0.000044629 | 0.000126086 |
| 0.04 | 0.000000594 | 0.000003964 | 0.000015117 | 0.000043237 |
| 0.03 | 0.000000142 | 0.000000960 | 0.000003707 | 0.000010732 |
| 0.02 | 0.000000019 | 0.000000129 | 0.000000504 | 0.000001478 |
| 0.01 | 0.000000001 | 0.000000004 | 0.000000016 | 0.000000048 |

TABLE 1:   CLUSTERING PROBABILITY  P(n;N,p)

n = 6

| P | N =10 | N =11 | N =12 | N =13 |
|------|------------|------------|------------|------------|
| 0.50 | 0.958984375 | 1 | 1 | 1 |
| 0.49 | 0.950004870 | 0.999927882 | 0.999999979 | 1.000000000 |
| 0.48 | 0.939330858 | 0.999424714 | 0.999998725 | 0.999999999 |
| 0.47 | 0.926775040 | 0.998067713 | 0.999986547 | 0.999999971 |
| 0.46 | 0.912185971 | 0.995450478 | 0.999930149 | 0.999999641 |
| 0.45 | 0.895450164 | 0.991190817 | 0.999754284 | 0.999997558 |
| 0.44 | 0.876493546 | 0.984938125 | 0.999324936 | 0.999988531 |
| 0.43 | 0.855282276 | 0.976380201 | 0.998437431 | 0.999958336 |
| 0.42 | 0.831822890 | 0.965249359 | 0.996811696 | 0.999874702 |
| 0.41 | 0.806161805 | 0.951327725 | 0.994095729 | 0.999674180 |
| 0.40 | 0.778384179 | 0.934451610 | 0.989877314 | 0.999245348 |
| 0.39 | 0.748612155 | 0.914514884 | 0.983703132 | 0.998410018 |
| 0.38 | 0.717002515 | 0.891471263 | 0.975103661 | 0.996905279 |
| 0.37 | 0.683743797 | 0.865335457 | 0.963621677 | 0.994370299 |
| 0.36 | 0.649052893 | 0.836183153 | 0.948841805 | 0.990342367 |
| 0.35 | 0.613171216 | 0.804149803 | 0.930418440 | 0.984266493 |
| 0.34 | 0.576360469 | 0.769428229 | 0.908099470 | 0.975521537 |
| 0.33 | 0.538898103 | 0.732265072 | 0.881743756 | 0.963462999 |
| 0.32 | 0.501072516 | 0.692956129 | 0.851331112 | 0.947477830 |
| 0.31 | 0.463178095 | 0.651840653 | 0.816964776 | 0.927040938 |
| 0.30 | 0.425510156 | 0.609294701 | 0.778867529 | 0.901762938 |
| 0.29 | 0.388359886 | 0.565723657 | 0.737372968 | 0.871424732 |
| 0.28 | 0.352009369 | 0.521554050 | 0.692913191 | 0.835998046 |
| 0.27 | 0.316726769 | 0.477224826 | 0.646003914 | 0.795652433 |
| 0.26 | 0.282761780 | 0.433178249 | 0.597227748 | 0.750750132 |
| 0.25 | 0.250341415 | 0.389850616 | 0.547216296 | 0.701830626 |
| 0.24 | 0.219666215 | 0.347662984 | 0.496631539 | 0.649587003 |
| 0.23 | 0.190906964 | 0.307012116 | 0.446146962 | 0.594836212 |
| 0.22 | 0.164201990 | 0.268261885 | 0.396428807 | 0.538485178 |
| 0.21 | 0.139655098 | 0.231735324 | 0.348117832 | 0.481494570 |
| 0.20 | 0.117334221 | 0.197707571 | 0.301811958 | 0.424841791 |
| 0.19 | 0.097270810 | 0.166399913 | 0.258050177 | 0.369484547 |
| 0.18 | 0.079460030 | 0.137975126 | 0.217298142 | 0.316326191 |
| 0.17 | 0.063861751 | 0.112534313 | 0.179935839 | 0.266183871 |
| 0.16 | 0.050402377 | 0.090115382 | 0.146247755 | 0.219760437 |
| 0.15 | 0.038977473 | 0.070693317 | 0.116415964 | 0.177620959 |
| 0.14 | 0.029455184 | 0.054182317 | 0.090516496 | 0.140174673 |
| 0.13 | 0.021680379 | 0.040439868 | 0.068519340 | 0.107663114 |
| 0.12 | 0.015479454 | 0.029272710 | 0.050292325 | 0.080155139 |
| 0.11 | 0.010665657 | 0.020444663 | 0.035609052 | 0.057549419 |
| 0.10 | 0.007045074 | 0.013686118 | 0.024160872 | 0.039584838 |
| 0.09 | 0.004422292 | 0.008704991 | 0.015572779 | 0.025858997 |
| 0.08 | 0.002606931 | 0.005198771 | 0.009422815 | 0.015854645 |
| 0.07 | 0.001419437 | 0.002867232 | 0.005264348 | 0.008973425 |
| 0.06 | 0.000696684 | 0.001425231 | 0.002650275 | 0.004575675 |
| 0.05 | 0.000296822 | 0.000614862 | 0.001157793 | 0.002024247 |
| 0.04 | 0.000103046 | 0.000216110 | 0.000412005 | 0.000729331 |
| 0.03 | 0.000025891 | 0.000054965 | 0.000106076 | 0.000190086 |
| 0.02 | 0.000003608 | 0.000007752 | 0.000015141 | 0.000027462 |
| 0.01 | 0.000000119 | 0.000000259 | 0.000000512 | 0.000000940 |

NEFF and NAUS

TABLE 1:  CLUSTERING PROBABILITY  P(n;N,p)

n = 7

| P | N = 7 | N = 8 | N = 9 | N =10 |
|---|---|---|---|---|
| 0.50 | 0.062500000 | 0.226562500 | 0.460937500 | 0.695312500 |
| 0.49 | 0.056195626 | 0.208253239 | 0.432945830 | 0.666648262 |
| 0.48 | 0.050390033 | 0.190836349 | 0.405302590 | 0.637004282 |
| 0.47 | 0.045057120 | 0.174323626 | 0.378126337 | 0.606540034 |
| 0.46 | 0.040171019 | 0.158721001 | 0.351529158 | 0.575427610 |
| 0.45 | 0.035706192 | 0.144028815 | 0.325615561 | 0.543848487 |
| 0.44 | 0.031637528 | 0.130242126 | 0.300481519 | 0.511990310 |
| 0.43 | 0.027940425 | 0.117351048 | 0.276213678 | 0.480043725 |
| 0.42 | 0.024590862 | 0.105341106 | 0.252888738 | 0.448199301 |
| 0.41 | 0.021565473 | 0.094193617 | 0.230572986 | 0.416644595 |
| 0.40 | 0.018841600 | 0.083886080 | 0.209321984 | 0.385561395 |
| 0.39 | 0.016397346 | 0.074392577 | 0.189180417 | 0.355123163 |
| 0.38 | 0.014211620 | 0.065684179 | 0.170182075 | 0.325492721 |
| 0.37 | 0.012264172 | 0.057729357 | 0.152349981 | 0.296820203 |
| 0.36 | 0.010535627 | 0.050494384 | 0.135696640 | 0.269241280 |
| 0.35 | 0.009007502 | 0.043943740 | 0.120224410 | 0.242875699 |
| 0.34 | 0.007662230 | 0.038040500 | 0.105925978 | 0.217826121 |
| 0.33 | 0.006483169 | 0.032746720 | 0.092784938 | 0.194177285 |
| 0.32 | 0.005454608 | 0.028023803 | 0.080776446 | 0.171995482 |
| 0.31 | 0.004561769 | 0.023832846 | 0.069867960 | 0.151328361 |
| 0.30 | 0.003790800 | 0.020134980 | 0.060020028 | 0.132205025 |
| 0.29 | 0.003128771 | 0.016891674 | 0.051187131 | 0.114636442 |
| 0.28 | 0.002563656 | 0.014065029 | 0.043318561 | 0.098616131 |
| 0.27 | 0.002084322 | 0.011618043 | 0.036359323 | 0.084121107 |
| 0.26 | 0.001680502 | 0.009514853 | 0.030251040 | 0.071113068 |
| 0.25 | 0.001342773 | 0.007720947 | 0.024932861 | 0.059539795 |
| 0.24 | 0.001062533 | 0.006203355 | 0.020342358 | 0.049336730 |
| 0.23 | 0.000831962 | 0.004930809 | 0.016416385 | 0.040428703 |
| 0.22 | 0.000643998 | 0.003873874 | 0.013091910 | 0.032731770 |
| 0.21 | 0.000492298 | 0.003005056 | 0.010306794 | 0.026155132 |
| 0.20 | 0.000371200 | 0.002298880 | 0.008000512 | 0.020603085 |
| 0.19 | 0.000275689 | 0.001731938 | 0.006114810 | 0.015976976 |
| 0.18 | 0.000201352 | 0.001282914 | 0.004594285 | 0.012177112 |
| 0.17 | 0.000144343 | 0.000932584 | 0.003386883 | 0.009104595 |
| 0.16 | 0.000101334 | 0.000663787 | 0.002444313 | 0.006663037 |
| 0.15 | 0.000069483 | 0.000461377 | 0.001722365 | 0.004760122 |
| 0.14 | 0.000046382 | 0.000312150 | 0.001181139 | 0.003308990 |
| 0.13 | 0.000030023 | 0.000204754 | 0.000785175 | 0.002229400 |
| 0.12 | 0.000018752 | 0.000129577 | 0.000503488 | 0.001448657 |
| 0.11 | 0.000011232 | 0.000078625 | 0.000309513 | 0.000902274 |
| 0.10 | 0.000006400 | 0.000045380 | 0.000180556 | 0.000534375 |
| 0.09 | 0.000003433 | 0.000024653 | 0.000099566 | 0.000297803 |
| 0.08 | 0.000001709 | 0.000012429 | 0.000050829 | 0.000153961 |
| 0.07 | 0.000000774 | 0.000005699 | 0.000023601 | 0.000072382 |
| 0.06 | 0.000000310 | 0.000002309 | 0.000009679 | 0.000030054 |
| 0.05 | 0.000000105 | 0.000000790 | 0.000003351 | 0.000010533 |
| 0.04 | 0.000000028 | 0.000000211 | 0.000000908 | 0.000002888 |
| 0.03 | 0.000000005 | 0.000000038 | 0.000000167 | 0.000000537 |
| 0.02 | 0.000000000 | 0.000000003 | 0.000000015 | 0.000000049 |
| 0.01 | 0.000000000 | 0.000000000 | 0.000000000 | 0.000000001 |

TABLE 1:   CLUSTERING PROBABILITY   P(n;N,p)

n = 7

| P | N = 11 | N = 12 | N = 13 | N = 14 |
|------|------|------|------|------|
| 0.50 | 0.871093750 | 0.967773438 | 1 | 1 |
| 0.49 | 0.850720680 | 0.959133397 | 0.999906269 | 0.999999952 |
| 0.48 | 0.828264535 | 0.948478001 | 0.999252845 | 0.999997237 |
| 0.47 | 0.803744690 | 0.935538696 | 0.997493443 | 0.999971566 |
| 0.46 | 0.777222654 | 0.920100849 | 0.994108266 | 0.999856034 |
| 0.45 | 0.748799798 | 0.902007638 | 0.988616457 | 0.999506331 |
| 0.44 | 0.718614433 | 0.881162658 | 0.980587611 | 0.998678272 |
| 0.43 | 0.686838285 | 0.857531182 | 0.969652078 | 0.997019270 |
| 0.42 | 0.653672465 | 0.831140105 | 0.955509794 | 0.994075654 |
| 0.41 | 0.619343021 | 0.802076556 | 0.937937432 | 0.989314863 |
| 0.40 | 0.584096154 | 0.770485256 | 0.916793695 | 0.982159473 |
| 0.39 | 0.548193203 | 0.736564066 | 0.892022604 | 0.972028658 |
| 0.38 | 0.511905497 | 0.700562046 | 0.863654705 | 0.958382162 |
| 0.37 | 0.475509168 | 0.662767529 | 0.831806145 | 0.940761876 |
| 0.36 | 0.439280017 | 0.623507348 | 0.796675640 | 0.918826855 |
| 0.35 | 0.403488539 | 0.583136387 | 0.758539391 | 0.892378708 |
| 0.34 | 0.368395186 | 0.542030203 | 0.717744074 | 0.861375811 |
| 0.33 | 0.334245952 | 0.500576721 | 0.674698060 | 0.825936334 |
| 0.32 | 0.301268366 | 0.459167780 | 0.629861095 | 0.786331505 |
| 0.31 | 0.269667953 | 0.418190723 | 0.583732682 | 0.742971380 |
| 0.30 | 0.239625216 | 0.378020229 | 0.536839482 | 0.696385524 |
| 0.29 | 0.211293213 | 0.339010570 | 0.489722041 | 0.647200478 |
| 0.28 | 0.184795733 | 0.301488473 | 0.442921222 | 0.596115399 |
| 0.27 | 0.160226127 | 0.265746756 | 0.396964699 | 0.543876876 |
| 0.26 | 0.137646784 | 0.232038895 | 0.352353889 | 0.491253685 |
| 0.25 | 0.117089272 | 0.200574636 | 0.309551716 | 0.439012110 |
| 0.24 | 0.098555106 | 0.171516784 | 0.268971556 | 0.387892407 |
| 0.23 | 0.082017149 | 0.144979214 | 0.230967704 | 0.338586908 |
| 0.22 | 0.067421572 | 0.121026194 | 0.195827676 | 0.291720314 |
| 0.21 | 0.054690345 | 0.099672999 | 0.163766597 | 0.247832653 |
| 0.20 | 0.043724186 | 0.080887816 | 0.134923873 | 0.207365408 |
| 0.19 | 0.034405881 | 0.064594885 | 0.109362293 | 0.170651239 |
| 0.18 | 0.026603914 | 0.050678771 | 0.087069600 | 0.137907701 |
| 0.17 | 0.020176280 | 0.038989648 | 0.067962521 | 0.109235205 |
| 0.16 | 0.014974406 | 0.029349432 | 0.051893127 | 0.084619401 |
| 0.15 | 0.010847054 | 0.021558554 | 0.038657318 | 0.063937966 |
| 0.14 | 0.007644103 | 0.015403154 | 0.028005120 | 0.046971606 |
| 0.13 | 0.005220104 | 0.010662445 | 0.019652408 | 0.033418886 |
| 0.12 | 0.003437484 | 0.007115967 | 0.013293563 | 0.022914285 |
| 0.11 | 0.002169321 | 0.004550459 | 0.008614506 | 0.015048645 |
| 0.10 | 0.001301575 | 0.002766069 | 0.005305506 | 0.009391004 |
| 0.09 | 0.000734715 | 0.001581617 | 0.003073100 | 0.005510595 |
| 0.08 | 0.000384679 | 0.000838681 | 0.001650472 | 0.002997695 |
| 0.07 | 0.000183125 | 0.000404288 | 0.000805684 | 0.001481919 |
| 0.06 | 0.000076980 | 0.000172067 | 0.000347187 | 0.000646589 |
| 0.05 | 0.000027310 | 0.000061794 | 0.000126222 | 0.000237976 |
| 0.04 | 0.000007579 | 0.000017357 | 0.000035886 | 0.000068483 |
| 0.03 | 0.000001427 | 0.000003308 | 0.000006920 | 0.000013366 |
| 0.02 | 0.000000132 | 0.000000311 | 0.000000658 | 0.000001285 |
| 0.01 | 0.000000002 | 0.000000005 | 0.000000011 | 0.000000022 |

NEFF and NAUS

## TABLE 1:   CLUSTERING PROBABILITY   P(n;N,p)

n = 8

| P | N = 8 | N = 9 | N = 10 | N = 11 |
|------|-------------|-------------|-------------|-------------|
| 0.50 | 0.03515€250 | 0.144531250 | 0.329101562 | 0.548828125 |
| 0.49 | 0.030994794 | 0.130269018 | 0.303189268 | 0.51€506244 |
| 0.48 | 0.027239971 | 0.117000372 | 0.278244469 | 0.484128153 |
| 0.47 | 0.02386194S | 0.104701774 | 0.25434441€ | 0.451894071 |
| 0.46 | 0.020832084 | 0.093345169 | 0.231553722 | 0.419999596 |
| 0.45 | 0.01€122968 | 0.082898568 | 0.209924229 | 0.38€632579 |
| 0.44 | 0.01570€468 | 0.073326619 | 0.189495109 | 0.357970316 |
| 0.43 | 0.013563749 | 0.064591169 | 0.170293107 | 0.32€177105 |
| 0.42 | 0.011665290 | 0.056651814 | 0.152332959 | 0.299402185 |
| 0.41 | 0.009990894 | 0.049466417 | 0.135617951 | 0.271778057 |
| 0.40 | 0.008519680 | 0.042991616 | 0.120140595 | 0.245419213 |
| 0.39 | 0.007232074 | 0.037183290 | 0.105883414 | 0.220421241 |
| 0.38 | 0.00€109792 | 0.031997004 | 0.092819806 | 0.19€860328 |
| 0.37 | 0.00€135815 | 0.02738€416 | 0.080914978 | 0.174793126 |
| 0.36 | 0.004294356 | 0.023313652 | 0.070126924 | 0.154256966 |
| 0.35 | 0.00357€831 | 0.019729645 | 0.060407442 | 0.135270396 |
| 0.34 | 0.002951812 | 0.016594437 | 0.05170315C | 0.117834022 |
| 0.33 | 0.002424989 | 0.013867445 | 0.043956512 | 0.101931600 |
| 0.32 | 0.001979121 | 0.011509688 | 0.037106846 | 0.087531364 |
| 0.31 | 0.001603985 | 0.009483983 | 0.031091292 | 0.074587544 |
| 0.30 | 0.001290330 | 0.007755102 | 0.025845747 | 0.063042025 |
| 0.29 | 0.001029818 | 0.006289891 | 0.021305745 | 0.052826131 |
| 0.28 | 0.00081497S | 0.005057365 | 0.017407268 | 0.04386247C |
| 0.27 | 0.00063912€ | 0.004028763 | 0.014087490 | 0.03€066820 |
| 0.26 | 0.00049€366 | 0.003177577 | 0.011285444 | 0.029349986 |
| 0.25 | 0.000381470 | 0.002479553 | 0.008942604 | 0.023619652 |
| 0.24 | 0.000289865 | 0.001912669 | 0.00700337S | 0.018782117 |
| 0.23 | 0.000217568 | 0.001457081 | 0.005415523 | 0.014743952 |
| 0.22 | 0.000161136 | 0.001095063 | 0.004130455 | 0.011413503 |
| 0.21 | 0.000117611 | 0.000810911 | 0.003103484 | 0.008702244 |
| 0.20 | 0.00008448C | 0.000590848 | 0.002293965 | 0.00€525952 |
| 0.19 | 0.000059621 | 0.000422903 | 0.00166535I | 0.004805683 |
| 0.18 | 0.000041264 | 0.000296790 | 0.00118519I | 0.003468552 |
| 0.17 | 0.000027944 | 0.000203768 | 0.000825038 | 0.002448302 |
| 0.16 | 0.00001€468 | 0.000136511 | 0.00056030S | 0.001685666 |
| 0.15 | 0.000011875 | 0.000088958 | 0.000370077 | 0.001128529 |
| 0.14 | 0.00000740C | 0.000056175 | 0.000236825 | 0.000731899 |
| 0.13 | 0.000004449 | 0.000034217 | 0.000146161 | 0.00C457703 |
| 0.12 | 0.000002566 | 0.00001998S | 0.000086500 | 0.000274426 |
| 0.11 | 0.000001409 | 0.000011118 | 0.000048735 | 0.000156616 |
| 0.10 | 0.00C000730 | 0.000005834 | 0.00002589€ | 0.000084291 |
| 0.09 | 0.000000353 | 0.000002853 | 0.000012822 | 0.000042260 |
| 0.08 | 0.000000156 | 0.000001278 | 0.000005817 | 0.000019413 |
| 0.07 | 0.000000062 | 0.000000513 | 0.000002363 | 0.000007983 |
| 0.06 | 0.00C000021 | 0.000000178 | 0.000000831 | 0.000002840 |
| 0.05 | 0.000000006 | 0.000000051 | 0.000000240 | 0.000000829 |
| 0.04 | 0.000000001 | 0.000000011 | 0.000000052 | 0.00000018C |
| 0.03 | 0.000000000 | 0.000000001 | 0.000000007 | 0.000000025 |
| 0.02 | 0.000000000 | 0.000000000 | 0.000000000 | 0.000000002 |
| 0.01 | 0.000000000 | 0.000000000 | 0.000000000 | 0.000000000 |

## TABLE 1: CLUSTERING PROBABILITY $P(n;N,p)$

$n = 8$

| P | N =12 | N =13 | N =14 | N =15 |
|---|---|---|---|---|
| 0.50 | 0.750244141 | 0.895263672 | 0.973815918 | 1 |
| 0.49 | 0.720396136 | 0.875138633 | 0.965457494 | 0.999882864 |
| 0.48 | 0.688975561 | 0.852428103 | 0.954771332 | 0.999066952 |
| 0.47 | 0.656179525 | 0.827120865 | 0.941398320 | 0.996873558 |
| 0.46 | 0.622232352 | 0.799270303 | 0.925055005 | 0.992663509 |
| 0.45 | 0.587379917 | 0.768991388 | 0.905539913 | 0.985855456 |
| 0.44 | 0.551883813 | 0.736456305 | 0.882737589 | 0.975942424 |
| 0.43 | 0.516015467 | 0.701888881 | 0.856620275 | 0.962506140 |
| 0.42 | 0.480050340 | 0.665557975 | 0.827247227 | 0.945228708 |
| 0.41 | 0.444262322 | 0.627770030 | 0.794761741 | 0.923901280 |
| 0.40 | 0.408918426 | 0.588860988 | 0.759385995 | 0.898429463 |
| 0.39 | 0.374273884 | 0.549187794 | 0.721413902 | 0.868835313 |
| 0.38 | 0.340567718 | 0.509119695 | 0.681202185 | 0.835255854 |
| 0.37 | 0.308018865 | 0.469029551 | 0.639159976 | 0.797938188 |
| 0.36 | 0.276822904 | 0.429285386 | 0.595737236 | 0.757231339 |
| 0.35 | 0.247149431 | 0.390242353 | 0.551412340 | 0.713575127 |
| 0.34 | 0.219140110 | 0.352235320 | 0.506679202 | 0.667486401 |
| 0.33 | 0.192907394 | 0.315572223 | 0.462034302 | 0.619543106 |
| 0.32 | 0.168533945 | 0.280528340 | 0.417963990 | 0.570366683 |
| 0.31 | 0.146072689 | 0.247341593 | 0.374932428 | 0.520603405 |
| 0.30 | 0.125547523 | 0.216208965 | 0.333370509 | 0.470905232 |
| 0.29 | 0.106954590 | 0.187284091 | 0.293666053 | 0.421910865 |
| 0.28 | 0.090264083 | 0.160676042 | 0.256155555 | 0.374227584 |
| 0.27 | 0.075422520 | 0.136449290 | 0.221117694 | 0.328414525 |
| 0.26 | 0.062355394 | 0.114624823 | 0.188768766 | 0.284967925 |
| 0.25 | 0.050970137 | 0.095182329 | 0.159260135 | 0.244308859 |
| 0.24 | 0.041159304 | 0.078063357 | 0.132677736 | 0.206773851 |
| 0.23 | 0.032803886 | 0.063175319 | 0.109043596 | 0.172608686 |
| 0.22 | 0.025776668 | 0.050396193 | 0.088319264 | 0.141965598 |
| 0.21 | 0.019945544 | 0.039579764 | 0.070410988 | 0.114903879 |
| 0.20 | 0.015176704 | 0.030561201 | 0.055176397 | 0.091393843 |
| 0.19 | 0.011337615 | 0.023162817 | 0.042432422 | 0.071323904 |
| 0.18 | 0.008299724 | 0.017199780 | 0.031964102 | 0.054510425 |
| 0.17 | 0.005940826 | 0.012485629 | 0.023533941 | 0.040709841 |
| 0.16 | 0.004147049 | 0.008837381 | 0.016891406 | 0.029632486 |
| 0.15 | 0.002814410 | 0.006080103 | 0.011782199 | 0.020957428 |
| 0.14 | 0.001849932 | 0.004050781 | 0.007956913 | 0.014347603 |
| 0.13 | 0.001172312 | 0.002601403 | 0.005178724 | 0.009464475 |
| 0.12 | 0.000712140 | 0.001601159 | 0.003229832 | 0.005981505 |
| 0.11 | 0.000411704 | 0.000937745 | 0.001916388 | 0.003595759 |
| 0.10 | 0.000224423 | 0.000517754 | 0.001071766 | 0.002037068 |
| 0.09 | 0.000113944 | 0.000266216 | 0.000558103 | 0.001074341 |
| 0.08 | 0.000052997 | 0.000125376 | 0.000266150 | 0.000518802 |
| 0.07 | 0.000022062 | 0.000052839 | 0.000113562 | 0.000224121 |
| 0.06 | 0.000007945 | 0.000019261 | 0.000041902 | 0.000083713 |
| 0.05 | 0.000002347 | 0.000005760 | 0.000012682 | 0.000025644 |
| 0.04 | 0.000000521 | 0.000001293 | 0.000002882 | 0.000005897 |
| 0.03 | 0.000000074 | 0.000000185 | 0.000000416 | 0.000000862 |
| 0.02 | 0.000000005 | 0.000000012 | 0.000000026 | 0.000000055 |
| 0.01 | 0.000000000 | 0.000000000 | 0.000000000 | 0.000000000 |

NEFF and NAUS

TABLE 1:  CLUSTERING PROBABILITY  P(n;N,p)

n = 9

| P | N = 9 | N = 10 | N = 11 | N = 12 |
|------|-------------|-------------|-------------|-------------|
| 0.50 | 0.019531250 | 0.089843750 | 0.226562500 | 0.414550781 |
| 0.49 | 0.016882329 | 0.079390812 | 0.204654269 | 0.382677678 |
| 0.48 | 0.014540509 | 0.069875598 | 0.184065256 | 0.351631107 |
| 0.47 | 0.012477114 | 0.061249296 | 0.164812654 | 0.321577830 |
| 0.46 | 0.010665330 | 0.053461368 | 0.146900242 | 0.292666183 |
| 0.45 | 0.009080168 | 0.046460191 | 0.130319324 | 0.265024558 |
| 0.44 | 0.007698427 | 0.040193654 | 0.115049782 | 0.238760390 |
| 0.43 | 0.006498639 | 0.034609696 | 0.101061217 | 0.213959642 |
| 0.42 | 0.005461016 | 0.029656801 | 0.088314152 | 0.190686729 |
| 0.41 | 0.004567377 | 0.025284425 | 0.076761266 | 0.168984871 |
| 0.40 | 0.003801088 | 0.021443379 | 0.066348646 | 0.148876820 |
| 0.39 | 0.003146981 | 0.018086152 | 0.057017026 | 0.130365921 |
| 0.38 | 0.002591284 | 0.015167186 | 0.048702994 | 0.113437164 |
| 0.37 | 0.002121538 | 0.012643100 | 0.041340159 | 0.098060219 |
| 0.36 | 0.001726519 | 0.010472863 | 0.034860252 | 0.084188297 |
| 0.35 | 0.001396163 | 0.008617927 | 0.029194157 | 0.071762948 |
| 0.34 | 0.001121479 | 0.007042314 | 0.024272854 | 0.060714596 |
| 0.33 | 0.000894476 | 0.005712663 | 0.020028273 | 0.050964851 |
| 0.32 | 0.000708085 | 0.004598246 | 0.016394053 | 0.042428519 |
| 0.31 | 0.000556085 | 0.003670945 | 0.013306194 | 0.035015578 |
| 0.30 | 0.000433026 | 0.002905211 | 0.010703615 | 0.028633057 |
| 0.29 | 0.000334165 | 0.002277982 | 0.008528600 | 0.023186806 |
| 0.28 | 0.000255394 | 0.001768597 | 0.006727148 | 0.018583122 |
| 0.27 | 0.000193182 | 0.001358678 | 0.005249235 | 0.014730209 |
| 0.26 | 0.000144508 | 0.001032007 | 0.004048966 | 0.011539454 |
| 0.25 | 0.000106812 | 0.000774384 | 0.003084660 | 0.008926511 |
| 0.24 | 0.000077933 | 0.000573484 | 0.002318841 | 0.006812183 |
| 0.23 | 0.000056071 | 0.000418704 | 0.001718175 | 0.005123102 |
| 0.22 | 0.000039730 | 0.000301010 | 0.001253329 | 0.003792211 |
| 0.21 | 0.000027686 | 0.000212781 | 0.000898788 | 0.002759054 |
| 0.20 | 0.000018944 | 0.000147661 | 0.000632627 | 0.001969889 |
| 0.19 | 0.000012704 | 0.000100409 | 0.000436247 | 0.001377637 |
| 0.18 | 0.000008331 | 0.000066760 | 0.000294086 | 0.000941684 |
| 0.17 | 0.000005329 | 0.000043291 | 0.000193320 | 0.000627562 |
| 0.16 | 0.000003316 | 0.000027297 | 0.000123549 | 0.000406530 |
| 0.15 | 0.000001999 | 0.000016677 | 0.000076492 | 0.000255073 |
| 0.14 | 0.000001163 | 0.000009829 | 0.000045680 | 0.000154348 |
| 0.13 | 0.000000649 | 0.000005560 | 0.000026175 | 0.000089600 |
| 0.12 | 0.000000346 | 0.000002998 | 0.000014297 | 0.000049574 |
| 0.11 | 0.000000174 | 0.000001529 | 0.000007383 | 0.000025926 |
| 0.10 | 0.000000082 | 0.000000729 | 0.000003566 | 0.000012681 |
| 0.09 | 0.000000036 | 0.000000321 | 0.000001589 | 0.000005720 |
| 0.08 | 0.000000014 | 0.000000128 | 0.000000641 | 0.000002335 |
| 0.07 | 0.000000005 | 0.000000045 | 0.000000228 | 0.000000840 |
| 0.06 | 0.000000001 | 0.000000013 | 0.000000069 | 0.000000256 |
| 0.05 | 0.000000000 | 0.000000003 | 0.000000016 | 0.000000062 |
| 0.04 | 0.000000000 | 0.000000001 | 0.000000003 | 0.000000011 |
| 0.03 | 0.000000000 | 0.000000000 | 0.000000000 | 0.000000001 |
| 0.02 | 0.000000000 | 0.000000000 | 0.000000000 | 0.000000000 |

TABLE 1:   CLUSTERING PROBABILITY   P(n;N,p)

n = 9

| P | N =13 | N =14 | N =15 | N =16 |
|------|------------|------------|------------|------------|
| 0.50 | 0.615966797 | 0.790527344 | 0.912719727 | 0.978179932 |
| 0.49 | 0.580681014 | 0.760054113 | 0.892896624 | 0.970055565 |
| 0.48 | 0.544832721 | 0.727410026 | 0.870004173 | 0.959296496 |
| 0.47 | 0.508704551 | 0.692820087 | 0.843990688 | 0.945444955 |
| 0.46 | 0.472581018 | 0.656556348 | 0.814896039 | 0.928144302 |
| 0.45 | 0.436742257 | 0.618929198 | 0.782847937 | 0.907148442 |
| 0.44 | 0.401458255 | 0.580278102 | 0.748055815 | 0.882327551 |
| 0.43 | 0.366983688 | 0.540962058 | 0.710802583 | 0.853669980 |
| 0.42 | 0.333553444 | 0.501350038 | 0.671434563 | 0.821280382 |
| 0.41 | 0.301378869 | 0.461811676 | 0.630349998 | 0.785374185 |
| 0.40 | 0.270644806 | 0.422708416 | 0.587986501 | 0.746268687 |
| 0.39 | 0.241507405 | 0.384385355 | 0.544807880 | 0.704371145 |
| 0.38 | 0.214092737 | 0.347163936 | 0.501290754 | 0.660164316 |
| 0.37 | 0.188496166 | 0.311335667 | 0.457911356 | 0.614190017 |
| 0.36 | 0.164782451 | 0.277156943 | 0.415132936 | 0.567031280 |
| 0.35 | 0.142986527 | 0.244845088 | 0.373394098 | 0.519293769 |
| 0.34 | 0.123114894 | 0.214575627 | 0.333098411 | 0.471587087 |
| 0.33 | 0.105147551 | 0.186480810 | 0.294605532 | 0.424506633 |
| 0.32 | 0.089040370 | 0.160649340 | 0.258224080 | 0.378616611 |
| 0.31 | 0.074727844 | 0.137127271 | 0.224206369 | 0.334434740 |
| 0.30 | 0.062126102 | 0.115919944 | 0.192745118 | 0.292419159 |
| 0.29 | 0.051136099 | 0.096994876 | 0.163972115 | 0.252957903 |
| 0.28 | 0.041646895 | 0.080285447 | 0.137958800 | 0.216361242 |
| 0.27 | 0.033538924 | 0.065695221 | 0.114718648 | 0.182857052 |
| 0.26 | 0.026687185 | 0.053102752 | 0.094211174 | 0.152589258 |
| 0.25 | 0.020964265 | 0.042366683 | 0.076347338 | 0.125619287 |
| 0.24 | 0.016243131 | 0.033330972 | 0.060996085 | 0.101930336 |
| 0.23 | 0.012399643 | 0.025830073 | 0.047991729 | 0.081434152 |
| 0.22 | 0.009314739 | 0.019693916 | 0.037141866 | 0.063979941 |
| 0.21 | 0.006876251 | 0.014752541 | 0.028235489 | 0.049364917 |
| 0.20 | 0.004980351 | 0.010840255 | 0.021051003 | 0.037345975 |
| 0.19 | 0.003532603 | 0.007799226 | 0.015363832 | 0.027651925 |
| 0.18 | 0.002448635 | 0.005482423 | 0.010953359 | 0.019995714 |
| 0.17 | 0.001654450 | 0.003755872 | 0.007608975 | 0.014086093 |
| 0.16 | 0.001086399 | 0.002500191 | 0.005135057 | 0.009638234 |
| 0.15 | 0.000690849 | 0.001611441 | 0.003354756 | 0.006382872 |
| 0.14 | 0.000423609 | 0.001001304 | 0.002112550 | 0.004073637 |
| 0.13 | 0.000249142 | 0.000596681 | 0.001275554 | 0.002492377 |
| 0.12 | 0.000139633 | 0.000338769 | 0.000733669 | 0.001452364 |
| 0.11 | 0.000073961 | 0.000181747 | 0.000398684 | 0.000799443 |
| 0.10 | 0.000036634 | 0.000091163 | 0.000202523 | 0.000411284 |
| 0.09 | 0.000016732 | 0.000042158 | 0.000094833 | 0.000195013 |
| 0.08 | 0.000006914 | 0.000017637 | 0.000040166 | 0.000083623 |
| 0.07 | 0.000002517 | 0.000006500 | 0.000014984 | 0.000031578 |
| 0.06 | 0.000000777 | 0.000002030 | 0.000004735 | 0.000010100 |
| 0.05 | 0.000000191 | 0.000000505 | 0.000001193 | 0.000002576 |
| 0.04 | 0.000000034 | 0.000000091 | 0.000000217 | 0.000000473 |
| 0.03 | 0.000000004 | 0.000000010 | 0.000000023 | 0.000000052 |
| 0.02 | 0.000000000 | 0.000000000 | 0.000000001 | 0.000000002 |
| 0.01 | 0.000000000 | 0.000000000 | 0.000000000 | 0.000000000 |

TABLE 1:   CLUSTERING PROBABILITY   P(n;N,p)

n = 9

| P | N =17 | N =18 | N =19 | N =20 |
|---|---|---|---|---|
| 0.50 | 1 | 1 | 1 | 1 |
| 0.49 | 0.999857798 | 0.999999834 | 1.000000000 | 1.000000000 |
| 0.48 | 0.998868100 | 0.999990860 | 0.999999972 | 1.000000000 |
| 0.47 | 0.996211786 | 0.999910418 | 0.999999136 | 0.999999996 |
| 0.46 | 0.991125462 | 0.999568012 | 0.999990721 | 0.999999908 |
| 0.45 | 0.982926823 | 0.998589182 | 0.999944494 | 0.999998942 |
| 0.44 | 0.971037105 | 0.996402165 | 0.999771145 | 0.999992575 |
| 0.43 | 0.954999841 | 0.992269855 | 0.999271249 | 0.999963122 |
| 0.42 | 0.934495255 | 0.985356716 | 0.998081818 | 0.999858078 |
| 0.41 | 0.909349801 | 0.974814604 | 0.995639559 | 0.999551535 |
| 0.40 | 0.879540539 | 0.959871024 | 0.991178842 | 0.998789434 |
| 0.39 | 0.845194213 | 0.939906207 | 0.983774232 | 0.997128869 |
| 0.38 | 0.806581129 | 0.914509961 | 0.972425323 | 0.993891312 |
| 0.37 | 0.764104098 | 0.883514308 | 0.956169602 | 0.988154493 |
| 0.36 | 0.718282909 | 0.847002305 | 0.934201288 | 0.978802710 |
| 0.35 | 0.669734969 | 0.805296651 | 0.905972935 | 0.964639634 |
| 0.34 | 0.619152879 | 0.758933312 | 0.871262340 | 0.944545384 |
| 0.33 | 0.567279847 | 0.708625518 | 0.830197481 | 0.917641008 |
| 0.32 | 0.514883895 | 0.655222656 | 0.783242778 | 0.883419926 |
| 0.31 | 0.462731894 | 0.599667278 | 0.731156611 | 0.841821230 |
| 0.30 | 0.411564408 | 0.542952324 | 0.674931246 | 0.793241804 |
| 0.29 | 0.362072346 | 0.486079984 | 0.615724425 | 0.738497577 |
| 0.28 | 0.314876280 | 0.430023279 | 0.554789263 | 0.678748882 |
| 0.27 | 0.270509213 | 0.375691318 | 0.493406990 | 0.615404965 |
| 0.26 | 0.229403385 | 0.323899133 | 0.432825455 | 0.550020586 |
| 0.25 | 0.191881564 | 0.275342915 | 0.374205324 | 0.484194583 |
| 0.24 | 0.158153016 | 0.230581385 | 0.318575202 | 0.419477282 |
| 0.23 | 0.128314192 | 0.190023820 | 0.266796585 | 0.357291138 |
| 0.22 | 0.102153887 | 0.153925061 | 0.219539197 | 0.298867150 |
| 0.21 | 0.080162487 | 0.122387487 | 0.177267076 | 0.245198285 |
| 0.20 | 0.061544679 | 0.095369640 | 0.140235393 | 0.197010328 |
| 0.19 | 0.046234870 | 0.072700847 | 0.108497645 | 0.154749898 |
| 0.18 | 0.033914421 | 0.054100843 | 0.081922395 | 0.118588838 |
| 0.17 | 0.024229756 | 0.039203158 | 0.060218261 | 0.088443580 |
| 0.16 | 0.016810356 | 0.027580791 | 0.042965376 | 0.064007474 |
| 0.15 | 0.011285703 | 0.018772596 | 0.029651168 | 0.044793410 |
| 0.14 | 0.007300325 | 0.012308766 | 0.019707993 | 0.030183449 |
| 0.13 | 0.004526238 | 0.007733916 | 0.012550070 | 0.019481701 |
| 0.12 | 0.002672277 | 0.004626461 | 0.007607221 | 0.011966476 |
| 0.11 | 0.001490036 | 0.002613291 | 0.004353211 | 0.006937756 |
| 0.10 | 0.000776384 | 0.001379148 | 0.002326994 | 0.003756530 |
| 0.09 | 0.000372777 | 0.000670578 | 0.001145818 | 0.001873294 |
| 0.08 | 0.000161840 | 0.000294766 | 0.000509573 | 0.000844220 |
| 0.07 | 0.000061865 | 0.000114065 | 0.000199780 | 0.000334812 |
| 0.06 | 0.000020027 | 0.000037374 | 0.000066256 | 0.000112393 |
| 0.05 | 0.000005169 | 0.000009761 | 0.000017512 | 0.000030064 |
| 0.04 | 0.000000961 | 0.000001837 | 0.000003334 | 0.000005792 |
| 0.03 | 0.000000107 | 0.000000206 | 0.000000378 | 0.000000665 |
| 0.02 | 0.000000005 | 0.000000009 | 0.000000017 | 0.000000030 |
| 0.01 | 0.000000000 | 0.000000000 | 0.000000000 | 0.000000000 |

TABLE 1a:   CLUSTERING PROBABILITY P(n;N,p)

|  | | | | $n = 3$ |
| --- | --- | --- | --- | --- |
| p | N = 6 | p | N = 6 | N = 7 |
| 0.350 | 0.996355000 | 0.300 | 0.979560000 | 0.999840400 |
| 0.349 | 0.996206748 | 0.299 | 0.978944240 | 0.999819410 |
| 0.348 | 0.996053505 | 0.298 | 0.978313596 | 0.999796361 |
| 0.347 | 0.995895138 | 0.297 | 0.977667827 | 0.999771117 |
| 0.346 | 0.995731510 | 0.296 | 0.977006687 | 0.999743534 |
| 0.345 | 0.995562482 | 0.295 | 0.976329933 | 0.999713466 |
| 0.344 | 0.995387912 | 0.294 | 0.975637319 | 0.999680758 |
| 0.343 | 0.995207657 | 0.293 | 0.974928601 | 0.999645253 |
| 0.342 | 0.995021569 | 0.292 | 0.974203532 | 0.999606785 |
| 0.341 | 0.994829498 | 0.291 | 0.973461866 | 0.999565187 |
| 0.340 | 0.994631291 | 0.290 | 0.972703358 | 0.999520283 |
| 0.339 | 0.994426792 | 0.289 | 0.971927763 | 0.999471893 |
| 0.338 | 0.994215843 | 0.288 | 0.971134832 | 0.999419832 |
| 0.337 | 0.993998282 | 0.287 | 0.970324321 | 0.999363908 |
| 0.336 | 0.993773943 | 0.286 | 0.969495984 | 0.999303924 |
| 0.335 | 0.993542660 | 0.285 | 0.968649574 | 0.999239678 |
| 0.334 | 0.993304261 | 0.284 | 0.967784847 | 0.999170961 |
| 0.333 | 0.993058572 | 0.283 | 0.966901556 | 0.999097560 |
| 0.332 | 0.992805416 | 0.282 | 0.965999458 | 0.999019255 |
| 0.331 | 0.992544613 | 0.281 | 0.965078308 | 0.998935822 |
| 0.330 | 0.992275978 | 0.280 | 0.964137861 | 0.998847028 |
| 0.329 | 0.991999326 | 0.279 | 0.963177875 | 0.998752637 |
| 0.328 | 0.991714467 | 0.278 | 0.962198107 | 0.998652407 |
| 0.327 | 0.991421209 | 0.277 | 0.961198315 | 0.998546090 |
| 0.326 | 0.991119356 | 0.276 | 0.960178258 | 0.998433431 |
| 0.325 | 0.990808711 | 0.275 | 0.959137695 | 0.998314171 |
| 0.324 | 0.990489073 | 0.274 | 0.958076388 | 0.998188043 |
| 0.323 | 0.990160240 | 0.273 | 0.956994097 | 0.998054778 |
| 0.322 | 0.989822005 | 0.272 | 0.955890585 | 0.997914096 |
| 0.321 | 0.989474161 | 0.271 | 0.954765616 | 0.997765715 |
| 0.320 | 0.989116498 | 0.270 | 0.953618955 | 0.997609347 |
| 0.319 | 0.988748803 | 0.269 | 0.952450368 | 0.997444697 |
| 0.318 | 0.988370862 | 0.268 | 0.951259622 | 0.997271464 |
| 0.317 | 0.987982457 | 0.267 | 0.950046486 | 0.997089343 |
| 0.316 | 0.987583370 | 0.266 | 0.948810730 | 0.996898022 |
| 0.315 | 0.987173381 | 0.265 | 0.947552126 | 0.996697183 |
| 0.314 | 0.986752266 | 0.264 | 0.946270447 | 0.996486503 |
| 0.313 | 0.986319802 | 0.263 | 0.944965467 | 0.996265654 |
| 0.312 | 0.985875762 | 0.262 | 0.943636964 | 0.996034300 |
| 0.311 | 0.985419918 | 0.261 | 0.942284715 | 0.995792103 |
| 0.310 | 0.984952041 | 0.260 | 0.940908500 | 0.995538716 |
| 0.309 | 0.984471900 | 0.259 | 0.939508102 | 0.995273789 |
| 0.308 | 0.983979263 | 0.258 | 0.938083304 | 0.994996966 |
| 0.307 | 0.983473895 | 0.257 | 0.936633891 | 0.994707883 |
| 0.306 | 0.982955563 | 0.256 | 0.935159652 | 0.994406176 |
| 0.305 | 0.982424029 | 0.255 | 0.933660377 | 0.994091470 |
| 0.304 | 0.981879057 | 0.254 | 0.932135856 | 0.993763388 |
| 0.303 | 0.981320408 | 0.253 | 0.930585886 | 0.993421547 |
| 0.302 | 0.980747843 | 0.252 | 0.929010262 | 0.993065561 |
| 0.301 | 0.980161120 | 0.251 | 0.927408783 | 0.992695034 |

TABLE 1a:   CLUSTERING PROBABILITY  P(n;N,p)

n = 3

| P | N = 6 | N = 7 | N = 8 | N = 9 |
|---|-------|-------|-------|-------|
| 0.250 | 0.925781250 | 0.992309570 | 0.999786377 | 1 |
| 0.249 | 0.924127468 | 0.991908766 | 0.999764967 | 1.000000000 |
| 0.248 | 0.922447242 | 0.991492214 | 0.999741746 | 1.000000000 |
| 0.247 | 0.920740381 | 0.991059502 | 0.999716518 | 1.000000000 |
| 0.246 | 0.919006697 | 0.990610213 | 0.999689159 | 0.999999999 |
| 0.245 | 0.917246005 | 0.990143925 | 0.999669518 | 0.999999998 |
| 0.244 | 0.915458120 | 0.989660212 | 0.999627433 | 0.999999995 |
| 0.243 | 0.913642863 | 0.989158645 | 0.999592734 | 0.999999990 |
| 0.242 | 0.911800056 | 0.988638788 | 0.999555241 | 0.999999980 |
| 0.241 | 0.909929525 | 0.988100203 | 0.999514765 | 0.999999963 |
| 0.240 | 0.908031099 | 0.987542447 | 0.999471107 | 0.999999935 |
| 0.239 | 0.906104608 | 0.986965074 | 0.999424057 | 0.999999893 |
| 0.238 | 0.904149889 | 0.986367632 | 0.999373394 | 0.999999832 |
| 0.237 | 0.902166778 | 0.985749669 | 0.999318888 | 0.999999743 |
| 0.236 | 0.900155116 | 0.985110726 | 0.999260296 | 0.999999619 |
| 0.235 | 0.898114748 | 0.984450343 | 0.999197364 | 0.999999450 |
| 0.234 | 0.896045522 | 0.983768054 | 0.999129828 | 0.999999225 |
| 0.233 | 0.893947288 | 0.983063392 | 0.999057410 | 0.999998927 |
| 0.232 | 0.891819901 | 0.982335887 | 0.998979822 | 0.999998542 |
| 0.231 | 0.889663219 | 0.981585065 | 0.998896760 | 0.999998049 |
| 0.230 | 0.887477104 | 0.980810450 | 0.998807913 | 0.999997426 |
| 0.229 | 0.885261420 | 0.980011563 | 0.998712954 | 0.999996648 |
| 0.228 | 0.883016036 | 0.979187923 | 0.998611543 | 0.999995685 |
| 0.227 | 0.880740825 | 0.978339045 | 0.998503328 | 0.999994503 |
| 0.226 | 0.878435663 | 0.977464444 | 0.998387943 | 0.999993065 |
| 0.225 | 0.876100430 | 0.976563633 | 0.998265011 | 0.999991330 |
| 0.224 | 0.873735009 | 0.975636121 | 0.998134139 | 0.999989248 |
| 0.223 | 0.871339289 | 0.974681418 | 0.997994922 | 0.999986769 |
| 0.222 | 0.868913162 | 0.973699029 | 0.997846941 | 0.999983833 |
| 0.221 | 0.866456523 | 0.972688462 | 0.997689763 | 0.999980377 |
| 0.220 | 0.863969272 | 0.971649220 | 0.997522940 | 0.999976328 |
| 0.219 | 0.861451314 | 0.970580807 | 0.997346014 | 0.999971608 |
| 0.218 | 0.858902556 | 0.969482726 | 0.997158509 | 0.999966132 |
| 0.217 | 0.856322911 | 0.968354478 | 0.996959937 | 0.999959805 |
| 0.216 | 0.853712295 | 0.967195566 | 0.996749795 | 0.999952526 |
| 0.215 | 0.851070631 | 0.966005490 | 0.996527567 | 0.999944183 |
| 0.214 | 0.848397842 | 0.964783752 | 0.996292723 | 0.999934655 |
| 0.213 | 0.845693861 | 0.963529852 | 0.996044717 | 0.999923812 |
| 0.212 | 0.842958619 | 0.962243292 | 0.995782991 | 0.999911513 |
| 0.211 | 0.840192058 | 0.960923574 | 0.995506973 | 0.999897605 |
| 0.210 | 0.837394120 | 0.959570201 | 0.995216075 | 0.999881924 |
| 0.209 | 0.834564754 | 0.958182675 | 0.994909698 | 0.999864296 |
| 0.208 | 0.831703912 | 0.956760502 | 0.994587225 | 0.999844530 |
| 0.207 | 0.828811552 | 0.955303185 | 0.994248029 | 0.999822426 |
| 0.206 | 0.825887637 | 0.953810234 | 0.993891468 | 0.999797767 |
| 0.205 | 0.822932134 | 0.952281155 | 0.993516886 | 0.999770324 |
| 0.204 | 0.819945015 | 0.950715460 | 0.993123613 | 0.999739851 |
| 0.203 | 0.816926257 | 0.949112661 | 0.992710967 | 0.999706088 |
| 0.202 | 0.813875843 | 0.947472273 | 0.992278251 | 0.999668759 |
| 0.201 | 0.810793760 | 0.945793813 | 0.991824756 | 0.999627569 |

TABLE 1a: CLUSTERING PROBABILITY P(n;N,p)

n = 3

| P | N = 6 | N = 7 | N = 8 | N = 9 |
|---|---|---|---|---|
| 0.200 | 0.807680000 | 0.944076800 | 0.991349760 | 0.999582208 |
| 0.199 | 0.804534560 | 0.942320758 | 0.990852528 | 0.999532347 |
| 0.198 | 0.801357443 | 0.940525212 | 0.990332312 | 0.999477639 |
| 0.197 | 0.798148658 | 0.938689692 | 0.989788352 | 0.999417718 |
| 0.196 | 0.794908216 | 0.936813730 | 0.989219875 | 0.999352197 |
| 0.195 | 0.791636136 | 0.934896862 | 0.988626098 | 0.999280671 |
| 0.194 | 0.788332443 | 0.932938630 | 0.988006223 | 0.999202712 |
| 0.193 | 0.784997165 | 0.930938577 | 0.987359443 | 0.999117872 |
| 0.192 | 0.781630338 | 0.928896252 | 0.986684939 | 0.999025680 |
| 0.191 | 0.778232000 | 0.926811210 | 0.985981882 | 0.998925645 |
| 0.190 | 0.774802199 | 0.924683009 | 0.985249429 | 0.998817251 |
| 0.189 | 0.771340986 | 0.922511212 | 0.984486730 | 0.998699958 |
| 0.188 | 0.767848417 | 0.920295389 | 0.983692923 | 0.998573205 |
| 0.187 | 0.764324555 | 0.918035114 | 0.982867138 | 0.998436403 |
| 0.186 | 0.760769468 | 0.915729967 | 0.982008493 | 0.998288941 |
| 0.185 | 0.757183232 | 0.913379534 | 0.981116099 | 0.998130181 |
| 0.184 | 0.753565926 | 0.910983409 | 0.980189057 | 0.997959462 |
| 0.183 | 0.749917635 | 0.908541191 | 0.979226460 | 0.997776092 |
| 0.182 | 0.746238452 | 0.906052485 | 0.978227393 | 0.997579358 |
| 0.181 | 0.742528475 | 0.903516904 | 0.977190933 | 0.997368518 |
| 0.180 | 0.738787807 | 0.900934069 | 0.976116151 | 0.997142801 |
| 0.179 | 0.735016558 | 0.898303606 | 0.975002109 | 0.996901413 |
| 0.178 | 0.731214844 | 0.895625151 | 0.973847864 | 0.996643528 |
| 0.177 | 0.727382787 | 0.892898347 | 0.972652467 | 0.996368294 |
| 0.176 | 0.723520516 | 0.890122845 | 0.971414964 | 0.996074832 |
| 0.175 | 0.719628164 | 0.887298305 | 0.970134395 | 0.995762234 |
| 0.174 | 0.715705873 | 0.884424396 | 0.968809795 | 0.995429563 |
| 0.173 | 0.711753789 | 0.881500794 | 0.967440197 | 0.995075853 |
| 0.172 | 0.707772067 | 0.878527185 | 0.966024628 | 0.994700112 |
| 0.171 | 0.703760865 | 0.875503267 | 0.964562113 | 0.994301316 |
| 0.170 | 0.699720350 | 0.872428743 | 0.963051674 | 0.993878415 |
| 0.169 | 0.695650694 | 0.869303330 | 0.961492333 | 0.993430329 |
| 0.168 | 0.691552076 | 0.866126752 | 0.959883108 | 0.992955951 |
| 0.167 | 0.687424683 | 0.862898746 | 0.958223016 | 0.992454142 |
| 0.166 | 0.683268707 | 0.859619057 | 0.956511076 | 0.991923739 |
| 0.165 | 0.679084345 | 0.856287443 | 0.954746306 | 0.991363548 |
| 0.164 | 0.674871805 | 0.852903673 | 0.952927723 | 0.990772347 |
| 0.163 | 0.670631298 | 0.849467524 | 0.951054349 | 0.990148887 |
| 0.162 | 0.666363043 | 0.845978790 | 0.949125204 | 0.989491892 |
| 0.161 | 0.662067266 | 0.842437273 | 0.947139314 | 0.988800058 |
| 0.160 | 0.657744200 | 0.838842788 | 0.945095706 | 0.988072053 |
| 0.159 | 0.653394083 | 0.835195163 | 0.942993412 | 0.987306522 |
| 0.158 | 0.649017164 | 0.831494237 | 0.940831469 | 0.986502080 |
| 0.157 | 0.644613694 | 0.827739864 | 0.938608917 | 0.985657319 |
| 0.156 | 0.640183934 | 0.823931909 | 0.936324804 | 0.984770804 |
| 0.155 | 0.635728151 | 0.820070252 | 0.933978183 | 0.983841078 |
| 0.154 | 0.631246620 | 0.816154786 | 0.931568115 | 0.982866658 |
| 0.153 | 0.626739622 | 0.812185417 | 0.929093670 | 0.981846039 |
| 0.152 | 0.622207446 | 0.808162067 | 0.926553924 | 0.980777692 |
| 0.151 | 0.617650388 | 0.804084669 | 0.923947964 | 0.979660068 |

TABLE 1a:   CLUSTERING PROBABILITY   P(n;N,p)

n = 3

| P | N = 6 | N = 7 | N = 8 | N = 9 |
|---|---|---|---|---|
| 0.150 | 0.613068750 | 0.799953173 | 0.921274887 | 0.978491595 |
| 0.149 | 0.608462843 | 0.795767545 | 0.918533800 | 0.977270683 |
| 0.148 | 0.603832983 | 0.791527763 | 0.915723821 | 0.975995720 |
| 0.147 | 0.599179497 | 0.787233822 | 0.912844083 | 0.974665078 |
| 0.146 | 0.594502715 | 0.782885732 | 0.909893728 | 0.973277109 |
| 0.145 | 0.589802978 | 0.778483519 | 0.906871915 | 0.971830151 |
| 0.144 | 0.585080632 | 0.774027225 | 0.903777816 | 0.970322524 |
| 0.143 | 0.580336031 | 0.769516908 | 0.900610618 | 0.968752536 |
| 0.142 | 0.575569538 | 0.764952643 | 0.897369525 | 0.967118479 |
| 0.141 | 0.570781522 | 0.760334521 | 0.894053757 | 0.965418634 |
| 0.140 | 0.565972360 | 0.755662651 | 0.890662551 | 0.963651273 |
| 0.139 | 0.561142435 | 0.750937158 | 0.887195165 | 0.961814654 |
| 0.138 | 0.556292141 | 0.746158185 | 0.883650873 | 0.959907031 |
| 0.137 | 0.551421877 | 0.741325893 | 0.880028972 | 0.957926648 |
| 0.136 | 0.546532050 | 0.736440460 | 0.876328777 | 0.955871744 |
| 0.135 | 0.541623076 | 0.731502083 | 0.872549627 | 0.953740555 |
| 0.134 | 0.536695377 | 0.726510978 | 0.868690881 | 0.951531312 |
| 0.133 | 0.531749385 | 0.721467377 | 0.864751925 | 0.949242246 |
| 0.132 | 0.526785537 | 0.716371534 | 0.860732166 | 0.946871589 |
| 0.131 | 0.521804281 | 0.711223720 | 0.856631036 | 0.944417573 |
| 0.130 | 0.516806070 | 0.706024226 | 0.852447996 | 0.941878435 |
| 0.129 | 0.511791366 | 0.700773363 | 0.848182529 | 0.939252416 |
| 0.128 | 0.506760640 | 0.695471459 | 0.843834150 | 0.936537765 |
| 0.127 | 0.501714369 | 0.690118867 | 0.839402400 | 0.933732739 |
| 0.126 | 0.496653041 | 0.684715955 | 0.834886849 | 0.930835605 |
| 0.125 | 0.491577148 | 0.679263115 | 0.830287099 | 0.927844644 |
| 0.124 | 0.486487194 | 0.673760758 | 0.825602781 | 0.924758149 |
| 0.123 | 0.481383688 | 0.668209316 | 0.820833559 | 0.921574430 |
| 0.122 | 0.476267149 | 0.662609243 | 0.815979129 | 0.918291817 |
| 0.121 | 0.471138104 | 0.656961013 | 0.811039222 | 0.914908656 |
| 0.120 | 0.465997087 | 0.651265123 | 0.806013602 | 0.911423319 |
| 0.119 | 0.460844641 | 0.645522089 | 0.800902069 | 0.907834199 |
| 0.118 | 0.455681318 | 0.639732452 | 0.795704459 | 0.904139718 |
| 0.117 | 0.450507677 | 0.633896774 | 0.790420646 | 0.900338323 |
| 0.116 | 0.445324285 | 0.628015638 | 0.785050540 | 0.896428496 |
| 0.115 | 0.440131720 | 0.622089651 | 0.779594091 | 0.892408747 |
| 0.114 | 0.434930566 | 0.616119444 | 0.774051291 | 0.888277625 |
| 0.113 | 0.429721415 | 0.610105668 | 0.768422168 | 0.884033714 |
| 0.112 | 0.424504869 | 0.604048998 | 0.762706794 | 0.879675638 |
| 0.111 | 0.419281538 | 0.597950134 | 0.756905286 | 0.875202062 |
| 0.110 | 0.414052039 | 0.591809798 | 0.751017799 | 0.870611697 |
| 0.109 | 0.408817001 | 0.585628736 | 0.745044535 | 0.865903300 |
| 0.108 | 0.403577057 | 0.579407718 | 0.738985742 | 0.861075676 |
| 0.107 | 0.398332852 | 0.573147538 | 0.732841711 | 0.856127682 |
| 0.106 | 0.393085039 | 0.566849015 | 0.726612783 | 0.851058229 |
| 0.105 | 0.387834278 | 0.560512991 | 0.720299343 | 0.845866285 |
| 0.104 | 0.382581239 | 0.554140334 | 0.713901828 | 0.840550876 |
| 0.103 | 0.377326600 | 0.547731937 | 0.707420722 | 0.835111089 |
| 0.102 | 0.372071049 | 0.541288716 | 0.700856559 | 0.829546077 |
| 0.101 | 0.366815281 | 0.534811615 | 0.694209926 | 0.823855056 |

TABLE 1a:   CLUSTERING PROBABILITY   P(n;N,p)

_n = 3_

| P | N = 6 | N = 7 | N = 8 | N = 9 |
|---|---|---|---|---|
| C.100 | 0.361560000 | 0.528301600 | 0.68748146C | 0.818037314 |
| 0.099 | 0.35630592C | 0.521759666 | 0.680671851 | 0.812092210 |
| 0.098 | 0.351053763 | 0.515186831 | 0.673781844 | 0.806019176 |
| 0.097 | 0.345804259 | 0.508584141 | 0.666812236 | 0.799817724 |
| 0.C96 | 0.340558148 | 0.501952667 | 0.659763881 | 0.793487440 |
| 0.095 | 0.335316179 | 0.495293504 | 0.652637688 | 0.787027999 |
| 0.C94 | 0.330079110 | 0.488607778 | 0.645434623 | 0.780439154 |
| 0.093 | 0.324847705 | 0.481896638 | 0.638155710 | 0.773720751 |
| 0.C92 | 0.319622741 | 0.475161259 | 0.630802031 | 0.766872723 |
| 0.C91 | 0.314405002 | 0.468402847 | 0.623374727 | 0.759895096 |
| C.090 | 0.309195280 | 0.461622630 | 0.615874999 | 0.752787993 |
| 0.089 | 0.303994378 | 0.454821866 | 0.608304109 | 0.745551634 |
| 0.C88 | 0.298803108 | 0.448001840 | 0.600663381 | 0.738186338 |
| 0.C87 | 0.293622289 | 0.441163863 | 0.592954199 | 0.730692531 |
| 0.C86 | 0.288452750 | 0.434309274 | 0.585178012 | 0.723070743 |
| 0.C85 | 0.283295330 | 0.427439441 | 0.577336332 | 0.715321612 |
| 0.C84 | 0.278150876 | 0.420555757 | 0.569430735 | 0.707445890 |
| 0.C83 | 0.273020245 | 0.413659645 | 0.561462864 | 0.699444439 |
| 0.082 | 0.267904303 | 0.406752556 | 0.553434425 | 0.691318242 |
| 0.081 | 0.262803924 | 0.399835968 | 0.545347192 | 0.683068397 |
| C.080 | 0.257719992 | 0.392911387 | 0.537203005 | 0.674696128 |
| 0.079 | 0.252653401 | 0.385980348 | 0.529003774 | 0.666202780 |
| 0.C78 | 0.247605054 | 0.379044414 | 0.520751475 | 0.657589826 |
| 0.C77 | 0.242575861 | 0.372105177 | 0.512448154 | 0.648858869 |
| 0.076 | 0.237566744 | 0.365164258 | 0.504095927 | 0.640011643 |
| 0.075 | 0.232578633 | 0.358223305 | 0.495696979 | 0.631050016 |
| 0.C74 | 0.227612468 | 0.351283996 | 0.487253568 | 0.621975996 |
| 0.073 | 0.222669197 | 0.344348038 | 0.478768022 | 0.612791725 |
| 0.072 | 0.217749780 | 0.337417166 | 0.470242740 | 0.603499491 |
| C.071 | 0.212855183 | 0.330493145 | 0.461680195 | 0.594101724 |
| 0.C70 | 0.207986384 | 0.323577769 | 0.453082933 | 0.584601001 |
| 0.069 | 0.203144369 | 0.316672860 | 0.444453572 | 0.575000047 |
| 0.068 | 0.198330133 | 0.309780270 | 0.435794805 | 0.565301738 |
| C.C67 | 0.193544683 | 0.302901882 | 0.427109400 | 0.555509102 |
| 0.066 | 0.188789032 | 0.296039605 | 0.418400199 | 0.545625323 |
| 0.065 | 0.184064205 | 0.289195381 | 0.409670120 | 0.535653744 |
| 0.C64 | 0.179371236 | 0.282371177 | 0.400922154 | 0.525597862 |
| 0.063 | 0.174711167 | 0.275568995 | 0.392159373 | 0.515461340 |
| 0.062 | 0.170085051 | 0.268790863 | 0.383384921 | 0.505248003 |
| 0.061 | 0.165493952 | 0.262038838 | 0.374602021 | 0.494961838 |
| 0.060 | 0.160938939 | 0.255315009 | 0.365813972 | 0.484607001 |
| 0.C59 | 0.156421095 | 0.248621494 | 0.357024152 | 0.474187817 |
| 0.C58 | 0.151941510 | 0.241960441 | 0.348236014 | 0.463708780 |
| 0.C57 | 0.147501285 | 0.235334025 | 0.339453092 | 0.453174555 |
| 0.C56 | 0.143101530 | 0.228744455 | 0.330678996 | 0.442589981 |
| 0.055 | 0.138743365 | 0.222193966 | 0.321917415 | 0.431960071 |
| 0.054 | 0.134427919 | 0.215684826 | 0.313172117 | 0.421290013 |
| 0.053 | 0.130156331 | 0.209219329 | 0.304446946 | 0.410585175 |
| 0.052 | 0.125929750 | 0.202799803 | 0.295745829 | 0.399851100 |
| 0.051 | 0.121749333 | 0.196428603 | 0.287072770 | 0.389093511 |

TABLE 1a:   CLUSTERING PROBABILITY   P(n;N,p)

n = 3

| P | N = 6 | N = 7 | N = 8 | N = 9 |
|---|---|---|---|---|
| 0.050 | 0.117616250 | 0.190108114 | 0.278431849 | 0.378318313 |
| 0.049 | 0.113531678 | 0.183840752 | 0.269827231 | 0.367531590 |
| 0.048 | 0.109496803 | 0.177628961 | 0.261263154 | 0.356739611 |
| 0.047 | 0.105512825 | 0.171475216 | 0.252743939 | 0.345948823 |
| 0.046 | 0.101580948 | 0.165382021 | 0.244273984 | 0.335165862 |
| 0.045 | 0.097702391 | 0.159351911 | 0.235857766 | 0.324397542 |
| 0.044 | 0.093878379 | 0.153387449 | 0.227499842 | 0.313650865 |
| 0.043 | 0.090110149 | 0.147491227 | 0.219204847 | 0.302933016 |
| 0.042 | 0.086398948 | 0.141665869 | 0.210977493 | 0.292251364 |
| 0.041 | 0.082746029 | 0.135914027 | 0.202822572 | 0.281613464 |
| 0.040 | 0.079152660 | 0.130238381 | 0.194744953 | 0.271027053 |
| 0.039 | 0.075620117 | 0.124641644 | 0.186749584 | 0.260500054 |
| 0.038 | 0.072149683 | 0.119126555 | 0.178841490 | 0.250040573 |
| 0.037 | 0.068742654 | 0.113695883 | 0.171025770 | 0.239656898 |
| 0.036 | 0.065400337 | 0.108352426 | 0.163307605 | 0.229357502 |
| 0.035 | 0.062124044 | 0.103099013 | 0.155692249 | 0.219151036 |
| 0.034 | 0.058915102 | 0.097938499 | 0.148185033 | 0.209046336 |
| 0.033 | 0.055774844 | 0.092873770 | 0.140791361 | 0.199052415 |
| 0.032 | 0.052704616 | 0.087907740 | 0.133516716 | 0.189178463 |
| 0.031 | 0.049705771 | 0.083043352 | 0.126366653 | 0.179433850 |
| 0.030 | 0.046779675 | 0.078283576 | 0.119346799 | 0.169828119 |
| 0.029 | 0.043927702 | 0.073631412 | 0.112462858 | 0.160370988 |
| 0.028 | 0.041151235 | 0.069089889 | 0.105720603 | 0.151072344 |
| 0.027 | 0.038451669 | 0.064662062 | 0.099125880 | 0.141942247 |
| 0.026 | 0.035830408 | 0.060351015 | 0.092684607 | 0.132990922 |
| 0.025 | 0.033288867 | 0.056159860 | 0.086402770 | 0.124228758 |
| 0.024 | 0.030828469 | 0.052091738 | 0.080286426 | 0.115666308 |
| 0.023 | 0.028450649 | 0.048149816 | 0.074341699 | 0.107314282 |
| 0.022 | 0.026156849 | 0.044337288 | 0.068574782 | 0.099183549 |
| 0.021 | 0.023948526 | 0.040657376 | 0.062991933 | 0.091285128 |
| 0.020 | 0.021827141 | 0.037113330 | 0.057599477 | 0.083630188 |
| 0.019 | 0.019794170 | 0.033708427 | 0.052403802 | 0.076230045 |
| 0.018 | 0.017851096 | 0.030445968 | 0.047411361 | 0.069096156 |
| 0.017 | 0.015999414 | 0.027329284 | 0.042628666 | 0.062240116 |
| 0.016 | 0.014240627 | 0.024361730 | 0.038062293 | 0.055673652 |
| 0.015 | 0.012576250 | 0.021546688 | 0.033718877 | 0.049408622 |
| 0.014 | 0.011007806 | 0.018887568 | 0.029605110 | 0.043457007 |
| 0.013 | 0.009536830 | 0.016387802 | 0.025727743 | 0.037830907 |
| 0.012 | 0.008164865 | 0.014050849 | 0.022093581 | 0.032542536 |
| 0.011 | 0.006893467 | 0.011880196 | 0.018709484 | 0.027604216 |
| 0.010 | 0.005724198 | 0.009879351 | 0.015582364 | 0.023028372 |
| 0.009 | 0.004658635 | 0.008051849 | 0.012719186 | 0.018827525 |
| 0.008 | 0.003698360 | 0.006401250 | 0.010126961 | 0.015014287 |
| 0.007 | 0.002844968 | 0.004931138 | 0.007812750 | 0.011601352 |
| 0.006 | 0.002100064 | 0.003645119 | 0.005783661 | 0.008601492 |
| 0.005 | 0.001465262 | 0.002546827 | 0.004046844 | 0.006027551 |
| 0.004 | 0.000942188 | 0.001639916 | 0.002609493 | 0.003892431 |
| 0.003 | 0.000532474 | 0.000928064 | 0.001478842 | 0.002209091 |
| 0.002 | 0.000237767 | 0.000414975 | 0.000662165 | 0.000990539 |
| 0.001 | 0.000059720 | 0.000104371 | 0.000166769 | 0.000249817 |

TABLE 1a:  CLUSTERING PROBABILITY  P(n;N,p)

n = 3

| P | N =10 | N =11 | N =12 | N =13 |
|---|---|---|---|---|
| 0.150 | 0.996420108 | 0.999715102 | 0.999992232 | 0.999999980 |
| 0.149 | 0.996091159 | 0.999671074 | 0.999990185 | 0.999999968 |
| 0.148 | 0.995737315 | 0.999621305 | 0.999987664 | 0.999999949 |
| 0.147 | 0.995357132 | 0.999565187 | 0.999984574 | 0.999999921 |
| 0.146 | 0.994949109 | 0.999502061 | 0.999980806 | 0.999999880 |
| 0.145 | 0.994511692 | 0.999431216 | 0.999976234 | 0.999999821 |
| 0.144 | 0.994043270 | 0.999351889 | 0.999970711 | 0.999999737 |
| 0.143 | 0.993542179 | 0.999263258 | 0.999964070 | 0.999999619 |
| 0.142 | 0.993006696 | 0.999164443 | 0.999956119 | 0.999999455 |
| 0.141 | 0.992435045 | 0.999054502 | 0.999946639 | 0.999999232 |
| 0.140 | 0.991825392 | 0.998932426 | 0.999935383 | 0.999998930 |
| 0.139 | 0.991175848 | 0.998797142 | 0.999922072 | 0.999998527 |
| 0.138 | 0.990484467 | 0.998647504 | 0.999906390 | 0.999997992 |
| 0.137 | 0.989749250 | 0.998482296 | 0.999887984 | 0.999997291 |
| 0.136 | 0.988968140 | 0.998300227 | 0.999866459 | 0.999996379 |
| 0.135 | 0.988139027 | 0.998099926 | 0.999841373 | 0.999995201 |
| 0.134 | 0.987259745 | 0.997879945 | 0.999812236 | 0.999993694 |
| 0.133 | 0.986328075 | 0.997638752 | 0.999778505 | 0.999991778 |
| 0.132 | 0.985341747 | 0.997374733 | 0.999739580 | 0.999989360 |
| 0.131 | 0.984298436 | 0.997086183 | 0.999694798 | 0.999986328 |
| 0.130 | 0.983195767 | 0.996771313 | 0.999643431 | 0.999982549 |
| 0.129 | 0.982031316 | 0.996428240 | 0.999584682 | 0.999977870 |
| 0.128 | 0.980802608 | 0.996054990 | 0.999517677 | 0.999972108 |
| 0.127 | 0.979507124 | 0.995649493 | 0.999441464 | 0.999965051 |
| 0.126 | 0.978142295 | 0.995209583 | 0.999355006 | 0.999956454 |
| 0.125 | 0.976705510 | 0.994732998 | 0.999257177 | 0.999946034 |
| 0.124 | 0.975194115 | 0.994217376 | 0.999146757 | 0.999933464 |
| 0.123 | 0.973605415 | 0.993660253 | 0.999022426 | 0.999918372 |
| 0.122 | 0.971936674 | 0.993059067 | 0.998882760 | 0.999900334 |
| 0.121 | 0.970185123 | 0.992411151 | 0.998726227 | 0.999878865 |
| 0.120 | 0.968347956 | 0.991713737 | 0.998551178 | 0.999853421 |
| 0.119 | 0.966422333 | 0.990963953 | 0.998355846 | 0.999823385 |
| 0.118 | 0.964405389 | 0.990158823 | 0.998138339 | 0.999788065 |
| 0.117 | 0.962294227 | 0.989295268 | 0.997896636 | 0.999746686 |
| 0.116 | 0.960085928 | 0.988370107 | 0.997628581 | 0.999698383 |
| 0.115 | 0.957777551 | 0.987380054 | 0.997331878 | 0.999642193 |
| 0.114 | 0.955366136 | 0.986321722 | 0.997004088 | 0.999577048 |
| 0.113 | 0.952848709 | 0.985191623 | 0.996642623 | 0.999501766 |
| 0.112 | 0.950222283 | 0.983986168 | 0.996244742 | 0.999415042 |
| 0.111 | 0.947483862 | 0.982701669 | 0.995807547 | 0.999315444 |
| 0.110 | 0.944630445 | 0.981334345 | 0.995327980 | 0.999201397 |
| 0.109 | 0.941659032 | 0.979880316 | 0.994802816 | 0.999071179 |
| 0.108 | 0.938566621 | 0.978335613 | 0.994228664 | 0.998922911 |
| 0.107 | 0.935350221 | 0.976696176 | 0.993601959 | 0.998754545 |
| 0.106 | 0.932006850 | 0.974957859 | 0.992918965 | 0.998563860 |
| 0.105 | 0.928533540 | 0.973116433 | 0.992175767 | 0.998348447 |
| 0.104 | 0.924927344 | 0.971167589 | 0.991368273 | 0.998105699 |
| 0.103 | 0.921185337 | 0.969106944 | 0.990492210 | 0.997832810 |
| 0.102 | 0.917304623 | 0.966930044 | 0.989543125 | 0.997526754 |
| 0.101 | 0.913282342 | 0.964632367 | 0.988516381 | 0.997184283 |

TABLE 1a:   CLUSTERING PROBABILITY  $P(n;N,p)$

$n = 3$

| P | N≡10 | N≡11 | N≡12 | N≡13 |
|---|---|---|---|---|
| 0.100 | 0.909115668 | 0.962209333 | 0.987407161 | 0.996801919 |
| 0.099 | 0.904801822 | 0.959656304 | 0.986210464 | 0.996375937 |
| 0.098 | 0.900338069 | 0.956968592 | 0.984921110 | 0.995902365 |
| 0.097 | 0.895721733 | 0.954141468 | 0.983533739 | 0.995376972 |
| 0.096 | 0.890950193 | 0.951170164 | 0.982042814 | 0.994795259 |
| 0.095 | 0.886020894 | 0.948049883 | 0.980442625 | 0.994152455 |
| 0.094 | 0.880931351 | 0.944775804 | 0.978727288 | 0.993443506 |
| 0.093 | 0.875679157 | 0.941343093 | 0.976890757 | 0.992663073 |
| 0.092 | 0.870261983 | 0.937746909 | 0.974926820 | 0.991805523 |
| 0.091 | 0.864677592 | 0.933982411 | 0.972829113 | 0.990864927 |
| 0.090 | 0.858923837 | 0.930044770 | 0.970591120 | 0.989835054 |
| 0.089 | 0.852998674 | 0.925929177 | 0.968206183 | 0.988709370 |
| 0.088 | 0.846900164 | 0.921630853 | 0.965667513 | 0.987481033 |
| 0.087 | 0.840626481 | 0.917145058 | 0.962968192 | 0.986142897 |
| 0.086 | 0.834175917 | 0.912467101 | 0.960101190 | 0.984687507 |
| 0.085 | 0.827546892 | 0.907592355 | 0.957059373 | 0.983107104 |
| 0.084 | 0.820737955 | 0.902516261 | 0.953835511 | 0.981393626 |
| 0.083 | 0.813747795 | 0.897234347 | 0.950422299 | 0.979538712 |
| 0.082 | 0.806575249 | 0.891742235 | 0.946812361 | 0.977533709 |
| 0.081 | 0.799219302 | 0.886035656 | 0.942998271 | 0.975369681 |
| 0.080 | 0.791679101 | 0.880110458 | 0.938972569 | 0.973037412 |
| 0.079 | 0.783953959 | 0.873962628 | 0.934727770 | 0.970527421 |
| 0.078 | 0.776043359 | 0.867588294 | 0.930256391 | 0.967829974 |
| 0.077 | 0.767946968 | 0.860983749 | 0.925550961 | 0.964935096 |
| 0.076 | 0.759664637 | 0.854145456 | 0.920604048 | 0.961832589 |
| 0.075 | 0.751196412 | 0.847070069 | 0.915408273 | 0.958512049 |
| 0.074 | 0.742542539 | 0.839754444 | 0.909956334 | 0.954962885 |
| 0.073 | 0.733703473 | 0.832195655 | 0.904241029 | 0.951174343 |
| 0.072 | 0.724679883 | 0.824391009 | 0.898255280 | 0.947135529 |
| 0.071 | 0.715472661 | 0.816338061 | 0.891992153 | 0.942835436 |
| 0.070 | 0.706082926 | 0.808034628 | 0.885444889 | 0.938262970 |
| 0.069 | 0.696512036 | 0.799478809 | 0.878606924 | 0.933406985 |
| 0.068 | 0.686761590 | 0.790668995 | 0.871471921 | 0.928256311 |
| 0.067 | 0.676833438 | 0.781603891 | 0.864033798 | 0.922799794 |
| 0.066 | 0.666729684 | 0.772282527 | 0.856286753 | 0.917026331 |
| 0.065 | 0.656452699 | 0.762704280 | 0.848225295 | 0.910924910 |
| 0.064 | 0.646005124 | 0.752868885 | 0.839844278 | 0.904484653 |
| 0.063 | 0.635389874 | 0.742776454 | 0.831138926 | 0.897694860 |
| 0.062 | 0.624610152 | 0.732427495 | 0.822104872 | 0.890545057 |
| 0.061 | 0.613669448 | 0.721822922 | 0.812738185 | 0.883025043 |
| 0.060 | 0.602571550 | 0.710964081 | 0.803035404 | 0.875124943 |
| 0.059 | 0.591320548 | 0.699852758 | 0.792993577 | 0.866835261 |
| 0.058 | 0.579920841 | 0.688491201 | 0.782610289 | 0.858146937 |
| 0.057 | 0.568377146 | 0.676882134 | 0.771883699 | 0.849051403 |
| 0.056 | 0.556694497 | 0.665028776 | 0.760812578 | 0.839540642 |
| 0.055 | 0.544878256 | 0.652934852 | 0.749396341 | 0.829607254 |
| 0.054 | 0.532934119 | 0.640604618 | 0.737635086 | 0.819244514 |
| 0.053 | 0.520868118 | 0.628042868 | 0.725529626 | 0.808446442 |
| 0.052 | 0.508686629 | 0.615254955 | 0.713081530 | 0.797207868 |
| 0.051 | 0.496396373 | 0.602246803 | 0.700293157 | 0.785524499 |

TABLE 1a:  CLUSTERING PROBABILITY  P(n;N,p)

n = 3

| P | N = 10 | N = 11 | N = 12 | N = 13 |
|---|---|---|---|---|
| 0.050 | 0.484004429 | 0.589024928 | 0.687167693 | 0.773392990 |
| 0.049 | 0.471518228 | 0.575596444 | 0.673709186 | 0.760811015 |
| 0.048 | 0.458945565 | 0.561969086 | 0.659922582 | 0.747777340 |
| 0.047 | 0.446294600 | 0.548151219 | 0.645813762 | 0.734291891 |
| 0.046 | 0.433573863 | 0.534151853 | 0.631389578 | 0.720355833 |
| 0.045 | 0.420792256 | 0.519980654 | 0.616657883 | 0.705971639 |
| 0.044 | 0.407959058 | 0.505647961 | 0.601627571 | 0.691143167 |
| 0.043 | 0.395083926 | 0.491164792 | 0.586308606 | 0.675875730 |
| 0.042 | 0.382176899 | 0.476542861 | 0.570712057 | 0.660176172 |
| 0.041 | 0.369248400 | 0.461794583 | 0.554850125 | 0.644052941 |
| 0.040 | 0.356309237 | 0.446933088 | 0.538736180 | 0.627516158 |
| 0.039 | 0.343370606 | 0.431972228 | 0.522384784 | 0.610577690 |
| 0.038 | 0.330444091 | 0.416926585 | 0.505811721 | 0.593251217 |
| 0.037 | 0.317541664 | 0.401811478 | 0.489034024 | 0.575552301 |
| 0.036 | 0.304675689 | 0.386642972 | 0.472069996 | 0.557498449 |
| 0.035 | 0.291858916 | 0.371437878 | 0.454939235 | 0.539109174 |
| 0.034 | 0.279104487 | 0.356213764 | 0.437662654 | 0.520406059 |
| 0.033 | 0.266425929 | 0.340988950 | 0.420262497 | 0.501412808 |
| 0.032 | 0.253837157 | 0.325782519 | 0.402762360 | 0.482155301 |
| 0.031 | 0.241352470 | 0.310614309 | 0.385187197 | 0.462661640 |
| 0.030 | 0.228986549 | 0.295504919 | 0.367563337 | 0.442962193 |
| 0.029 | 0.216754452 | 0.280475703 | 0.349918488 | 0.423089632 |
| 0.028 | 0.204671613 | 0.265548768 | 0.332281745 | 0.403078966 |
| 0.027 | 0.192753840 | 0.250746971 | 0.314683588 | 0.382967566 |
| 0.026 | 0.181017303 | 0.236093910 | 0.297155881 | 0.362795188 |
| 0.025 | 0.169478536 | 0.221613916 | 0.279731869 | 0.342603981 |
| 0.024 | 0.158154427 | 0.207332045 | 0.262446167 | 0.322438499 |
| 0.023 | 0.147062212 | 0.193274069 | 0.245334745 | 0.302345693 |
| 0.022 | 0.136219470 | 0.179466455 | 0.228434914 | 0.282374900 |
| 0.021 | 0.125644112 | 0.165936359 | 0.211785302 | 0.262577826 |
| 0.020 | 0.115354374 | 0.152711604 | 0.195425827 | 0.243008515 |
| 0.019 | 0.105368809 | 0.139820663 | 0.179397670 | 0.223723304 |
| 0.018 | 0.095706274 | 0.127292635 | 0.163743232 | 0.204780778 |
| 0.017 | 0.086385921 | 0.115157228 | 0.148506099 | 0.186241706 |
| 0.016 | 0.077427186 | 0.103444729 | 0.133730990 | 0.168168962 |
| 0.015 | 0.068849776 | 0.092185975 | 0.119463710 | 0.150627443 |
| 0.014 | 0.060673653 | 0.081412331 | 0.105751086 | 0.133683967 |
| 0.013 | 0.052919027 | 0.071155648 | 0.092640906 | 0.117407159 |
| 0.012 | 0.045606333 | 0.061448234 | 0.080181849 | 0.101867322 |
| 0.011 | 0.038756222 | 0.052322818 | 0.068423404 | 0.087136296 |
| 0.010 | 0.032389539 | 0.043812505 | 0.057415789 | 0.073287297 |
| 0.009 | 0.026527311 | 0.035950736 | 0.047209860 | 0.060394745 |
| 0.008 | 0.021190723 | 0.028771243 | 0.037857013 | 0.048534069 |
| 0.007 | 0.016401101 | 0.022308002 | 0.029409077 | 0.037781503 |
| 0.006 | 0.012179894 | 0.016595178 | 0.021918203 | 0.028213859 |
| 0.005 | 0.008548648 | 0.011667072 | 0.015436741 | 0.019908281 |
| 0.004 | 0.005528987 | 0.007558067 | 0.010017115 | 0.012941983 |
| 0.003 | 0.003142587 | 0.004302564 | 0.005711680 | 0.007391968 |
| 0.002 | 0.001411154 | 0.001934919 | 0.002572583 | 0.003334726 |
| 0.001 | 0.000356397 | 0.000489377 | 0.000651605 | 0.000845911 |

TABLE 1a:   CLUSTERING PROBABILITY  P(n;N,p)

n = 3

| P | N = 14 | N = 15 | N = 16 | N = 17 |
|---|--------|--------|--------|--------|
| C.100 | 0.999421701 | 0.999932406 | 0.999995562 | 0.999999870 |
| 0.099 | 0.999313259 | 0.999914299 | 0.999993819 | 0.999999791 |
| 0.098 | 0.999187408 | 0.999891982 | 0.999991477 | 0.999999670 |
| 0.097 | 0.999041837 | 0.999864627 | 0.999988357 | 0.999999488 |
| 0.096 | 0.998873995 | 0.999831269 | 0.999984239 | 0.999999218 |
| 0.095 | 0.998681078 | 0.999790795 | 0.999978847 | 0.999998822 |
| 0.094 | 0.998460009 | 0.999741927 | 0.999971843 | 0.999998252 |
| 0.093 | 0.998207421 | 0.999683195 | 0.999962815 | 0.999997439 |
| 0.092 | 0.997919644 | 0.999612928 | 0.999951259 | 0.999996295 |
| 0.091 | 0.997592682 | 0.999529224 | 0.999936573 | 0.999994703 |
| 0.090 | 0.997222201 | 0.999429931 | 0.999918031 | 0.999992510 |
| 0.089 | 0.996803509 | 0.999312623 | 0.999894770 | 0.999989521 |
| 0.088 | 0.996331540 | 0.999174575 | 0.999865768 | 0.999985486 |
| 0.087 | 0.995800839 | 0.999012735 | 0.999829818 | 0.999980092 |
| 0.086 | 0.995205544 | 0.998823699 | 0.999785510 | 0.999972942 |
| 0.085 | 0.994539372 | 0.998603686 | 0.999731195 | 0.999963545 |
| 0.084 | 0.993795608 | 0.998348502 | 0.999664963 | 0.999951294 |
| 0.083 | 0.992967085 | 0.998053517 | 0.999584605 | 0.999935446 |
| 0.082 | 0.992046179 | 0.997713635 | 0.999487583 | 0.999915096 |
| 0.081 | 0.991024793 | 0.997323261 | 0.999370991 | 0.999889148 |
| C.080 | 0.989894352 | 0.996876272 | 0.999231518 | 0.999856286 |
| 0.079 | 0.988645791 | 0.996365993 | 0.999065405 | 0.999814940 |
| 0.078 | 0.987269553 | 0.995785159 | 0.998868405 | 0.999763243 |
| 0.077 | 0.985755585 | 0.995125894 | 0.998635734 | 0.999698993 |
| 0.076 | 0.984093335 | 0.994379679 | 0.998362029 | 0.999619604 |
| 0.075 | 0.982271753 | 0.993537330 | 0.998041297 | 0.999522054 |
| 0.074 | 0.980279296 | 0.992588970 | 0.997666867 | 0.999402833 |
| 0.073 | 0.978103934 | 0.991524010 | 0.997231342 | 0.999257884 |
| 0.072 | 0.975733159 | 0.990331126 | 0.996726547 | 0.999082536 |
| 0.071 | 0.973153999 | 0.988988245 | 0.996143481 | 0.998871444 |
| 0.070 | 0.970353032 | 0.987512530 | 0.995472264 | 0.998618510 |
| 0.069 | 0.967316408 | 0.985860368 | 0.994702095 | 0.998316820 |
| 0.068 | 0.964029869 | 0.984027366 | 0.993821199 | 0.997958558 |
| 0.067 | 0.960478781 | 0.981998349 | 0.992816788 | 0.997534933 |
| 0.066 | 0.956648161 | 0.979757362 | 0.991675017 | 0.997036097 |
| 0.065 | 0.952522717 | 0.977287678 | 0.990380949 | 0.996451065 |
| 0.064 | 0.948086884 | 0.974571815 | 0.988918519 | 0.995767633 |
| 0.063 | 0.943324870 | 0.971591553 | 0.987270511 | 0.994972294 |
| 0.062 | 0.938220708 | 0.968327963 | 0.985418532 | 0.994050167 |
| 0.061 | 0.932758306 | 0.964761443 | 0.983343004 | 0.992984915 |
| 0.060 | 0.926921509 | 0.960871756 | 0.981023154 | 0.991758675 |
| 0.059 | 0.920694161 | 0.956638081 | 0.978437017 | 0.990351995 |
| 0.058 | 0.914060176 | 0.952039075 | 0.975561452 | 0.988743775 |
| 0.057 | 0.907003610 | 0.947052932 | 0.972372164 | 0.986911214 |
| 0.056 | 0.899508742 | 0.941657470 | 0.968843741 | 0.984829777 |
| 0.055 | 0.891560156 | 0.935830205 | 0.964949703 | 0.982473165 |
| 0.054 | 0.883142834 | 0.929548454 | 0.960662566 | 0.979813305 |
| 0.053 | 0.874242246 | 0.922789439 | 0.955953917 | 0.976820350 |
| 0.052 | 0.864844452 | 0.915530398 | 0.950794511 | 0.973462709 |
| 0.051 | 0.854936205 | 0.907748716 | 0.945154380 | 0.969707083 |

TABLE 1a:   CLUSTERING PROBABILITY   P(n;N,p)

n = 3

| P | N = 14 | N = 15 | N = 16 | N = 17 |
|---|---|---|---|---|
| 0.050 | 0.844505059 | 0.899422059 | 0.939002961 | 0.965518532 |
| 0.049 | 0.833539485 | 0.890528520 | 0.932309239 | 0.960860566 |
| 0.048 | 0.822028984 | 0.881046777 | 0.925041919 | 0.955695259 |
| 0.047 | 0.809964215 | 0.870956262 | 0.917169604 | 0.949983395 |
| 0.046 | 0.797337112 | 0.860237334 | 0.908661001 | 0.943684640 |
| 0.045 | 0.784141021 | 0.848871472 | 0.899485148 | 0.936757751 |
| 0.044 | 0.770370825 | 0.836841471 | 0.889611655 | 0.929160815 |
| 0.043 | 0.756023084 | 0.824131649 | 0.879010975 | 0.920851522 |
| 0.042 | 0.741096168 | 0.810728060 | 0.867654684 | 0.911787482 |
| 0.041 | 0.725590398 | 0.796618722 | 0.855515796 | 0.901926570 |
| 0.040 | 0.709508187 | 0.781793847 | 0.842569080 | 0.891227314 |
| 0.039 | 0.692854176 | 0.766246078 | 0.828791413 | 0.879649321 |
| 0.038 | 0.675635380 | 0.749970733 | 0.814162138 | 0.867153743 |
| 0.037 | 0.657861324 | 0.732966058 | 0.798663449 | 0.853703777 |
| 0.036 | 0.639544187 | 0.715233474 | 0.782280786 | 0.839265207 |
| 0.035 | 0.620698935 | 0.696777836 | 0.765003246 | 0.823806978 |
| 0.034 | 0.601343456 | 0.677607688 | 0.746824002 | 0.807301811 |
| 0.033 | 0.581498693 | 0.657735517 | 0.727740742 | 0.789726841 |
| 0.032 | 0.561188768 | 0.637178010 | 0.707756106 | 0.771064291 |
| 0.031 | 0.540441103 | 0.615956302 | 0.686878130 | 0.751302169 |
| 0.030 | 0.519286537 | 0.594096220 | 0.665120693 | 0.730434988 |
| 0.029 | 0.497759428 | 0.571628518 | 0.642503964 | 0.708464497 |
| 0.028 | 0.475897760 | 0.548589109 | 0.619054835 | 0.685400430 |
| 0.027 | 0.453743221 | 0.525019274 | 0.594807358 | 0.661261252 |
| 0.026 | 0.431341293 | 0.500965863 | 0.569803153 | 0.636074909 |
| 0.025 | 0.408741309 | 0.476481483 | 0.544091809 | 0.609879562 |
| 0.024 | 0.385996512 | 0.451624655 | 0.517731251 | 0.582724301 |
| 0.023 | 0.363164093 | 0.426459958 | 0.490788088 | 0.554669833 |
| 0.022 | 0.340305212 | 0.401058146 | 0.463337918 | 0.525789129 |
| 0.021 | 0.317485007 | 0.375496234 | 0.435465596 | 0.496168018 |
| 0.020 | 0.294772581 | 0.349857556 | 0.407265449 | 0.465905724 |
| 0.019 | 0.272240972 | 0.324231786 | 0.378841448 | 0.435115320 |
| 0.018 | 0.249967095 | 0.298714923 | 0.350307308 | 0.403924101 |
| 0.017 | 0.228031674 | 0.273409237 | 0.321786529 | 0.372473851 |
| 0.016 | 0.206519135 | 0.248423166 | 0.293412355 | 0.340920996 |
| 0.015 | 0.185517488 | 0.223871171 | 0.265327658 | 0.309436622 |
| 0.014 | 0.165118171 | 0.199873535 | 0.237684730 | 0.278206353 |
| 0.013 | 0.145415870 | 0.176556114 | 0.210644973 | 0.247430062 |
| 0.012 | 0.126508314 | 0.154050024 | 0.184378499 | 0.217321410 |
| 0.011 | 0.108496027 | 0.132491271 | 0.159063599 | 0.188107187 |
| 0.010 | 0.091482058 | 0.112020317 | 0.134886114 | 0.160026450 |
| 0.009 | 0.075571677 | 0.092781577 | 0.112038661 | 0.133329438 |
| 0.008 | 0.060872024 | 0.074922844 | 0.090719746 | 0.108276253 |
| 0.007 | 0.047491735 | 0.058594648 | 0.071132720 | 0.085135282 |
| 0.006 | 0.035540526 | 0.043949528 | 0.053484602 | 0.064181372 |
| 0.005 | 0.025128729 | 0.031141235 | 0.037984749 | 0.045693716 |
| 0.004 | 0.016366805 | 0.020323849 | 0.024843372 | 0.029953470 |
| 0.003 | 0.009364797 | 0.011650812 | 0.014269888 | 0.017241071 |
| 0.002 | 0.004231753 | 0.005273879 | 0.006471115 | 0.007833262 |
| 0.001 | 0.001075106 | 0.001341981 | 0.001649304 | 0.001999824 |

TABLE 1a:  CLUSTERING PROBABILITY  P(n;N,p)

n = 3

| P | N =18 | P | N =18 | N =19 |
|-------|-------------|-------|-------------|-------------|
| 0.100 | 0.99999999 | 0.050 | 0.98194839 2 | 0.991311223 |
| 0.099 | 0.99999998 | 0.049 | 0.97896539 8 | 0.989562903 |
| 0.098 | 0.99999995 | 0.048 | 0.97557516 5 | 0.987518457 |
| 0.097 | 0.99999991 | 0.047 | 0.97173555 6 | 0.985138155 |
| 0.096 | 0.99999984 | 0.046 | 0.967401961 | 0.982378673 |
| 0.095 | 0.99999971 | 0.045 | 0.962527408 | 0.979193027 |
| 0.094 | 0.99999949 | 0.044 | 0.957062719 | 0.975530561 |
| 0.093 | 0.99999914 | 0.043 | 0.950956707 | 0.971336977 |
| 0.092 | 0.99999856 | 0.042 | 0.94415643 8 | 0.966554433 |
| 0.091 | 0.99999765 | 0.041 | 0.93660753 3 | 0.961121704 |
| 0.090 | 0.99999622 | 0.040 | 0.92825453 7 | 0.954974415 |
| 0.089 | 0.99999403 | 0.039 | 0.91904135 7 | 0.948045357 |
| 0.088 | 0.99999073 | 0.038 | 0.908911748 | 0.940264904 |
| 0.087 | 0.99998582 | 0.037 | 0.897809884 | 0.931561508 |
| 0.086 | 0.99997862 | 0.036 | 0.88563809 82 | 0.921862312 |
| 0.085 | 0.99996822 | 0.035 | 0.87247201 0 | 0.911093872 |
| 0.084 | 0.99995336 | 0.034 | 0.85813245 3 | 0.899182984 |
| 0.083 | 0.99999324 0 | 0.033 | 0.84261515 2 | 0.886057639 |
| 0.082 | 0.99999031 6 | 0.032 | 0.82587721 2 | 0.871648092 |
| 0.081 | 0.99998628 2 | 0.031 | 0.807880970 | 0.855888057 |
| 0.080 | 0.99998077 4 | 0.030 | 0.78859502 6 | 0.838716019 |
| 0.079 | 0.99997332 4 | 0.029 | 0.76799532 5 | 0.820076664 |
| 0.078 | 0.99996334 4 | 0.028 | 0.74606628 5 | 0.799922417 |
| 0.077 | 0.99995009 3 | 0.027 | 0.722801970 | 0.778215089 |
| 0.076 | 0.99993264 7 | 0.026 | 0.69820728 3 | 0.754927606 |
| 0.075 | 0.99990986 6 | 0.025 | 0.67299918 5 | 0.730045827 |
| 0.074 | 0.99988035 0 | 0.024 | 0.645107908 | 0.703570410 |
| 0.073 | 0.99984239 1 | 0.023 | 0.616678163 | 0.675518723 |
| 0.072 | 0.99979392 3 | 0.022 | 0.587070318 | 0.645926769 |
| 0.071 | 0.99973246 4 | 0.021 | 0.556361528 | 0.614851095 |
| 0.070 | 0.99965504 6 | 0.020 | 0.52464679 5 | 0.582370654 |
| 0.069 | 0.99955814 2 | 0.019 | 0.492039945 | 0.548588586 |
| 0.068 | 0.99943759 4 | 0.018 | 0.45867448 6 | 0.513633876 |
| 0.067 | 0.99928851 5 | 0.017 | 0.424704324 | 0.477662846 |
| 0.066 | 0.99910520 5 | 0.016 | 0.39030432 5 | 0.440860439 |
| 0.065 | 0.99888104 4 | 0.015 | 0.35567066 9 | 0.403441233 |
| 0.064 | 0.99860838 9 | 0.014 | 0.32102099 4 | 0.365650146 |
| 0.063 | 0.99827845 7 | 0.013 | 0.28659427 5 | 0.327762771 |
| 0.062 | 0.99788121 0 | 0.012 | 0.25265043 2 | 0.290085277 |
| 0.061 | 0.99740523 0 | 0.011 | 0.21946961 6 | 0.252953833 |
| 0.060 | 0.99683759 4 | 0.010 | 0.187351151 | 0.216733485 |
| 0.059 | 0.99616374 5 | 0.009 | 0.15661210 9 | 0.181816424 |
| 0.058 | 0.99536736 2 | 0.008 | 0.127854 71 | 0.148619612 |
| 0.057 | 0.99443023 3 | 0.007 | 0.100617869 | 0.117581682 |
| 0.056 | 0.99333213 1 | 0.006 | 0.076066868 | 0.089159086 |
| 0.055 | 0.99205069 5 | 0.005 | 0.05429776 9 | 0.063821439 |
| 0.054 | 0.99056131 6 | 0.004 | 0.035679919 | 0.042046025 |
| 0.053 | 0.988837041 | 0.003 | 0.02058251 7 | 0.024311436 |
| 0.052 | 0.98684848 2 | 0.002 | 0.009369892 | 0.011090334 |
| 0.051 | 0.98456374 6 | 0.001 | 0.002396266 | 0.002841330 |

TABLE 1a:  CLUSTERING PROBABILITY  P(n;N,p)

n = 4

| p | N = 8 | N = 9 | p | N =12 |
|---|---|---|---|---|
| 0.350 | 0.98412967 0 | 0.999173314 | 0.250 | 0.999972463 |
| 0.349 | 0.98355430 3 | 0.999122369 | 0.249 | 0.999968225 |
| 0.348 | 0.982962914 | 0.999068653 | 0.248 | 0.999963397 |
| 0.347 | 0.982355211 | 0.999012034 | 0.247 | 0.999957904 |
| 0.346 | 0.981730902 | 0.998952375 | 0.246 | 0.999951666 |
| 0.345 | 0.981089693 | 0.998889536 | 0.245 | 0.999944590 |
| 0.344 | 0.98043128 8 | 0.998823368 | 0.244 | 0.999936577 |
| 0.343 | 0.979755391 | 0.998753719 | 0.243 | 0.999927516 |
| 0.342 | 0.979061707 | 0.998680428 | 0.242 | 0.99991728 4 |
| 0.341 | 0.978349937 | 0.998603331 | 0.241 | 0.999905748 |
| 0.340 | 0.977619784 | 0.998522256 | 0.240 | 0.999892760 |
| 0.339 | 0.976870951 | 0.998437025 | 0.239 | 0.999878161 |
| 0.338 | 0.976103140 | 0.998347451 | 0.238 | 0.999861775 |
| 0.337 | 0.975316051 | 0.998253343 | 0.237 | 0.999843411 |
| 0.336 | 0.97450938 9 | 0.998154501 | 0.236 | 0.999822863 |
| 0.335 | 0.973682853 | 0.998050718 | 0.235 | 0.9997999' |
| 0.334 | 0.972836148 | 0.997941779 | 0.234 | 0.99977429 |
| 0.333 | 0.971968977 | 0.997827461 | 0.233 | 0.999745780 |
| 0.332 | 0.971081043 | 0.997707535 | 0.232 | 0.99971406 |
| 0.331 | 0.970172051 | 0.997581759 | 0.231 | 0.999678862 |
| 0.330 | 0.969241708 | 0.997449887 | 0.230 | 0.999639836 |
| 0.329 | 0.968289721 | 0.997311663 | 0.229 | 0.999596646 |
| 0.328 | 0.967315797 | 0.997166823 | 0.228 | 0.999548922 |
| 0.327 | 0.966319647 | 0.997015094 | 0.227 | 0.99949626 9 |
| 0.326 | 0.96530098 3 | 0.996856196 | 0.226 | 0.999438268 |
| 0.325 | 0.964259518 | 0.996689840 | 0.225 | 0.999374475 |
| 0.324 | 0.963194967 | 0.996515731 | 0.224 | 0.999304417 |
| 0.323 | 0.962107048 | 0.996333562 | 0.223 | 0.999227594 |
| 0.322 | 0.96099548 1 | 0.996143022 | 0.222 | 0.999143479 |
| 0.321 | 0.959859988 | 0.995943791 | 0.221 | 0.999051515 |
| 0.320 | 0.95870029 3 | 0.995735540 | 0.220 | 0.998951114 |
| 0.319 | 0.957516124 | 0.995517935 | 0.219 | 0.998841660 |
| 0.318 | 0.956307211 | 0.995290633 | 0.218 | 0.99872250 6 |
| 0.317 | 0.95507328 7 | 0.995053282 | 0.217 | 0.998592970 |
| 0.316 | 0.953814088 | 0.994805526 | 0.216 | 0.998452341 |
| 0.315 | 0.952529352 | 0.994547001 | 0.215 | 0.998299875 |
| 0.314 | 0.951218824 | 0.994277333 | 0.214 | 0.998134794 |
| 0.313 | 0.94988224 8 | 0.993996146 | 0.213 | 0.99795628 7 |
| 0.312 | 0.94851937 4 | 0.993703053 | 0.212 | 0.997763510 |
| 0.311 | 0.947129954 | 0.993397663 | 0.211 | 0.997555584 |
| 0.310 | 0.945713747 | 0.993079578 | 0.210 | 0.997331595 |
| 0.309 | 0.94427051 1 | 0.992748393 | 0.209 | 0.997090596 |
| 0.308 | 0.942800012 | 0.992403698 | 0.208 | 0.996831605 |
| 0.307 | 0.941302019 | 0.992045076 | 0.207 | 0.99655360 4 |
| 0.306 | 0.93977630 5 | 0.991672105 | 0.206 | 0.996255543 |
| 0.305 | 0.938222645 | 0.991284358 | 0.205 | 0.995936335 |
| 0.304 | 0.936640823 | 0.990881401 | 0.204 | 0.995594860 |
| 0.303 | 0.935030623 | 0.990462795 | 0.203 | 0.995229963 |
| 0.302 | 0.933391837 | 0.990028098 | 0.202 | 0.99484045 4 |
| 0.301 | 0.93172425 9 | 0.989576862 | 0.201 | 0.99442511 2 |

TABLE 1a:   CLUSTERING PROBABILITY   P(n;N,p)

n = 4

| P | N = 8 | N = 9 | N = 10 | N = 11 |
|---|---|---|---|---|
| 0.300 | 0.930027690 | 0.989108632 | 0.999782580 | 0.999999168 |
| 0.299 | 0.928301934 | 0.988622953 | 0.999752739 | 0.999998948 |
| 0.298 | 0.926546801 | 0.988119362 | 0.999719786 | 0.999998679 |
| 0.297 | 0.924762106 | 0.987597394 | 0.999683492 | 0.999998352 |
| 0.296 | 0.922947668 | 0.987056580 | 0.999643618 | 0.999997956 |
| 0.295 | 0.921103313 | 0.986496448 | 0.999599913 | 0.999997480 |
| 0.294 | 0.919228870 | 0.985916520 | 0.999552118 | 0.999996910 |
| 0.293 | 0.917324176 | 0.985316318 | 0.999499962 | 0.999996231 |
| 0.292 | 0.915389071 | 0.984695360 | 0.999443163 | 0.999995425 |
| 0.291 | 0.913423401 | 0.984053161 | 0.999381429 | 0.999994473 |
| 0.290 | 0.911427019 | 0.983389233 | 0.999314455 | 0.999993353 |
| 0.289 | 0.909399783 | 0.982703088 | 0.999241926 | 0.999992041 |
| 0.288 | 0.907341555 | 0.981994234 | 0.999163516 | 0.999990509 |
| 0.287 | 0.905252205 | 0.981262178 | 0.999078888 | 0.999988728 |
| 0.286 | 0.903131607 | 0.980506426 | 0.998987692 | 0.999986662 |
| 0.285 | 0.900979642 | 0.979726481 | 0.998889568 | 0.999984275 |
| 0.284 | 0.898796198 | 0.978921848 | 0.998784142 | 0.999981525 |
| 0.283 | 0.896581165 | 0.978092029 | 0.998671030 | 0.999978368 |
| 0.282 | 0.894334445 | 0.977236526 | 0.998549838 | 0.999974752 |
| 0.281 | 0.892055940 | 0.976354842 | 0.998420156 | 0.999970623 |
| 0.280 | 0.889745562 | 0.975446478 | 0.998281566 | 0.999965920 |
| 0.279 | 0.887403229 | 0.974510937 | 0.998133636 | 0.999960579 |
| 0.278 | 0.885028863 | 0.973547722 | 0.997975923 | 0.999954527 |
| 0.277 | 0.882622395 | 0.972556337 | 0.997807971 | 0.999947687 |
| 0.276 | 0.880183760 | 0.971536286 | 0.997629314 | 0.999939973 |
| 0.275 | 0.877712900 | 0.970487077 | 0.997439474 | 0.999931294 |
| 0.274 | 0.875209766 | 0.969408217 | 0.997237959 | 0.999921551 |
| 0.273 | 0.872674312 | 0.968299216 | 0.997024268 | 0.999910635 |
| 0.272 | 0.870106500 | 0.967159585 | 0.996797886 | 0.999898432 |
| 0.271 | 0.867506298 | 0.965988841 | 0.996558290 | 0.999884816 |
| 0.270 | 0.864873681 | 0.964786498 | 0.996304941 | 0.999869653 |
| 0.269 | 0.862208632 | 0.963552078 | 0.996037293 | 0.999852799 |
| 0.268 | 0.859511138 | 0.962285104 | 0.995754785 | 0.999834100 |
| 0.267 | 0.856781193 | 0.960985102 | 0.995456848 | 0.999813389 |
| 0.266 | 0.854018801 | 0.959651603 | 0.995142900 | 0.999790491 |
| 0.265 | 0.851223968 | 0.958284142 | 0.994812349 | 0.999765217 |
| 0.264 | 0.848396711 | 0.956882257 | 0.994464592 | 0.999737364 |
| 0.263 | 0.845537050 | 0.955445491 | 0.994099018 | 0.999706720 |
| 0.262 | 0.842645016 | 0.953973392 | 0.993715002 | 0.999673057 |
| 0.261 | 0.839720642 | 0.952465514 | 0.993311911 | 0.999636132 |
| 0.260 | 0.836763972 | 0.950921414 | 0.992889103 | 0.999595690 |
| 0.259 | 0.833775054 | 0.949340658 | 0.992445924 | 0.999551460 |
| 0.258 | 0.830753945 | 0.947722813 | 0.991981714 | 0.999503153 |
| 0.257 | 0.827700707 | 0.946067457 | 0.991495801 | 0.999450468 |
| 0.256 | 0.824615411 | 0.944374171 | 0.990987507 | 0.999393084 |
| 0.255 | 0.821498133 | 0.942642544 | 0.990456143 | 0.999330665 |
| 0.254 | 0.818348956 | 0.940872171 | 0.989901014 | 0.999262855 |
| 0.253 | 0.815167972 | 0.939062655 | 0.989321417 | 0.999189282 |
| 0.252 | 0.811955277 | 0.937213606 | 0.988716639 | 0.999109555 |
| 0.251 | 0.808710976 | 0.935324641 | 0.988085963 | 0.999023263 |

TABLE 1a:   CLUSTERING PROBABILITY   P(n;N,p)

$n = 4$

| P | N = 8 | N = 9 | N = 10 | N = 11 |
|---|---|---|---|---|
| 0.250 | 0.805435181 | 0.933395386 | 0.987428665 | 0.998929977 |
| 0.249 | 0.802128009 | 0.931425473 | 0.986744013 | 0.998829247 |
| 0.248 | 0.798789586 | 0.929414543 | 0.986031271 | 0.998720604 |
| 0.247 | 0.795420044 | 0.927362248 | 0.985289695 | 0.998603556 |
| 0.246 | 0.792019522 | 0.925268245 | 0.984518538 | 0.998477593 |
| 0.245 | 0.788588166 | 0.923132202 | 0.983717048 | 0.998342183 |
| 0.244 | 0.785126129 | 0.920953797 | 0.982884467 | 0.998196772 |
| 0.243 | 0.781633571 | 0.918732714 | 0.982020036 | 0.998040787 |
| 0.242 | 0.778110658 | 0.916468650 | 0.981122991 | 0.997873630 |
| 0.241 | 0.774557565 | 0.914161310 | 0.980192563 | 0.997694684 |
| 0.240 | 0.770974471 | 0.911810410 | 0.979227986 | 0.997503308 |
| 0.239 | 0.767361564 | 0.909415676 | 0.978228485 | 0.997298841 |
| 0.238 | 0.763719039 | 0.906976844 | 0.977193290 | 0.997080601 |
| 0.237 | 0.760047096 | 0.904493661 | 0.976121627 | 0.996847880 |
| 0.236 | 0.756345943 | 0.901965885 | 0.975012719 | 0.996599954 |
| 0.235 | 0.752615796 | 0.899393284 | 0.973865795 | 0.996336072 |
| 0.234 | 0.748856875 | 0.896775638 | 0.972680078 | 0.996055466 |
| 0.233 | 0.745069410 | 0.894112740 | 0.971454798 | 0.995757343 |
| 0.232 | 0.741253634 | 0.891404391 | 0.970189181 | 0.995440891 |
| 0.231 | 0.737409790 | 0.888650407 | 0.968882459 | 0.995105277 |
| 0.230 | 0.733538126 | 0.885850614 | 0.967533866 | 0.994749646 |
| 0.229 | 0.729638898 | 0.883004851 | 0.966142637 | 0.994373124 |
| 0.228 | 0.725712367 | 0.880112969 | 0.964708012 | 0.993974817 |
| 0.227 | 0.721758801 | 0.877174832 | 0.963229237 | 0.993553810 |
| 0.226 | 0.717778476 | 0.874190315 | 0.961705560 | 0.993109170 |
| 0.225 | 0.713771673 | 0.871159308 | 0.960136235 | 0.992639946 |
| 0.224 | 0.709738680 | 0.868081711 | 0.958520522 | 0.992145166 |
| 0.223 | 0.705679792 | 0.864957441 | 0.956857689 | 0.991623843 |
| 0.222 | 0.701595310 | 0.861786424 | 0.955147009 | 0.991074970 |
| 0.221 | 0.697485540 | 0.858568602 | 0.953387763 | 0.990497525 |
| 0.220 | 0.693350798 | 0.855303930 | 0.951579240 | 0.989890471 |
| 0.219 | 0.689191402 | 0.851992375 | 0.949720739 | 0.989252752 |
| 0.218 | 0.685007680 | 0.848633919 | 0.947811565 | 0.988583299 |
| 0.217 | 0.680799964 | 0.845228559 | 0.945851037 | 0.987881030 |
| 0.216 | 0.676568592 | 0.841776303 | 0.943838481 | 0.987144847 |
| 0.215 | 0.672313911 | 0.838277175 | 0.941773234 | 0.986373641 |
| 0.214 | 0.668036271 | 0.834731213 | 0.939654648 | 0.985566289 |
| 0.213 | 0.663736030 | 0.831138469 | 0.937482083 | 0.984721660 |
| 0.212 | 0.659413550 | 0.827499008 | 0.935254912 | 0.983838609 |
| 0.211 | 0.655069201 | 0.823812911 | 0.932972524 | 0.982915983 |
| 0.210 | 0.650703358 | 0.820080274 | 0.930634318 | 0.981952620 |
| 0.209 | 0.646316403 | 0.816301206 | 0.928239710 | 0.980947351 |
| 0.208 | 0.641908721 | 0.812475830 | 0.925788128 | 0.979898997 |
| 0.207 | 0.637480706 | 0.808604286 | 0.923279018 | 0.978806376 |
| 0.206 | 0.633032757 | 0.804686728 | 0.920711840 | 0.977668300 |
| 0.205 | 0.628565276 | 0.800723323 | 0.918086070 | 0.976483575 |
| 0.204 | 0.624078675 | 0.796714255 | 0.915401202 | 0.975251007 |
| 0.203 | 0.619573367 | 0.792659721 | 0.912656747 | 0.973969396 |
| 0.202 | 0.615049773 | 0.788559936 | 0.909852233 | 0.972637544 |
| 0.201 | 0.610508321 | 0.784415127 | 0.906987207 | 0.971254250 |

TABLE 1a:   CLUSTERING PROBABILITY  P(n;N,p)

n = 4

| P | N = 8 | N = 9 | N = 10 | N = 11 |
|---|---|---|---|---|
| 0.200 | 0.605949440 | 0.780225536 | 0.904061235 | 0.969818317 |
| 0.199 | 0.601373568 | 0.775991422 | 0.901073902 | 0.968328546 |
| 0.198 | 0.596781147 | 0.771713059 | 0.898024812 | 0.966783744 |
| 0.197 | 0.592172623 | 0.767390733 | 0.894913591 | 0.965182720 |
| 0.196 | 0.587548449 | 0.763024749 | 0.891739884 | 0.963524291 |
| 0.195 | 0.582909082 | 0.758615424 | 0.888503357 | 0.961807276 |
| 0.194 | 0.578254985 | 0.754163092 | 0.885203700 | 0.960030504 |
| 0.193 | 0.573586623 | 0.749668102 | 0.881840622 | 0.958192813 |
| 0.192 | 0.568904470 | 0.745130816 | 0.878413857 | 0.956293048 |
| 0.191 | 0.564209001 | 0.740551613 | 0.874923159 | 0.954330067 |
| 0.190 | 0.559500699 | 0.735930887 | 0.871368309 | 0.952302740 |
| 0.189 | 0.554780048 | 0.731269047 | 0.867749109 | 0.950209947 |
| 0.188 | 0.550047539 | 0.726566515 | 0.864065386 | 0.948050586 |
| 0.187 | 0.545303666 | 0.721823730 | 0.860316991 | 0.945823568 |
| 0.186 | 0.540548930 | 0.717041145 | 0.856503800 | 0.943527820 |
| 0.185 | 0.535783833 | 0.712219229 | 0.852625714 | 0.941162289 |
| 0.184 | 0.531008883 | 0.707358464 | 0.848682661 | 0.938725938 |
| 0.183 | 0.526224591 | 0.702459347 | 0.844674593 | 0.936217753 |
| 0.182 | 0.521431473 | 0.697522392 | 0.840601488 | 0.933636737 |
| 0.181 | 0.516630050 | 0.692548125 | 0.836463354 | 0.930981920 |
| 0.180 | 0.511820843 | 0.687537086 | 0.832260220 | 0.928252352 |
| 0.179 | 0.507004381 | 0.682489833 | 0.827992148 | 0.925447107 |
| 0.178 | 0.502181194 | 0.677406934 | 0.823659224 | 0.922565289 |
| 0.177 | 0.497351817 | 0.672288975 | 0.819261561 | 0.919606024 |
| 0.176 | 0.492516788 | 0.667136553 | 0.814799303 | 0.916568468 |
| 0.175 | 0.487676648 | 0.661950282 | 0.810272619 | 0.913451807 |
| 0.174 | 0.482831941 | 0.656730788 | 0.805681708 | 0.910255256 |
| 0.173 | 0.477983216 | 0.651478710 | 0.801026798 | 0.906978060 |
| 0.172 | 0.473131024 | 0.646194704 | 0.796308144 | 0.903619498 |
| 0.171 | 0.468275918 | 0.640879436 | 0.791526032 | 0.900178883 |
| 0.170 | 0.463418456 | 0.635533588 | 0.786680774 | 0.896655561 |
| 0.169 | 0.458559197 | 0.630157855 | 0.781772715 | 0.893048913 |
| 0.168 | 0.453698703 | 0.624752944 | 0.776802228 | 0.889358358 |
| 0.167 | 0.448837540 | 0.619319575 | 0.771769713 | 0.885583352 |
| 0.166 | 0.443976276 | 0.613858482 | 0.766675603 | 0.881723390 |
| 0.165 | 0.439115479 | 0.608370413 | 0.761520360 | 0.877778004 |
| 0.164 | 0.434255722 | 0.602856124 | 0.756304474 | 0.873746770 |
| 0.163 | 0.429397580 | 0.597316390 | 0.751028467 | 0.869629302 |
| 0.162 | 0.424541629 | 0.591751991 | 0.745692889 | 0.865425259 |
| 0.161 | 0.419688448 | 0.586163726 | 0.740298322 | 0.861134342 |
| 0.160 | 0.414838617 | 0.580552401 | 0.734845376 | 0.856756295 |
| 0.159 | 0.409992718 | 0.574918835 | 0.729334692 | 0.852290909 |
| 0.158 | 0.405151334 | 0.569263860 | 0.723766942 | 0.847738018 |
| 0.157 | 0.400315051 | 0.563588319 | 0.718142825 | 0.843097505 |
| 0.156 | 0.395484456 | 0.557893064 | 0.712463072 | 0.838369298 |
| 0.155 | 0.390660137 | 0.552178961 | 0.706728444 | 0.833553376 |
| 0.154 | 0.385842681 | 0.546446883 | 0.700939730 | 0.828649762 |
| 0.153 | 0.381032681 | 0.540697718 | 0.695097751 | 0.823658533 |
| 0.152 | 0.376230726 | 0.534932361 | 0.689203356 | 0.818579813 |
| 0.151 | 0.371437408 | 0.529151719 | 0.683257423 | 0.813413778 |

TABLE 1a: CLUSTERING PROBABILITY P(n;N,p)

n = 4

| P | N = 8 | N = 9 | N = 10 | N = 11 |
|---|---|---|---|---|
| 0.150 | 0.366653321 | 0.523356707 | 0.677260860 | 0.808160654 |
| 0.149 | 0.361879056 | 0.517548251 | 0.671214604 | 0.802820720 |
| 0.148 | 0.357115208 | 0.511727287 | 0.665119621 | 0.797394307 |
| 0.147 | 0.352362371 | 0.505894758 | 0.658976907 | 0.791881798 |
| 0.146 | 0.347621138 | 0.500051617 | 0.652787484 | 0.786283630 |
| 0.145 | 0.342892103 | 0.494198827 | 0.646552403 | 0.780600292 |
| 0.144 | 0.338175860 | 0.488337358 | 0.640272745 | 0.774832330 |
| 0.143 | 0.333473004 | 0.482468188 | 0.633949617 | 0.768980341 |
| 0.142 | 0.328784127 | 0.476592303 | 0.627584153 | 0.763044979 |
| 0.141 | 0.324109822 | 0.470710697 | 0.621177517 | 0.757026953 |
| 0.140 | 0.319450681 | 0.464824372 | 0.614730897 | 0.750927025 |
| 0.139 | 0.314807297 | 0.458934335 | 0.608245509 | 0.744746014 |
| 0.138 | 0.310180259 | 0.453041602 | 0.601722596 | 0.738484794 |
| 0.137 | 0.305570157 | 0.447147194 | 0.595163424 | 0.732144297 |
| 0.136 | 0.300977580 | 0.441252138 | 0.588569289 | 0.725725506 |
| 0.135 | 0.296403113 | 0.435357469 | 0.581941509 | 0.719229464 |
| 0.134 | 0.291847344 | 0.429464225 | 0.575281427 | 0.712657267 |
| 0.133 | 0.287310856 | 0.423573451 | 0.568590411 | 0.706010070 |
| 0.132 | 0.282794230 | 0.417686197 | 0.561869854 | 0.699289079 |
| 0.131 | 0.278298048 | 0.411803515 | 0.555121170 | 0.692495561 |
| 0.130 | 0.273822888 | 0.405926466 | 0.548345798 | 0.685630833 |
| 0.129 | 0.269369324 | 0.400056110 | 0.541545199 | 0.678696271 |
| 0.128 | 0.264937932 | 0.394193515 | 0.534720855 | 0.671693305 |
| 0.127 | 0.260529281 | 0.388339750 | 0.527874272 | 0.664623418 |
| 0.126 | 0.256143941 | 0.382495886 | 0.521006973 | 0.657488149 |
| 0.125 | 0.251782477 | 0.376662999 | 0.514120506 | 0.650289091 |
| 0.124 | 0.247445451 | 0.370842167 | 0.507216435 | 0.643027890 |
| 0.123 | 0.243133423 | 0.365034469 | 0.500296346 | 0.635706244 |
| 0.122 | 0.238846949 | 0.359240986 | 0.493361842 | 0.628325905 |
| 0.121 | 0.234586581 | 0.353462800 | 0.486414545 | 0.620888677 |
| 0.120 | 0.230352869 | 0.347700993 | 0.479456093 | 0.613396414 |
| 0.119 | 0.226146357 | 0.341956651 | 0.472488142 | 0.605851021 |
| 0.118 | 0.221967587 | 0.336230856 | 0.465512366 | 0.598254454 |
| 0.117 | 0.217817096 | 0.330524692 | 0.458530450 | 0.590608719 |
| 0.116 | 0.213695416 | 0.324839242 | 0.451544097 | 0.582915867 |
| 0.115 | 0.209603075 | 0.319175588 | 0.444555024 | 0.575178001 |
| 0.114 | 0.205540598 | 0.313534810 | 0.437564961 | 0.567397267 |
| 0.113 | 0.201508502 | 0.307917987 | 0.430575648 | 0.559575860 |
| 0.112 | 0.197507302 | 0.302326196 | 0.423588842 | 0.551716018 |
| 0.111 | 0.193537506 | 0.296760510 | 0.416606306 | 0.543820024 |
| 0.110 | 0.189599617 | 0.291222000 | 0.409629817 | 0.535890203 |
| 0.109 | 0.185694132 | 0.285711734 | 0.402661160 | 0.527928923 |
| 0.108 | 0.181821544 | 0.280230775 | 0.395702128 | 0.519938592 |
| 0.107 | 0.177982338 | 0.274780183 | 0.388754524 | 0.511921656 |
| 0.106 | 0.174176995 | 0.269361013 | 0.381820157 | 0.503880602 |
| 0.105 | 0.170405987 | 0.263974314 | 0.374900841 | 0.495817954 |
| 0.104 | 0.166669782 | 0.258621130 | 0.367998398 | 0.487736268 |
| 0.103 | 0.162968839 | 0.253302499 | 0.361114652 | 0.479638139 |
| 0.102 | 0.159303614 | 0.248019453 | 0.354251433 | 0.471526191 |
| 0.101 | 0.155674551 | 0.242773017 | 0.347410571 | 0.463403083 |

TABLE 1a:   CLUSTERING PROBABILITY   P(n;N,p)

n = 4

| P | N = 8 | N = 9 | N = 10 | N = 11 |
|------|------------|------------|------------|------------|
| 0.100 | 0.152082090 | 0.237564208 | 0.340593900 | 0.455271502 |
| 0.099 | 0.148526663 | 0.232394036 | 0.333803254 | 0.447134164 |
| 0.098 | 0.145008694 | 0.227263502 | 0.327040467 | 0.438993812 |
| 0.097 | 0.141528599 | 0.222173600 | 0.320307372 | 0.430853214 |
| 0.096 | 0.138086786 | 0.217125312 | 0.313605801 | 0.422715162 |
| 0.095 | 0.134683655 | 0.212119612 | 0.306937581 | 0.414582471 |
| 0.094 | 0.131319598 | 0.207157465 | 0.300304537 | 0.406457975 |
| 0.093 | 0.127994997 | 0.202239824 | 0.293708487 | 0.398344527 |
| 0.092 | 0.124710226 | 0.197367630 | 0.287151246 | 0.390244997 |
| 0.091 | 0.121465650 | 0.192541815 | 0.280634618 | 0.382162269 |
| 0.090 | 0.118261624 | 0.187763297 | 0.274160403 | 0.374099242 |
| 0.089 | 0.115098494 | 0.183032981 | 0.267730389 | 0.366058824 |
| 0.088 | 0.111976597 | 0.178351761 | 0.261346355 | 0.358043933 |
| 0.087 | 0.108896259 | 0.173720516 | 0.255010069 | 0.350057495 |
| 0.086 | 0.105857795 | 0.169140111 | 0.248723286 | 0.342102440 |
| 0.085 | 0.102861512 | 0.164611396 | 0.242487748 | 0.334181701 |
| 0.084 | 0.099907704 | 0.160135208 | 0.236305182 | 0.326298213 |
| 0.083 | 0.096996656 | 0.155712365 | 0.230177300 | 0.318454910 |
| 0.082 | 0.094128641 | 0.151343671 | 0.224105796 | 0.310654720 |
| 0.081 | 0.091303920 | 0.147029913 | 0.218092347 | 0.302900569 |
| 0.080 | 0.088522744 | 0.142771861 | 0.212138610 | 0.295195372 |
| 0.079 | 0.085785351 | 0.138570266 | 0.206246224 | 0.287542038 |
| 0.078 | 0.083091968 | 0.134425862 | 0.200416804 | 0.279943460 |
| 0.077 | 0.080442809 | 0.130339363 | 0.194651942 | 0.272402518 |
| 0.076 | 0.077838076 | 0.126311465 | 0.188953208 | 0.264922074 |
| 0.075 | 0.075277957 | 0.122342841 | 0.183322146 | 0.257504972 |
| 0.074 | 0.072762630 | 0.118434147 | 0.177760275 | 0.250154033 |
| 0.073 | 0.070292255 | 0.114586015 | 0.172269083 | 0.242872054 |
| 0.072 | 0.067866984 | 0.110799057 | 0.166850032 | 0.235661805 |
| 0.071 | 0.065486952 | 0.107073862 | 0.161504554 | 0.228526027 |
| 0.070 | 0.063152279 | 0.103410995 | 0.156234049 | 0.221467429 |
| 0.069 | 0.060863075 | 0.099811000 | 0.151039882 | 0.214488686 |
| 0.068 | 0.058619431 | 0.096274395 | 0.145923389 | 0.207592436 |
| 0.067 | 0.056421427 | 0.092801673 | 0.140885867 | 0.200781275 |
| 0.066 | 0.054269124 | 0.089393303 | 0.135928576 | 0.194057761 |
| 0.065 | 0.052162572 | 0.086049727 | 0.131052741 | 0.187424404 |
| 0.064 | 0.050101803 | 0.082771360 | 0.126259546 | 0.180883667 |
| 0.063 | 0.048086834 | 0.079558591 | 0.121550135 | 0.174437962 |
| 0.062 | 0.046117664 | 0.076411781 | 0.116925609 | 0.168089648 |
| 0.061 | 0.044194280 | 0.073331262 | 0.112387026 | 0.161841030 |
| 0.060 | 0.042316647 | 0.070317337 | 0.107935400 | 0.155694353 |
| 0.059 | 0.040484718 | 0.067370279 | 0.103571699 | 0.149651799 |
| 0.058 | 0.038698425 | 0.064490330 | 0.099296843 | 0.143715486 |
| 0.057 | 0.036957686 | 0.061677703 | 0.095111702 | 0.137887468 |
| 0.056 | 0.035262399 | 0.058932578 | 0.091017098 | 0.132169725 |
| 0.055 | 0.033612443 | 0.056255101 | 0.087013799 | 0.126564165 |
| 0.054 | 0.032007683 | 0.053645387 | 0.083102520 | 0.121072621 |
| 0.053 | 0.030447961 | 0.051103517 | 0.079283923 | 0.115696845 |
| 0.052 | 0.028933102 | 0.048629537 | 0.075558611 | 0.110438510 |
| 0.051 | 0.027462912 | 0.046223457 | 0.071927133 | 0.105299200 |

TABLE 1a:   CLUSTERING PROBABILITY   P(n;N,p)

n = 4

| P | N ≡ 8 | N ≡ 9 | N ≡ 10 | N ≡ 11 |
|---|---|---|---|---|
| 0.050 | 0.026037177 | 0.043885252 | 0.068389974 | 0.100280416 |
| 0.049 | 0.024655664 | 0.041614861 | 0.064947563 | 0.095383562 |
| 0.048 | 0.023318119 | 0.039412184 | 0.061600262 | 0.090609954 |
| 0.047 | 0.022024268 | 0.037277085 | 0.058348374 | 0.085960805 |
| 0.046 | 0.020773817 | 0.035209386 | 0.055192134 | 0.081437232 |
| 0.045 | 0.019566450 | 0.033208874 | 0.052131708 | 0.077040246 |
| 0.044 | 0.018401830 | 0.031275292 | 0.049167198 | 0.072770753 |
| 0.043 | 0.017279599 | 0.029408343 | 0.046298633 | 0.068629547 |
| 0.042 | 0.016199376 | 0.027607689 | 0.043525970 | 0.064617310 |
| 0.041 | 0.015160760 | 0.025872948 | 0.040849093 | 0.060734609 |
| 0.040 | 0.014163324 | 0.024203696 | 0.038267812 | 0.056981889 |
| 0.039 | 0.013206621 | 0.022599465 | 0.035781859 | 0.053359474 |
| 0.038 | 0.012290180 | 0.021059740 | 0.033390889 | 0.049867560 |
| 0.037 | 0.011413506 | 0.019583962 | 0.031094473 | 0.046506214 |
| 0.036 | 0.010576080 | 0.018171526 | 0.028892106 | 0.043275371 |
| 0.035 | 0.009777359 | 0.016821778 | 0.026783194 | 0.040174829 |
| 0.034 | 0.009016777 | 0.015534017 | 0.024767062 | 0.037204246 |
| 0.033 | 0.008293741 | 0.014307493 | 0.022842944 | 0.034363136 |
| 0.032 | 0.007607634 | 0.013141405 | 0.021009987 | 0.031650868 |
| 0.031 | 0.006957812 | 0.012034903 | 0.019267249 | 0.029066658 |
| 0.030 | 0.006343608 | 0.010987084 | 0.017613692 | 0.026609572 |
| 0.029 | 0.005764325 | 0.009996993 | 0.016048186 | 0.024278514 |
| 0.028 | 0.005219241 | 0.009063622 | 0.014569503 | 0.022072231 |
| 0.027 | 0.004707610 | 0.008185909 | 0.013176318 | 0.019989301 |
| 0.026 | 0.004228653 | 0.007362736 | 0.011867207 | 0.018028137 |
| 0.025 | 0.003781568 | 0.006592930 | 0.010640640 | 0.016186978 |
| 0.024 | 0.003365522 | 0.005875260 | 0.009494988 | 0.014463887 |
| 0.023 | 0.002979656 | 0.005208439 | 0.008428512 | 0.012856746 |
| 0.022 | 0.002623081 | 0.004591120 | 0.007439367 | 0.011363256 |
| 0.021 | 0.002294878 | 0.004021895 | 0.006525597 | 0.009980926 |
| 0.020 | 0.001994099 | 0.003499298 | 0.005685135 | 0.008707076 |
| 0.019 | 0.001719767 | 0.003021801 | 0.004915799 | 0.007538829 |
| 0.018 | 0.001470875 | 0.002587810 | 0.004215290 | 0.006473108 |
| 0.017 | 0.001246382 | 0.002195672 | 0.003581191 | 0.005506632 |
| 0.016 | 0.001045220 | 0.001843667 | 0.003010966 | 0.004635910 |
| 0.015 | 0.000866288 | 0.001530010 | 0.002501953 | 0.003857239 |
| 0.014 | 0.000708452 | 0.001252848 | 0.002051368 | 0.003166699 |
| 0.013 | 0.000570547 | 0.001010263 | 0.001656296 | 0.002560147 |
| 0.012 | 0.000451374 | 0.000800265 | 0.001313696 | 0.002033215 |
| 0.011 | 0.000349704 | 0.000620798 | 0.001020392 | 0.001581302 |
| 0.010 | 0.000264270 | 0.000469733 | 0.000773075 | 0.001199574 |
| 0.009 | 0.000193776 | 0.000344868 | 0.000568299 | 0.000882953 |
| 0.008 | 0.000136887 | 0.000243930 | 0.000402477 | 0.000626119 |
| 0.007 | 0.000092237 | 0.000164572 | 0.000271884 | 0.000423497 |
| 0.006 | 0.000058422 | 0.000104371 | 0.000172645 | 0.000269261 |
| 0.005 | 0.000034005 | 0.000060827 | 0.000100744 | 0.000157321 |
| 0.004 | 0.000017512 | 0.000031363 | 0.000052011 | 0.000081322 |
| 0.003 | 0.000007430 | 0.000013325 | 0.000022125 | 0.000034637 |
| 0.002 | 0.000002214 | 0.000003976 | 0.000006610 | 0.000010361 |
| 0.001 | 0.000000278 | 0.000000500 | 0.000000833 | 0.000001308 |

TABLE 1a:   CLUSTERING PROBABILITY   P(n;N,p)

n = 4

| P | N ≡12 | N ≡13 | N ≡14 | N ≡15 |
|---|---|---|---|---|
| 0.200 | 0.993982681 | 0.999415443 | 0.999978281 | 0.999999803 |
| 0.199 | 0.993511872 | 0.999338377 | 0.999973570 | 0.999999735 |
| 0.198 | 0.993011363 | 0.999252917 | 0.999967971 | 0.999999646 |
| 0.197 | 0.992479804 | 0.999158344 | 0.999961342 | 0.999999528 |
| 0.196 | 0.991915809 | 0.999053896 | 0.999953522 | 0.999999376 |
| 0.195 | 0.991317964 | 0.998938765 | 0.999944331 | 0.999999178 |
| 0.194 | 0.990684827 | 0.998812099 | 0.999933565 | 0.999998923 |
| 0.193 | 0.990014925 | 0.998672995 | 0.999920999 | 0.999998596 |
| 0.192 | 0.989306758 | 0.998520507 | 0.999906378 | 0.999998178 |
| 0.191 | 0.988558800 | 0.998353636 | 0.999889422 | 0.999997646 |
| 0.190 | 0.987769499 | 0.998171333 | 0.999869820 | 0.999996974 |
| 0.189 | 0.986937276 | 0.997972500 | 0.999847227 | 0.999996127 |
| 0.188 | 0.986060534 | 0.997755987 | 0.999821264 | 0.999995066 |
| 0.187 | 0.985137648 | 0.997520591 | 0.999791516 | 0.999993742 |
| 0.186 | 0.984166977 | 0.997265055 | 0.999757528 | 0.999992099 |
| 0.185 | 0.983146858 | 0.996988071 | 0.999718801 | 0.999990068 |
| 0.184 | 0.982075610 | 0.996688275 | 0.999674794 | 0.999987569 |
| 0.183 | 0.980951537 | 0.996364251 | 0.999624919 | 0.999984510 |
| 0.182 | 0.979772928 | 0.996014527 | 0.999568538 | 0.999980781 |
| 0.181 | 0.978538056 | 0.995637579 | 0.999504961 | 0.999976257 |
| 0.180 | 0.977245185 | 0.995231827 | 0.999433447 | 0.999970790 |
| 0.179 | 0.975892570 | 0.994795639 | 0.999353194 | 0.999964214 |
| 0.178 | 0.974478454 | 0.994327328 | 0.999263345 | 0.999956338 |
| 0.177 | 0.973001077 | 0.993825155 | 0.999162981 | 0.999946941 |
| 0.176 | 0.971458674 | 0.993287327 | 0.999051118 | 0.999935778 |
| 0.175 | 0.969849475 | 0.992712002 | 0.998926709 | 0.999922567 |
| 0.174 | 0.968171711 | 0.992097286 | 0.998788639 | 0.999906994 |
| 0.173 | 0.966423617 | 0.991441233 | 0.998635721 | 0.999888705 |
| 0.172 | 0.964603425 | 0.990741853 | 0.998466701 | 0.999867305 |
| 0.171 | 0.962709379 | 0.989997103 | 0.998280248 | 0.999842353 |
| 0.170 | 0.960739726 | 0.989204899 | 0.998074957 | 0.999813361 |
| 0.169 | 0.958692724 | 0.988363109 | 0.997849349 | 0.999779788 |
| 0.168 | 0.956566643 | 0.987469560 | 0.997601865 | 0.999741039 |
| 0.167 | 0.954359767 | 0.986522037 | 0.997330867 | 0.999696456 |
| 0.166 | 0.952070396 | 0.985518287 | 0.997034639 | 0.999645322 |
| 0.165 | 0.949696847 | 0.984456020 | 0.996711382 | 0.999586850 |
| 0.164 | 0.947237459 | 0.983332908 | 0.996359216 | 0.999520184 |
| 0.163 | 0.944690595 | 0.982146594 | 0.995976180 | 0.999444392 |
| 0.162 | 0.942054639 | 0.980894690 | 0.995560228 | 0.999358464 |
| 0.161 | 0.939328007 | 0.979574780 | 0.995109236 | 0.999261308 |
| 0.160 | 0.936509142 | 0.978184424 | 0.994620993 | 0.999151745 |
| 0.159 | 0.933596520 | 0.976721157 | 0.994093208 | 0.999028507 |
| 0.158 | 0.930588649 | 0.975182499 | 0.993523509 | 0.998890232 |
| 0.157 | 0.927484077 | 0.973565951 | 0.992909442 | 0.998735462 |
| 0.156 | 0.924281389 | 0.971869003 | 0.992248473 | 0.998562637 |
| 0.155 | 0.920979210 | 0.970089134 | 0.991537992 | 0.998370096 |
| 0.154 | 0.917576211 | 0.968223817 | 0.990775310 | 0.998156069 |
| 0.153 | 0.914071108 | 0.966270522 | 0.989957662 | 0.997918678 |
| 0.152 | 0.910462662 | 0.964226720 | 0.989082213 | 0.997655934 |
| 0.151 | 0.906749688 | 0.962089886 | 0.988146053 | 0.997365731 |

TABLE la: CLUSTERING PROBABILITY P(n;N,p)

n = 4

| P | N =12 | N =13 | N =14 | N =15 |
|---|---|---|---|---|
| 0.150 | 0.902931052 | 0.959857504 | 0.987146208 | 0.997045849 |
| 0.149 | 0.899005674 | 0.957527068 | 0.986079634 | 0.996693950 |
| 0.148 | 0.894972532 | 0.955096088 | 0.984943229 | 0.996307573 |
| 0.147 | 0.890830661 | 0.952562097 | 0.983733827 | 0.995884140 |
| 0.146 | 0.886579160 | 0.949922647 | 0.982448209 | 0.995420949 |
| 0.145 | 0.882217188 | 0.947175320 | 0.981083105 | 0.994915175 |
| 0.144 | 0.877743971 | 0.944317730 | 0.979635199 | 0.994363871 |
| 0.143 | 0.873158802 | 0.941347526 | 0.978101116 | 0.993763968 |
| 0.142 | 0.868461043 | 0.938262398 | 0.976477465 | 0.993112275 |
| 0.141 | 0.863650128 | 0.935060081 | 0.974760808 | 0.992405480 |
| 0.140 | 0.858725562 | 0.931738358 | 0.972947678 | 0.991640152 |
| 0.139 | 0.853686927 | 0.928295063 | 0.971034586 | 0.990812742 |
| 0.138 | 0.848533879 | 0.924728091 | 0.969018023 | 0.989919586 |
| 0.137 | 0.843266157 | 0.921035397 | 0.966894469 | 0.988956908 |
| 0.136 | 0.837883574 | 0.917215001 | 0.964660397 | 0.987920822 |
| 0.135 | 0.832386030 | 0.913264995 | 0.962312276 | 0.986807338 |
| 0.134 | 0.826773504 | 0.909183543 | 0.959846584 | 0.985612362 |
| 0.133 | 0.821046064 | 0.904968890 | 0.957259810 | 0.984331707 |
| 0.132 | 0.815203860 | 0.900619364 | 0.954548459 | 0.982961093 |
| 0.131 | 0.809247132 | 0.896133380 | 0.951709064 | 0.981496152 |
| 0.130 | 0.803176208 | 0.891509443 | 0.948738190 | 0.979932440 |
| 0.129 | 0.796991505 | 0.886746155 | 0.945632439 | 0.978265436 |
| 0.128 | 0.790693534 | 0.881842218 | 0.942388462 | 0.976490557 |
| 0.127 | 0.784282896 | 0.876796436 | 0.939002963 | 0.974603158 |
| 0.126 | 0.777760284 | 0.871607723 | 0.935472706 | 0.972598544 |
| 0.125 | 0.771126489 | 0.866275102 | 0.931794525 | 0.970471978 |
| 0.124 | 0.764382393 | 0.860797713 | 0.927965331 | 0.968218690 |
| 0.123 | 0.757528975 | 0.855174814 | 0.923982116 | 0.965833885 |
| 0.122 | 0.750567312 | 0.849405786 | 0.919841966 | 0.963312753 |
| 0.121 | 0.743498575 | 0.843490137 | 0.915542066 | 0.960650482 |
| 0.120 | 0.736324034 | 0.837427502 | 0.911079708 | 0.957842263 |
| 0.119 | 0.729045056 | 0.831217652 | 0.906452296 | 0.954883305 |
| 0.118 | 0.721663107 | 0.824860490 | 0.901657361 | 0.951768845 |
| 0.117 | 0.714179751 | 0.818356063 | 0.896692561 | 0.948494160 |
| 0.116 | 0.706596649 | 0.811704555 | 0.891555692 | 0.945054576 |
| 0.115 | 0.698915560 | 0.804906296 | 0.886244695 | 0.941445486 |
| 0.114 | 0.691138344 | 0.797961766 | 0.880757666 | 0.937662355 |
| 0.113 | 0.683266957 | 0.790871590 | 0.875092858 | 0.933700737 |
| 0.112 | 0.675303451 | 0.783636549 | 0.869248693 | 0.929556289 |
| 0.111 | 0.667249978 | 0.776257575 | 0.863223766 | 0.925224779 |
| 0.110 | 0.659108786 | 0.768735757 | 0.857016856 | 0.920702102 |
| 0.109 | 0.650882219 | 0.761072343 | 0.850626928 | 0.915984295 |
| 0.108 | 0.642572715 | 0.753268738 | 0.844053141 | 0.911067546 |
| 0.107 | 0.634182808 | 0.745326509 | 0.837294859 | 0.905948211 |
| 0.106 | 0.625715126 | 0.737247385 | 0.830351649 | 0.900622823 |
| 0.105 | 0.617172389 | 0.729033257 | 0.823223294 | 0.895088111 |
| 0.104 | 0.608557407 | 0.720686182 | 0.815909796 | 0.889341008 |
| 0.103 | 0.599873082 | 0.712208377 | 0.808411381 | 0.883378668 |
| 0.102 | 0.591122404 | 0.703602228 | 0.800728505 | 0.877198474 |
| 0.101 | 0.582308451 | 0.694870283 | 0.792861859 | 0.870798058 |

TABLE 1a:   CLUSTERING PROBABILITY  P(n;N,p)

n = 4

| P | N ≡12 | N ≡13 | N ≡14 | N ≡15 |
|---|---|---|---|---|
| 0.100 | 0.573434385 | 0.686015257 | 0.784812372 | 0.864175307 |
| 0.099 | 0.564503455 | 0.677040027 | 0.776581217 | 0.857328378 |
| 0.098 | 0.555518989 | 0.667947635 | 0.768169817 | 0.850255711 |
| 0.097 | 0.546484397 | 0.658741288 | 0.759579842 | 0.842956038 |
| 0.096 | 0.537403167 | 0.649424353 | 0.750813219 | 0.835428399 |
| 0.095 | 0.528278863 | 0.640000359 | 0.741872130 | 0.827672149 |
| 0.094 | 0.519115122 | 0.630472994 | 0.732759018 | 0.819686968 |
| 0.093 | 0.509915655 | 0.620846105 | 0.723476587 | 0.811472876 |
| 0.092 | 0.500684240 | 0.611123694 | 0.714027804 | 0.803030238 |
| 0.091 | 0.491424722 | 0.601309918 | 0.704415900 | 0.794359775 |
| 0.090 | 0.482141009 | 0.591409084 | 0.694644370 | 0.785462572 |
| 0.089 | 0.472837071 | 0.581425649 | 0.684716976 | 0.776340088 |
| 0.088 | 0.463516937 | 0.571364216 | 0.674637745 | 0.766994160 |
| 0.087 | 0.454184690 | 0.561229530 | 0.664410966 | 0.757427011 |
| 0.086 | 0.444844464 | 0.551026475 | 0.654041193 | 0.747641259 |
| 0.085 | 0.435500444 | 0.540760073 | 0.643533243 | 0.737639915 |
| 0.084 | 0.426156860 | 0.530435474 | 0.632892188 | 0.727426397 |
| 0.083 | 0.416817982 | 0.520057959 | 0.622123362 | 0.717004525 |
| 0.082 | 0.407488122 | 0.509632931 | 0.611232350 | 0.706378529 |
| 0.081 | 0.398171626 | 0.499165911 | 0.600224986 | 0.695553047 |
| 0.080 | 0.388872869 | 0.488662536 | 0.589107354 | 0.684533134 |
| 0.079 | 0.379596256 | 0.478128548 | 0.577885775 | 0.673324250 |
| 0.078 | 0.370346215 | 0.467569797 | 0.566566808 | 0.661932271 |
| 0.077 | 0.361127195 | 0.456992226 | 0.555157244 | 0.650363482 |
| 0.076 | 0.351943658 | 0.446401872 | 0.543664097 | 0.638624572 |
| 0.075 | 0.342800079 | 0.435804859 | 0.532094596 | 0.626722636 |
| 0.074 | 0.333700941 | 0.425207388 | 0.520456184 | 0.614665167 |
| 0.073 | 0.324650728 | 0.414615735 | 0.508756504 | 0.602460051 |
| 0.072 | 0.315653924 | 0.404036240 | 0.497003395 | 0.590115562 |
| 0.071 | 0.306715006 | 0.393475304 | 0.485204880 | 0.577640351 |
| 0.070 | 0.297838441 | 0.382939379 | 0.473369159 | 0.565043442 |
| 0.069 | 0.289028682 | 0.372434963 | 0.461504595 | 0.552334218 |
| 0.068 | 0.280290158 | 0.361968588 | 0.449619711 | 0.539522413 |
| 0.067 | 0.271627278 | 0.351546818 | 0.437723173 | 0.526618099 |
| 0.066 | 0.263044418 | 0.341176237 | 0.425823779 | 0.513631674 |
| 0.065 | 0.254545921 | 0.330863441 | 0.413930451 | 0.500573848 |
| 0.064 | 0.246136089 | 0.320615033 | 0.402052219 | 0.487455628 |
| 0.063 | 0.237819180 | 0.310437609 | 0.390198211 | 0.474288301 |
| 0.062 | 0.229599403 | 0.300337753 | 0.378377637 | 0.461083421 |
| 0.061 | 0.221480911 | 0.290322028 | 0.366599777 | 0.447852785 |
| 0.060 | 0.213467795 | 0.280396964 | 0.354873966 | 0.434608421 |
| 0.059 | 0.205564084 | 0.270569054 | 0.343209582 | 0.421362560 |
| 0.058 | 0.197773731 | 0.260844737 | 0.331616027 | 0.408127625 |
| 0.057 | 0.190100617 | 0.251230396 | 0.320102715 | 0.394916200 |
| 0.056 | 0.182548539 | 0.241732344 | 0.308679054 | 0.381741012 |
| 0.055 | 0.175121204 | 0.232356814 | 0.297354431 | 0.368614909 |
| 0.054 | 0.167822230 | 0.223109951 | 0.286138197 | 0.355550832 |
| 0.053 | 0.160655132 | 0.213997800 | 0.275039645 | 0.342561788 |
| 0.052 | 0.153623324 | 0.205026298 | 0.264068001 | 0.329660832 |
| 0.051 | 0.146730106 | 0.196201260 | 0.253232398 | 0.316861031 |

TABLE 1a:   CLUSTERING PROBABILITY   $P(n;N,p)$

$n = 4$

| P | N =12 | N =13 | N =14 | N =15 |
|---|---|---|---|---|
| 0.050 | 0.139978665 | 0.187528373 | 0.242541865 | 0.304175442 |
| 0.049 | 0.133372064 | 0.179013181 | 0.232005305 | 0.291617081 |
| 0.048 | 0.126913238 | 0.170661078 | 0.221631477 | 0.279198895 |
| 0.047 | 0.120604989 | 0.162477296 | 0.211428980 | 0.266933734 |
| 0.046 | 0.114449978 | 0.154466890 | 0.201406232 | 0.254834316 |
| 0.045 | 0.108450721 | 0.146634736 | 0.191571453 | 0.242913201 |
| 0.044 | 0.102609583 | 0.138985511 | 0.181932641 | 0.231182756 |
| 0.043 | 0.096928769 | 0.131523686 | 0.172497559 | 0.219655127 |
| 0.042 | 0.091410320 | 0.124253517 | 0.163273712 | 0.208342202 |
| 0.041 | 0.086056110 | 0.117179028 | 0.154268328 | 0.197255582 |
| 0.040 | 0.080867831 | 0.110304006 | 0.145488337 | 0.186406547 |
| 0.039 | 0.075846998 | 0.103631982 | 0.136940354 | 0.175806020 |
| 0.038 | 0.070994933 | 0.097166230 | 0.128630657 | 0.165464537 |
| 0.037 | 0.066312765 | 0.090909745 | 0.120565165 | 0.155392212 |
| 0.036 | 0.061801421 | 0.084865239 | 0.112749421 | 0.145598702 |
| 0.035 | 0.057461619 | 0.079035125 | 0.105188573 | 0.136093171 |
| 0.034 | 0.053293863 | 0.073421508 | 0.097887348 | 0.126884261 |
| 0.033 | 0.049298437 | 0.068026172 | 0.090850037 | 0.117980051 |
| 0.032 | 0.045475398 | 0.062850570 | 0.084080472 | 0.109388028 |
| 0.031 | 0.041824567 | 0.057895809 | 0.077582007 | 0.101115047 |
| 0.030 | 0.038345527 | 0.053162643 | 0.071357496 | 0.093167303 |
| 0.029 | 0.035037612 | 0.048651456 | 0.065409274 | 0.085550290 |
| 0.028 | 0.031899903 | 0.044362254 | 0.059739136 | 0.078268771 |
| 0.027 | 0.028931220 | 0.040294652 | 0.054348316 | 0.071326740 |
| 0.026 | 0.026130118 | 0.036447863 | 0.049237468 | 0.064727390 |
| 0.025 | 0.023494873 | 0.032820681 | 0.044406644 | 0.058473081 |
| 0.024 | 0.021023484 | 0.029411476 | 0.039855275 | 0.052565300 |
| 0.023 | 0.018713661 | 0.026218179 | 0.035582148 | 0.047004631 |
| 0.022 | 0.016562818 | 0.023238267 | 0.031585390 | 0.041790722 |
| 0.021 | 0.014568066 | 0.020468755 | 0.027862440 | 0.036922247 |
| 0.020 | 0.012726207 | 0.017906180 | 0.024410036 | 0.032396877 |
| 0.019 | 0.011033726 | 0.015546590 | 0.021224191 | 0.028211244 |
| 0.018 | 0.009486784 | 0.013385534 | 0.018300170 | 0.024360905 |
| 0.017 | 0.008081209 | 0.011418042 | 0.015632471 | 0.020840313 |
| 0.016 | 0.006812490 | 0.009638619 | 0.013214805 | 0.017642782 |
| 0.015 | 0.005675768 | 0.008041229 | 0.011040073 | 0.014760449 |
| 0.014 | 0.004665828 | 0.006619281 | 0.009100342 | 0.012184246 |
| 0.013 | 0.003777092 | 0.005365616 | 0.007386828 | 0.009903860 |
| 0.012 | 0.003003611 | 0.004272493 | 0.005889868 | 0.007907702 |
| 0.011 | 0.002339053 | 0.003331575 | 0.004598903 | 0.006182870 |
| 0.010 | 0.001776701 | 0.002533913 | 0.003502449 | 0.004715114 |
| 0.009 | 0.001309437 | 0.001869934 | 0.002588076 | 0.003488797 |
| 0.008 | 0.000929737 | 0.001329423 | 0.001842383 | 0.002486859 |
| 0.007 | 0.000629664 | 0.000901508 | 0.001250975 | 0.001690779 |
| 0.006 | 0.000400852 | 0.000574646 | 0.000798431 | 0.001080533 |
| 0.005 | 0.000234503 | 0.000336602 | 0.000468283 | 0.000634552 |
| 0.004 | 0.000121372 | 0.000174436 | 0.000242985 | 0.000329680 |
| 0.003 | 0.000051760 | 0.000074484 | 0.000103885 | 0.000141129 |
| 0.002 | 0.000015503 | 0.000022337 | 0.000031193 | 0.000042429 |
| 0.001 | 0.000001959 | 0.000002826 | 0.000003951 | 0.000005381 |

TABLE 1a:   CLUSTERING PROBABILITY   P(n;N,p)

n = 4

| P | N ≡16 | N ≡17 | N ≡18 | N ≡19 |
|---|---|---|---|---|
| 0.150 | 0.999565414 | 0.999965043 | 0.999998820 | 0.999999992 |
| 0.149 | 0.999489754 | 0.999955978 | 0.999998347 | 0.999999985 |
| 0.148 | 0.999402966 | 0.999944874 | 0.999997705 | 0.999999975 |
| 0.147 | 0.999303723 | 0.999931348 | 0.999996843 | 0.999999959 |
| 0.146 | 0.999190587 | 0.999914954 | 0.999995697 | 0.999999933 |
| 0.145 | 0.999061992 | 0.999895184 | 0.999994183 | 0.999999893 |
| 0.144 | 0.998916247 | 0.999871458 | 0.999992202 | 0.999999832 |
| 0.143 | 0.998751525 | 0.999843116 | 0.999989631 | 0.999999741 |
| 0.142 | 0.998565861 | 0.999809412 | 0.999986317 | 0.999999608 |
| 0.141 | 0.998357144 | 0.999769507 | 0.999982080 | 0.999999415 |
| 0.140 | 0.998123112 | 0.999722457 | 0.999976699 | 0.999999140 |
| 0.139 | 0.997861350 | 0.999667210 | 0.999969912 | 0.999998752 |
| 0.138 | 0.997569284 | 0.999602592 | 0.999961408 | 0.999998211 |
| 0.137 | 0.997244175 | 0.999527304 | 0.999950820 | 0.999997466 |
| 0.136 | 0.996883120 | 0.999439906 | 0.999937717 | 0.999996449 |
| 0.135 | 0.996483043 | 0.999338815 | 0.999921597 | 0.999995078 |
| 0.134 | 0.996040697 | 0.999222292 | 0.999901877 | 0.999993246 |
| 0.133 | 0.995552659 | 0.999088431 | 0.999877884 | 0.999990819 |
| 0.132 | 0.995015330 | 0.998935157 | 0.999848845 | 0.999987634 |
| 0.131 | 0.994424935 | 0.998760209 | 0.999813877 | 0.999983490 |
| 0.130 | 0.993777517 | 0.998561137 | 0.999771973 | 0.999978138 |
| 0.129 | 0.993068945 | 0.998335292 | 0.999721996 | 0.999971282 |
| 0.128 | 0.992294908 | 0.998079817 | 0.999662661 | 0.999962562 |
| 0.127 | 0.991450923 | 0.997791640 | 0.999592526 | 0.999951550 |
| 0.126 | 0.990532332 | 0.997467468 | 0.999509976 | 0.999937738 |
| 0.125 | 0.989534308 | 0.997103778 | 0.999413214 | 0.999920526 |
| 0.124 | 0.988451857 | 0.996696813 | 0.999300245 | 0.999899210 |
| 0.123 | 0.987279825 | 0.996242576 | 0.999168861 | 0.999872972 |
| 0.122 | 0.986012903 | 0.995736827 | 0.999016632 | 0.999840861 |
| 0.121 | 0.984645633 | 0.995175076 | 0.998840889 | 0.999801781 |
| 0.120 | 0.983172414 | 0.994552583 | 0.998638709 | 0.999754473 |
| 0.119 | 0.981587513 | 0.993864356 | 0.998406909 | 0.999697503 |
| 0.118 | 0.979885071 | 0.993105149 | 0.998142025 | 0.999629237 |
| 0.117 | 0.978059115 | 0.992269463 | 0.997840306 | 0.999547829 |
| 0.116 | 0.976103567 | 0.991351550 | 0.997497701 | 0.999451202 |
| 0.115 | 0.974012258 | 0.990345410 | 0.997109845 | 0.999337024 |
| 0.114 | 0.971778937 | 0.989244802 | 0.996672055 | 0.999202693 |
| 0.113 | 0.969397288 | 0.988043243 | 0.996179316 | 0.999045315 |
| 0.112 | 0.966860941 | 0.986734020 | 0.995626278 | 0.998861686 |
| 0.111 | 0.964163487 | 0.985310195 | 0.995007246 | 0.998648270 |
| 0.110 | 0.961298497 | 0.983764614 | 0.994316176 | 0.998401181 |
| 0.109 | 0.958259534 | 0.982089924 | 0.993546673 | 0.998116165 |
| 0.108 | 0.955040172 | 0.980278575 | 0.992691990 | 0.997788579 |
| 0.107 | 0.951634014 | 0.978322845 | 0.991745025 | 0.997413376 |
| 0.106 | 0.948034710 | 0.976214848 | 0.990698326 | 0.996985088 |
| 0.105 | 0.944235977 | 0.973946554 | 0.989544092 | 0.996497811 |
| 0.104 | 0.940231615 | 0.971509808 | 0.988274184 | 0.995945192 |
| 0.103 | 0.936015532 | 0.968896349 | 0.986880128 | 0.995320417 |
| 0.102 | 0.931581761 | 0.966097829 | 0.985353130 | 0.994616200 |
| 0.101 | 0.926924481 | 0.963105842 | 0.983684086 | 0.993824779 |

TABLE 1a:   CLUSTERING PROBABILITY   P(n;N,p)

n = 4

| P | N = 16 | N = 17 | N = 18 | N = 19 |
|---|---|---|---|---|
| 0.100 | 0.922038041 | 0.959911941 | 0.981863599 | 0.992937909 |
| 0.099 | 0.916916978 | 0.956507671 | 0.979881995 | 0.991946860 |
| 0.098 | 0.911556041 | 0.952684589 | 0.977729347 | 0.990842416 |
| 0.097 | 0.905950213 | 0.949034296 | 0.975395494 | 0.989614882 |
| 0.096 | 0.900094731 | 0.944948464 | 0.972870068 | 0.988254089 |
| 0.095 | 0.893985110 | 0.940618870 | 0.970142524 | 0.986749405 |
| 0.094 | 0.887617164 | 0.936037423 | 0.967202167 | 0.985089749 |
| 0.093 | 0.880987029 | 0.931196199 | 0.964038185 | 0.983263608 |
| 0.092 | 0.874091183 | 0.926087472 | 0.960639687 | 0.981259061 |
| 0.091 | 0.866926468 | 0.920703750 | 0.956995741 | 0.979063802 |
| 0.090 | 0.859490114 | 0.915037807 | 0.953095409 | 0.976665172 |
| 0.089 | 0.851779753 | 0.909082719 | 0.948927796 | 0.974050191 |
| 0.088 | 0.843793446 | 0.902831899 | 0.944482088 | 0.971205597 |
| 0.087 | 0.835529700 | 0.896279133 | 0.939747603 | 0.968117890 |
| 0.086 | 0.826987485 | 0.889418613 | 0.934713836 | 0.964773375 |
| 0.085 | 0.818166255 | 0.882244976 | 0.929370509 | 0.961158215 |
| 0.084 | 0.809065964 | 0.874753335 | 0.923707624 | 0.957258484 |
| 0.083 | 0.799687085 | 0.866939319 | 0.917715514 | 0.953060225 |
| 0.082 | 0.790030620 | 0.858799102 | 0.911384894 | 0.948549516 |
| 0.081 | 0.780098121 | 0.850329441 | 0.904706920 | 0.943712529 |
| 0.080 | 0.769891699 | 0.841527704 | 0.897673241 | 0.938535604 |
| 0.079 | 0.759414040 | 0.832391910 | 0.890276055 | 0.933005319 |
| 0.078 | 0.748668410 | 0.822920751 | 0.882508163 | 0.927108567 |
| 0.077 | 0.737658671 | 0.813113627 | 0.874363027 | 0.920832628 |
| 0.076 | 0.726389287 | 0.802970674 | 0.865834821 | 0.914165256 |
| 0.075 | 0.714865329 | 0.792492787 | 0.856918491 | 0.907094755 |
| 0.074 | 0.703092483 | 0.781681651 | 0.847609802 | 0.899610062 |
| 0.073 | 0.691077054 | 0.770539757 | 0.837905392 | 0.891700833 |
| 0.072 | 0.678825962 | 0.759070428 | 0.827802826 | 0.883357527 |
| 0.071 | 0.666346754 | 0.747277837 | 0.817300638 | 0.874571490 |
| 0.070 | 0.653647589 | 0.735167023 | 0.806398382 | 0.865335037 |
| 0.069 | 0.640737248 | 0.722743903 | 0.795096676 | 0.855641539 |
| 0.068 | 0.627625119 | 0.710015287 | 0.783397241 | 0.845485500 |
| 0.067 | 0.614321195 | 0.696988885 | 0.771302938 | 0.834862643 |
| 0.066 | 0.600836066 | 0.683673314 | 0.758817807 | 0.823769980 |
| 0.065 | 0.587180906 | 0.670078099 | 0.745947094 | 0.812205894 |
| 0.064 | 0.573367463 | 0.656213675 | 0.732697281 | 0.800170201 |
| 0.063 | 0.559408042 | 0.642091382 | 0.719076106 | 0.787664225 |
| 0.062 | 0.545315490 | 0.627723460 | 0.705092585 | 0.774690853 |
| 0.061 | 0.531103180 | 0.613123040 | 0.690757023 | 0.761254596 |
| 0.060 | 0.516784985 | 0.598304127 | 0.676081024 | 0.747361637 |
| 0.059 | 0.502375262 | 0.583281590 | 0.661077495 | 0.733019878 |
| 0.058 | 0.487888823 | 0.568071134 | 0.645760643 | 0.718238975 |
| 0.057 | 0.473340913 | 0.552689284 | 0.630145972 | 0.703030369 |
| 0.056 | 0.458747178 | 0.537153351 | 0.614250265 | 0.687407309 |
| 0.055 | 0.444123636 | 0.521481409 | 0.598091567 | 0.671384865 |
| 0.054 | 0.429486648 | 0.505692252 | 0.581689161 | 0.654979938 |
| 0.053 | 0.414852880 | 0.489805363 | 0.565063540 | 0.638211248 |
| 0.052 | 0.400239271 | 0.473840871 | 0.548236365 | 0.621099332 |
| 0.051 | 0.385662995 | 0.457819502 | 0.531230424 | 0.603666510 |

TABLE 1a:   CLUSTERING PROBABILITY   P(n;N,p)

n = 4

| P | N ≡16 | N ≡17 | N ≡18 | N ≡19 |
|---|---|---|---|---|
| 0.050 | 0.371141418 | 0.441762539 | 0.514069585 | 0.585936864 |
| 0.049 | 0.356692064 | 0.425691760 | 0.496778739 | 0.567936186 |
| 0.048 | 0.342332568 | 0.409629393 | 0.479383736 | 0.549691933 |
| 0.047 | 0.328080634 | 0.393598049 | 0.461911321 | 0.531233155 |
| 0.046 | 0.313953988 | 0.377620666 | 0.444389058 | 0.512590432 |
| 0.045 | 0.299970335 | 0.361720445 | 0.426845251 | 0.493795781 |
| 0.044 | 0.286147307 | 0.345920778 | 0.409308858 | 0.474882566 |
| 0.043 | 0.272502416 | 0.330245181 | 0.391809400 | 0.455885389 |
| 0.042 | 0.259053004 | 0.314717220 | 0.374376862 | 0.436839981 |
| 0.041 | 0.245816190 | 0.299360438 | 0.357041595 | 0.417783073 |
| 0.040 | 0.232808820 | 0.284198273 | 0.339834208 | 0.398752258 |
| 0.039 | 0.220047410 | 0.269253983 | 0.322785451 | 0.379785855 |
| 0.038 | 0.207548097 | 0.254550560 | 0.305926107 | 0.360922748 |
| 0.037 | 0.195326578 | 0.240110648 | 0.289286863 | 0.342202228 |
| 0.036 | 0.183398063 | 0.225956457 | 0.272898194 | 0.323663824 |
| 0.035 | 0.171777210 | 0.212109680 | 0.256790228 | 0.305347121 |
| 0.034 | 0.160478075 | 0.198591399 | 0.240992619 | 0.287291582 |
| 0.033 | 0.149514053 | 0.185421999 | 0.225534414 | 0.269536352 |
| 0.032 | 0.138897823 | 0.172621080 | 0.210443912 | 0.252120063 |
| 0.031 | 0.128641289 | 0.160207363 | 0.195748533 | 0.235080635 |
| 0.030 | 0.118755523 | 0.148198602 | 0.181474672 | 0.218455068 |
| 0.029 | 0.109250710 | 0.136611490 | 0.167647563 | 0.202279232 |
| 0.028 | 0.100136089 | 0.125461572 | 0.154291135 | 0.186587658 |
| 0.027 | 0.091419899 | 0.114763149 | 0.141427870 | 0.171413325 |
| 0.026 | 0.083109321 | 0.104529193 | 0.129078664 | 0.156787439 |
| 0.025 | 0.075210419 | 0.094771253 | 0.117262682 | 0.142739226 |
| 0.024 | 0.067728090 | 0.085499364 | 0.105997222 | 0.129295712 |
| 0.023 | 0.060666004 | 0.076721964 | 0.095297572 | 0.116481510 |
| 0.022 | 0.054026553 | 0.068445802 | 0.085176873 | 0.104318611 |
| 0.021 | 0.047810791 | 0.060675851 | 0.075645996 | 0.092826171 |
| 0.020 | 0.042018386 | 0.053415226 | 0.066713355 | 0.082020310 |
| 0.019 | 0.036647562 | 0.046665095 | 0.058384874 | 0.071913904 |
| 0.018 | 0.031695047 | 0.040424599 | 0.050663763 | 0.062516392 |
| 0.017 | 0.027156021 | 0.034690766 | 0.043550438 | 0.053833581 |
| 0.016 | 0.023024063 | 0.029458436 | 0.037042385 | 0.045867458 |
| 0.015 | 0.019291096 | 0.024720172 | 0.031134042 | 0.038616004 |
| 0.014 | 0.015947337 | 0.020466188 | 0.025816676 | 0.032073019 |
| 0.013 | 0.012981246 | 0.016684267 | 0.021078268 | 0.026227945 |
| 0.012 | 0.010379470 | 0.013359683 | 0.016903398 | 0.021065701 |
| 0.011 | 0.008126789 | 0.010475120 | 0.013273126 | 0.016566514 |
| 0.010 | 0.006206069 | 0.008010598 | 0.010164884 | 0.012705758 |
| 0.009 | 0.004598197 | 0.005943391 | 0.007552360 | 0.009453795 |
| 0.008 | 0.003282035 | 0.004247945 | 0.005405385 | 0.006775813 |
| 0.007 | 0.002234354 | 0.002895797 | 0.003689815 | 0.004631670 |
| 0.006 | 0.001429782 | 0.001855489 | 0.002367412 | 0.002975728 |
| 0.005 | 0.000840737 | 0.001092481 | 0.001395724 | 0.001756687 |
| 0.004 | 0.000437363 | 0.000569055 | 0.000727952 | 0.000917411 |
| 0.003 | 0.000187464 | 0.000244221 | 0.000312815 | 0.000394736 |
| 0.002 | 0.000056431 | 0.000073609 | 0.000094403 | 0.000119278 |
| 0.001 | 0.000007166 | 0.000009359 | 0.000012018 | 0.000015204 |

TABLE 1a: CLUSTERING PROBABILITY P(n;N,p)

n = 5

| P | N =10 | P | N =10 | N =11 |
|---|---|---|---|---|
| 0.400 | 0.995201843 | 0.350 | 0.961764078 | 0.994897871 |
| 0.399 | 0.994939948 | 0.349 | 0.960518120 | 0.994623834 |
| 0.398 | 0.994667014 | 0.348 | 0.959243404 | 0.994337578 |
| 0.397 | 0.994382725 | 0.347 | 0.957939612 | 0.994038685 |
| 0.396 | 0.994086759 | 0.346 | 0.956606430 | 0.993726727 |
| 0.395 | 0.993778793 | 0.345 | 0.955243552 | 0.993401269 |
| 0.394 | 0.993458497 | 0.344 | 0.953850673 | 0.993061868 |
| 0.393 | 0.993125540 | 0.343 | 0.952427498 | 0.992708071 |
| 0.392 | 0.992779587 | 0.342 | 0.950973733 | 0.992339419 |
| 0.391 | 0.992420299 | 0.341 | 0.949489094 | 0.991955442 |
| 0.390 | 0.992047334 | 0.340 | 0.947973300 | 0.991555666 |
| 0.389 | 0.991660347 | 0.339 | 0.946426077 | 0.991139607 |
| 0.388 | 0.991258992 | 0.338 | 0.944847157 | C.990706773 |
| 0.387 | 0.990842917 | 0.337 | 0.943236281 | 0.990256667 |
| 0.386 | 0.990411770 | 0.336 | 0.941593192 | 0.989788783 |
| 0.385 | 0.989965194 | 0.335 | 0.939917643 | 0.989302610 |
| 0.384 | 0.989502833 | 0.334 | 0.938209394 | 0.988797630 |
| 0.383 | 0.989024326 | 0.333 | 0.936468211 | 0.988273319 |
| 0.382 | 0.988529311 | 0.332 | 0.934693867 | 0.987729148 |
| 0.381 | 0.988017424 | 0.331 | 0.932886144 | 0.987164581 |
| 0.380 | 0.987488300 | 0.330 | 0.931044830 | 0.986579079 |
| 0.379 | 0.986941571 | 0.329 | 0.929169721 | 0.985972098 |
| 0.378 | 0.986376869 | 0.328 | 0.927260622 | 0.985343090 |
| 0.377 | 0.985793823 | 0.327 | 0.925317343 | 0.984691501 |
| 0.376 | 0.985192064 | 0.326 | 0.923339706 | 0.984016779 |
| 0.375 | 0.984571218 | 0.325 | 0.921327537 | 0.983318363 |
| 0.374 | 0.983930915 | 0.324 | 0.919280673 | 0.982595694 |
| 0.373 | 0.983270780 | 0.323 | 0.917198958 | 0.981848210 |
| 0.372 | 0.982590441 | 0.322 | 0.915082246 | 0.981075345 |
| 0.371 | 0.981889524 | 0.321 | 0.912930397 | 0.980276535 |
| 0.370 | 0.981167656 | 0.320 | 0.910743282 | 0.979451214 |
| 0.369 | 0.980424463 | 0.319 | 0.908520778 | 0.978598815 |
| 0.368 | 0.979659573 | 0.318 | 0.906262774 | 0.977718771 |
| 0.367 | 0.978872614 | 0.317 | 0.903969166 | 0.976810517 |
| 0.366 | 0.978063214 | 0.316 | 0.901639858 | 0.975873487 |
| 0.365 | 0.977231003 | 0.315 | 0.899274764 | 0.974907117 |
| 0.364 | 0.976375612 | 0.314 | 0.896873807 | 0.973910845 |
| 0.363 | 0.975496672 | 0.313 | 0.894436918 | 0.972884112 |
| 0.362 | 0.974593819 | 0.312 | 0.891964039 | 0.971826359 |
| 0.361 | 0.973666688 | 0.311 | 0.889455120 | 0.970737032 |
| 0.360 | 0.972714915 | 0.310 | 0.886910119 | 0.969615580 |
| 0.359 | 0.971738142 | 0.309 | 0.884329005 | 0.968461455 |
| 0.358 | 0.970736010 | 0.308 | 0.881711755 | 0.967274113 |
| 0.357 | 0.969708165 | 0.307 | 0.879058355 | 0.966053017 |
| 0.356 | 0.968654253 | 0.306 | 0.876368803 | 0.964797631 |
| 0.355 | 0.967573926 | 0.305 | 0.873643103 | 0.963507427 |
| 0.354 | 0.966466838 | 0.304 | 0.870881269 | 0.962181882 |
| 0.353 | 0.965332647 | 0.303 | 0.868083325 | 0.960820480 |
| 0.352 | 0.964171012 | 0.302 | 0.865249305 | 0.959422710 |
| 0.351 | 0.962981599 | 0.301 | 0.862379250 | 0.957988068 |

NEFF and NAUS

## TABLE 1a:  CLUSTERING PROBABILITY  P(n;N,p)

n = 5

| P | N = 10 | N = 11 | N = 12 | N = 13 |
|---|---|---|---|---|
| 0.300 | 0.859473212 | 0.956516058 | 0.992495077 | 0.999718955 |
| 0.299 | 0.856531253 | 0.955006192 | 0.992059962 | 0.999678866 |
| 0.298 | 0.853553443 | 0.953457988 | 0.991603990 | 0.999634368 |
| 0.297 | 0.850539860 | 0.951870973 | 0.991126451 | 0.999585112 |
| 0.296 | 0.847490594 | 0.950244685 | 0.990626628 | 0.999530728 |
| 0.295 | 0.844406744 | 0.948578667 | 0.990103791 | 0.999470833 |
| 0.294 | 0.841285416 | 0.946872473 | 0.989557207 | 0.999405021 |
| 0.293 | 0.838129726 | 0.945125669 | 0.988986132 | 0.999332871 |
| 0.292 | 0.834938802 | 0.943337827 | 0.988389814 | 0.999253942 |
| 0.291 | 0.831712777 | 0.941508532 | 0.987767497 | 0.999167774 |
| 0.290 | 0.828451796 | 0.939637378 | 0.987118417 | 0.999073887 |
| 0.289 | 0.825156012 | 0.937723971 | 0.986441804 | 0.998971783 |
| 0.288 | 0.821825586 | 0.935767927 | 0.985736884 | 0.998860944 |
| 0.287 | 0.818460692 | 0.933768875 | 0.985002877 | 0.998740831 |
| 0.286 | 0.815061508 | 0.931726454 | 0.984238999 | 0.998610887 |
| 0.285 | 0.811628223 | 0.929640315 | 0.983444464 | 0.998470535 |
| 0.284 | 0.808161037 | 0.927510124 | 0.982618481 | 0.998319174 |
| 0.283 | 0.804660156 | 0.925335556 | 0.981760257 | 0.998156189 |
| 0.282 | 0.801125796 | 0.923116301 | 0.980868999 | 0.997980940 |
| 0.281 | 0.797558181 | 0.920852061 | 0.979943911 | 0.997792770 |
| 0.280 | 0.793957545 | 0.918542552 | 0.978984196 | 0.997591000 |
| 0.279 | 0.790324129 | 0.916187503 | 0.977989059 | 0.997374931 |
| 0.278 | 0.786658184 | 0.913786657 | 0.976957703 | 0.997143845 |
| 0.277 | 0.782959969 | 0.911339769 | 0.975889336 | 0.996897005 |
| 0.276 | 0.779229751 | 0.908846612 | 0.974783164 | 0.996633653 |
| 0.275 | 0.775467807 | 0.906306970 | 0.973638398 | 0.996353012 |
| 0.274 | 0.771674419 | 0.903720642 | 0.972454250 | 0.996054286 |
| 0.273 | 0.767849881 | 0.901087443 | 0.971229939 | 0.995736660 |
| 0.272 | 0.763994494 | 0.898407202 | 0.969964686 | 0.995399301 |
| 0.271 | 0.760108565 | 0.895679763 | 0.968657717 | 0.995041358 |
| 0.270 | 0.756192411 | 0.892904984 | 0.967308265 | 0.994661961 |
| 0.269 | 0.752246357 | 0.890082741 | 0.965915569 | 0.994260224 |
| 0.268 | 0.748270736 | 0.887212924 | 0.964478875 | 0.993835243 |
| 0.267 | 0.744265887 | 0.884295438 | 0.962997437 | 0.993386098 |
| 0.266 | 0.740232159 | 0.881330204 | 0.961470515 | 0.992911854 |
| 0.265 | 0.736169907 | 0.878317158 | 0.959897382 | 0.992411560 |
| 0.264 | 0.732079493 | 0.875256255 | 0.958277317 | 0.991884249 |
| 0.263 | 0.727961289 | 0.872147463 | 0.956609612 | 0.991328941 |
| 0.262 | 0.723815673 | 0.868990767 | 0.954893567 | 0.990744642 |
| 0.261 | 0.719643028 | 0.865786167 | 0.953128496 | 0.990130348 |
| 0.260 | 0.715443748 | 0.862533682 | 0.951313723 | 0.989485038 |
| 0.259 | 0.711218232 | 0.859233346 | 0.949448587 | 0.988807682 |
| 0.258 | 0.706966885 | 0.855885207 | 0.947532439 | 0.988097242 |
| 0.257 | 0.702690121 | 0.852489333 | 0.945564644 | 0.987352666 |
| 0.256 | 0.698388360 | 0.849045807 | 0.943544580 | 0.986572895 |
| 0.255 | 0.694062028 | 0.845554727 | 0.941471644 | 0.985756862 |
| 0.254 | 0.689711558 | 0.842016211 | 0.939345244 | 0.984903492 |
| 0.253 | 0.685337389 | 0.838430390 | 0.937164808 | 0.984011705 |
| 0.252 | 0.680939967 | 0.834797413 | 0.934929778 | 0.983080414 |
| 0.251 | 0.676519744 | 0.831117446 | 0.932639616 | 0.982108530 |

## TABLE 1a:   CLUSTERING PROBABILITY   P(n;N,p)

n = 5

| P | N = 10 | N = 11 | N = 12 | N = 13 |
|---|---|---|---|---|
| 0.250 | 0.672077179 | 0.827390671 | 0.930293798 | 0.981094956 |
| 0.249 | 0.667612735 | 0.823617286 | 0.927891824 | 0.980038598 |
| 0.248 | 0.663126882 | 0.819797507 | 0.925433207 | 0.978938357 |
| 0.247 | 0.658620095 | 0.815931564 | 0.922917483 | 0.977793135 |
| 0.246 | 0.654092857 | 0.812019706 | 0.920344206 | 0.976601832 |
| 0.245 | 0.649545655 | 0.808062197 | 0.917712953 | 0.975363354 |
| 0.244 | 0.644978980 | 0.804059318 | 0.915023319 | 0.974076606 |
| 0.243 | 0.640393331 | 0.800011365 | 0.912274922 | 0.972740499 |
| 0.242 | 0.635789209 | 0.795918651 | 0.909467399 | 0.971353946 |
| 0.241 | 0.631167124 | 0.791781506 | 0.906600414 | 0.969915869 |
| 0.240 | 0.626527589 | 0.787600274 | 0.903673648 | 0.968425195 |
| 0.239 | 0.621871120 | 0.783375317 | 0.900686809 | 0.966880860 |
| 0.238 | 0.617198240 | 0.779107013 | 0.897639627 | 0.965281810 |
| 0.237 | 0.612509477 | 0.774795752 | 0.894531855 | 0.963626998 |
| 0.236 | 0.607805361 | 0.770441945 | 0.891363271 | 0.961915392 |
| 0.235 | 0.603086430 | 0.766046014 | 0.888133676 | 0.960145970 |
| 0.234 | 0.598353222 | 0.761608400 | 0.884842897 | 0.958317726 |
| 0.233 | 0.593606281 | 0.757129556 | 0.881490785 | 0.956429666 |
| 0.232 | 0.588846157 | 0.752609953 | 0.878077218 | 0.954480813 |
| 0.231 | 0.584073400 | 0.748050075 | 0.874602097 | 0.952470208 |
| 0.230 | 0.579288566 | 0.743450422 | 0.871065351 | 0.950396908 |
| 0.229 | 0.574492214 | 0.738811508 | 0.867466933 | 0.948259989 |
| 0.228 | 0.569684907 | 0.734133862 | 0.863806824 | 0.946058548 |
| 0.227 | 0.564867210 | 0.729418027 | 0.860085030 | 0.943791704 |
| 0.226 | 0.560039693 | 0.724664562 | 0.856301585 | 0.941458596 |
| 0.225 | 0.555202926 | 0.719874036 | 0.852456549 | 0.939058387 |
| 0.224 | 0.550357486 | 0.715047036 | 0.848550008 | 0.936590266 |
| 0.223 | 0.545503949 | 0.710184161 | 0.844582077 | 0.934053443 |
| 0.222 | 0.540642896 | 0.705286022 | 0.840552897 | 0.931447159 |
| 0.221 | 0.535774909 | 0.700353247 | 0.836462637 | 0.928770678 |
| 0.220 | 0.530900573 | 0.695386474 | 0.832311492 | 0.926023295 |
| 0.219 | 0.526020475 | 0.690386354 | 0.828099686 | 0.923204333 |
| 0.218 | 0.521135205 | 0.685353553 | 0.823827470 | 0.920313143 |
| 0.217 | 0.516245353 | 0.680288746 | 0.819495122 | 0.917349111 |
| 0.216 | 0.511351512 | 0.675192623 | 0.815102948 | 0.914311650 |
| 0.215 | 0.506454277 | 0.670065885 | 0.810651282 | 0.911200209 |
| 0.214 | 0.501554243 | 0.664909245 | 0.806140485 | 0.908014269 |
| 0.213 | 0.496652009 | 0.659723427 | 0.801570945 | 0.904753345 |
| 0.212 | 0.491748171 | 0.654509167 | 0.796943078 | 0.901416986 |
| 0.211 | 0.486843331 | 0.649267212 | 0.792257328 | 0.898004779 |
| 0.210 | 0.481938087 | 0.643998319 | 0.787514166 | 0.894516345 |
| 0.209 | 0.477033042 | 0.638703255 | 0.782714088 | 0.890951343 |
| 0.208 | 0.472128797 | 0.633382800 | 0.777857621 | 0.887309470 |
| 0.207 | 0.467225954 | 0.628037742 | 0.772945316 | 0.883590460 |
| 0.206 | 0.462325117 | 0.622668878 | 0.767977752 | 0.879794089 |
| 0.205 | 0.457426887 | 0.617277017 | 0.762955535 | 0.875920168 |
| 0.204 | 0.452531867 | 0.611862975 | 0.757879295 | 0.871968552 |
| 0.203 | 0.447640661 | 0.606427578 | 0.752749691 | 0.867939134 |
| 0.202 | 0.442753870 | 0.600971661 | 0.747567408 | 0.863831850 |
| 0.201 | 0.437872097 | 0.595496066 | 0.742333154 | 0.859646675 |

TABLE 1a:   CLUSTERING PROBABILITY   $P(n;N,p)$

n = 5

| P | N ≡ 10 | N ≡ 11 | N ≡ 12 | N ≡ 13 |
|---|---|---|---|---|
| 0.200 | 0.432995942 | 0.590001644 | 0.737047667 | 0.855383628 |
| 0.199 | 0.428126008 | 0.584489253 | 0.731711705 | 0.851042770 |
| 0.198 | 0.423262893 | 0.578959762 | 0.726326056 | 0.846624204 |
| 0.197 | 0.418407197 | 0.573414041 | 0.720891530 | 0.842128076 |
| 0.196 | 0.413559518 | 0.567852974 | 0.715408962 | 0.837554576 |
| 0.195 | 0.408720451 | 0.562277445 | 0.709879211 | 0.832903936 |
| 0.194 | 0.403890592 | 0.556688349 | 0.704303159 | 0.828176433 |
| 0.193 | 0.399070534 | 0.551086586 | 0.698681715 | 0.823372387 |
| 0.192 | 0.394260870 | 0.545473060 | 0.693015807 | 0.818492164 |
| 0.191 | 0.389462188 | 0.539848683 | 0.687306387 | 0.813536171 |
| 0.190 | 0.384675077 | 0.534214369 | 0.681554430 | 0.808504862 |
| 0.189 | 0.379900122 | 0.528571040 | 0.675760933 | 0.803398733 |
| 0.188 | 0.375137906 | 0.522919621 | 0.669926913 | 0.798218326 |
| 0.187 | 0.370389011 | 0.517261040 | 0.664053411 | 0.792964227 |
| 0.186 | 0.365654014 | 0.511596231 | 0.658141485 | 0.787637065 |
| 0.185 | 0.360933491 | 0.505926132 | 0.652192217 | 0.782237515 |
| 0.184 | 0.356228014 | 0.500251680 | 0.646206706 | 0.776766294 |
| 0.183 | 0.351538153 | 0.494573820 | 0.640186071 | 0.771224164 |
| 0.182 | 0.346864473 | 0.488893497 | 0.634131451 | 0.765611930 |
| 0.181 | 0.342207538 | 0.483211657 | 0.628044002 | 0.759930440 |
| 0.180 | 0.337567907 | 0.477529252 | 0.621924898 | 0.754180587 |
| 0.179 | 0.332946135 | 0.471847231 | 0.615775330 | 0.748363306 |
| 0.178 | 0.328342774 | 0.466166548 | 0.609596508 | 0.742479572 |
| 0.177 | 0.323758373 | 0.460488155 | 0.603389656 | 0.736530405 |
| 0.176 | 0.319193474 | 0.454813006 | 0.597156014 | 0.730516867 |
| 0.175 | 0.314648618 | 0.449142056 | 0.590896837 | 0.724440058 |
| 0.174 | 0.310124339 | 0.443476260 | 0.584613397 | 0.718301122 |
| 0.173 | 0.305621169 | 0.437816569 | 0.578306977 | 0.712101242 |
| 0.172 | 0.301139632 | 0.432163939 | 0.571978875 | 0.705841640 |
| 0.171 | 0.296680251 | 0.426519320 | 0.565630402 | 0.699523579 |
| 0.170 | 0.292243541 | 0.420883664 | 0.559262880 | 0.693148360 |
| 0.169 | 0.287830013 | 0.415257919 | 0.552877645 | 0.686717321 |
| 0.168 | 0.283440174 | 0.409643032 | 0.546476041 | 0.680231838 |
| 0.167 | 0.279074523 | 0.404039948 | 0.540059426 | 0.673693324 |
| 0.166 | 0.274733555 | 0.398449608 | 0.533629165 | 0.667103228 |
| 0.165 | 0.270417760 | 0.392872952 | 0.527186633 | 0.660463033 |
| 0.164 | 0.266127621 | 0.387310914 | 0.520733216 | 0.653774257 |
| 0.163 | 0.261863616 | 0.381764427 | 0.514270303 | 0.647038453 |
| 0.162 | 0.257626215 | 0.376234417 | 0.507799296 | 0.640257204 |
| 0.161 | 0.253415885 | 0.370721809 | 0.501321599 | 0.633432129 |
| 0.160 | 0.249233083 | 0.365227521 | 0.494838624 | 0.626564874 |
| 0.159 | 0.245078263 | 0.359752467 | 0.488351790 | 0.619657118 |
| 0.158 | 0.240951870 | 0.354297553 | 0.481862518 | 0.612710567 |
| 0.157 | 0.236854342 | 0.348863684 | 0.475372235 | 0.605726958 |
| 0.156 | 0.232786113 | 0.343451756 | 0.468882369 | 0.598708053 |
| 0.155 | 0.228747606 | 0.338062658 | 0.462394353 | 0.591655641 |
| 0.154 | 0.224739241 | 0.332697274 | 0.455909621 | 0.584571537 |
| 0.153 | 0.220761427 | 0.327356482 | 0.449429610 | 0.577457581 |
| 0.152 | 0.216814567 | 0.322041149 | 0.442955755 | 0.570315634 |
| 0.151 | 0.212899059 | 0.316752137 | 0.436489493 | 0.563147581 |

TABLE 1a:   CLUSTERING PROBABILITY   P(n;N,p)

n = 5

| P | N = 10 | N = 11 | N = 12 | N = 13 |
|------|------------|------------|------------|------------|
| 0.150 | 0.209015289 | 0.311490301 | 0.430032260 | 0.555955327 |
| 0.149 | 0.205163638 | 0.306256486 | 0.423585489 | 0.548740797 |
| 0.148 | 0.201344478 | 0.301051527 | 0.417150615 | 0.541505937 |
| 0.147 | 0.197558174 | 0.295876254 | 0.410729065 | 0.534252708 |
| 0.146 | 0.193805082 | 0.290731484 | 0.404322266 | 0.526983089 |
| 0.145 | 0.190085550 | 0.285618028 | 0.397931642 | 0.519699074 |
| 0.144 | 0.186399918 | 0.280536683 | 0.391558608 | 0.512402671 |
| 0.143 | 0.182748516 | 0.275488240 | 0.385204577 | 0.505095901 |
| 0.142 | 0.179131668 | 0.270473478 | 0.378870955 | 0.497780797 |
| 0.141 | 0.175549687 | 0.265493164 | 0.372559141 | 0.490459403 |
| 0.140 | 0.172002878 | 0.260548055 | 0.366270526 | 0.483133772 |
| 0.139 | 0.168491538 | 0.255638896 | 0.360006494 | 0.475805964 |
| 0.138 | 0.165015953 | 0.250766423 | 0.353768420 | 0.468478047 |
| 0.137 | 0.161576402 | 0.245931357 | 0.347557668 | 0.461152094 |
| 0.136 | 0.158173153 | 0.241134408 | 0.341375593 | 0.453830184 |
| 0.135 | 0.154806465 | 0.236376273 | 0.335223540 | 0.446514395 |
| 0.134 | 0.151476590 | 0.231657638 | 0.329102840 | 0.439206812 |
| 0.133 | 0.148183766 | 0.226979173 | 0.323014815 | 0.431909515 |
| 0.132 | 0.144928225 | 0.222341537 | 0.316960771 | 0.424624587 |
| 0.131 | 0.141710188 | 0.217745374 | 0.310942002 | 0.417354107 |
| 0.130 | 0.138529867 | 0.213191316 | 0.304959788 | 0.410100152 |
| 0.129 | 0.135387462 | 0.208679980 | 0.299015394 | 0.402864792 |
| 0.128 | 0.132283166 | 0.204211966 | 0.293110069 | 0.395650093 |
| 0.127 | 0.129217159 | 0.199787864 | 0.287245048 | 0.388458112 |
| 0.126 | 0.126189613 | 0.195408245 | 0.281421546 | 0.381290898 |
| 0.125 | 0.123200689 | 0.191073668 | 0.275640763 | 0.374150490 |
| 0.124 | 0.120250537 | 0.186784675 | 0.269903881 | 0.367038916 |
| 0.123 | 0.117339299 | 0.182541791 | 0.264212062 | 0.359958191 |
| 0.122 | 0.114467103 | 0.178345529 | 0.258566449 | 0.352910203 |
| 0.121 | 0.111634070 | 0.174196381 | 0.252968167 | 0.345897275 |
| 0.120 | 0.108840309 | 0.170094828 | 0.247418320 | 0.338921040 |
| 0.119 | 0.106085917 | 0.166041329 | 0.241917989 | 0.331983562 |
| 0.118 | 0.103370982 | 0.162036331 | 0.236468236 | 0.325086772 |
| 0.117 | 0.100695582 | 0.158080261 | 0.231070100 | 0.318232584 |
| 0.116 | 0.098059781 | 0.154173529 | 0.225724596 | 0.311422887 |
| 0.115 | 0.095463637 | 0.150316529 | 0.220432718 | 0.304659549 |
| 0.114 | 0.092907192 | 0.146509637 | 0.215195435 | 0.297944415 |
| 0.113 | 0.090390480 | 0.142753211 | 0.210013692 | 0.291279302 |
| 0.112 | 0.087913525 | 0.139047590 | 0.204888408 | 0.284666002 |
| 0.111 | 0.085476336 | 0.135393097 | 0.199820479 | 0.278106279 |
| 0.110 | 0.083078915 | 0.131790034 | 0.194810773 | 0.271601869 |
| 0.109 | 0.080721251 | 0.128238687 | 0.189860133 | 0.265154476 |
| 0.108 | 0.078403323 | 0.124739321 | 0.184969374 | 0.258765775 |
| 0.107 | 0.076125097 | 0.121292185 | 0.180139284 | 0.252437406 |
| 0.106 | 0.073886529 | 0.117897507 | 0.175370624 | 0.246170978 |
| 0.105 | 0.071687565 | 0.114555496 | 0.170664126 | 0.239968063 |
| 0.104 | 0.069528138 | 0.111266342 | 0.166020494 | 0.233830200 |
| 0.103 | 0.067408171 | 0.108030216 | 0.161440401 | 0.227758888 |
| 0.102 | 0.065327575 | 0.104847269 | 0.156924493 | 0.221755590 |
| 0.101 | 0.063286251 | 0.101717633 | 0.152473384 | 0.215821730 |

TABLE 1a:   CLUSTERING PROBABILITY   P(n;N,p)

n = 5

| P | N = 10 | N = 11 | N = 12 | N = 13 |
|------|-----------|-----------|-----------|-----------|
| 0.100 | 0.061284088 | 0.098641420 | 0.148087659 | 0.209958692 |
| 0.099 | 0.059320963 | 0.095618721 | 0.143767871 | 0.204167819 |
| 0.098 | 0.057396743 | 0.092649610 | 0.139514543 | 0.198450410 |
| 0.097 | 0.055511285 | 0.089734137 | 0.135328167 | 0.192807725 |
| 0.096 | 0.053664433 | 0.086872336 | 0.131209201 | 0.187240976 |
| 0.095 | 0.051856020 | 0.084064218 | 0.127158073 | 0.181751334 |
| 0.094 | 0.050085870 | 0.081309774 | 0.123175178 | 0.176339921 |
| 0.093 | 0.048353793 | 0.078608977 | 0.119260877 | 0.171007816 |
| 0.092 | 0.046659591 | 0.075961778 | 0.115415500 | 0.165756047 |
| 0.091 | 0.045003053 | 0.073368107 | 0.111639343 | 0.160585596 |
| 0.090 | 0.043383958 | 0.070827875 | 0.107932667 | 0.155497396 |
| 0.089 | 0.041802074 | 0.068340972 | 0.104295702 | 0.150492330 |
| 0.088 | 0.040257159 | 0.065907268 | 0.100728642 | 0.145571231 |
| 0.087 | 0.038748960 | 0.063526612 | 0.097231647 | 0.140734879 |
| 0.086 | 0.037277212 | 0.061198834 | 0.093804844 | 0.135984005 |
| 0.085 | 0.035841641 | 0.058923743 | 0.090448323 | 0.131319286 |
| 0.084 | 0.034441963 | 0.056701126 | 0.087162142 | 0.126741345 |
| 0.083 | 0.033077883 | 0.054530752 | 0.083946323 | 0.122250755 |
| 0.082 | 0.031749094 | 0.052412369 | 0.080800854 | 0.117848031 |
| 0.081 | 0.030455283 | 0.050345706 | 0.077725686 | 0.113533636 |
| 0.080 | 0.029196123 | 0.048330470 | 0.074720737 | 0.109307976 |
| 0.079 | 0.027971279 | 0.046366349 | 0.071785889 | 0.105171403 |
| 0.078 | 0.026780406 | 0.044453012 | 0.068920989 | 0.101124212 |
| 0.077 | 0.025623149 | 0.042590106 | 0.066125849 | 0.097166644 |
| 0.076 | 0.024499145 | 0.040777262 | 0.063400247 | 0.093298880 |
| 0.075 | 0.023408018 | 0.039014088 | 0.060743924 | 0.089521047 |
| 0.074 | 0.022349386 | 0.037300175 | 0.058156588 | 0.085833214 |
| 0.073 | 0.021322858 | 0.035635093 | 0.055637909 | 0.082235392 |
| 0.072 | 0.020328031 | 0.034018396 | 0.053187527 | 0.078727536 |
| 0.071 | 0.019364498 | 0.032449615 | 0.050805044 | 0.075309542 |
| 0.070 | 0.018431838 | 0.030928267 | 0.048490027 | 0.071981250 |
| 0.069 | 0.017529625 | 0.029453847 | 0.046242012 | 0.068742440 |
| 0.068 | 0.016657426 | 0.028025835 | 0.044060498 | 0.065592838 |
| 0.067 | 0.015814795 | 0.026643690 | 0.041944951 | 0.062532109 |
| 0.066 | 0.015001283 | 0.025306857 | 0.039894805 | 0.059559863 |
| 0.065 | 0.014216431 | 0.024014759 | 0.037909458 | 0.056675650 |
| 0.064 | 0.013459773 | 0.022766808 | 0.035988278 | 0.053878967 |
| 0.063 | 0.012730837 | 0.021562394 | 0.034130598 | 0.051169251 |
| 0.062 | 0.012029141 | 0.020400893 | 0.032335721 | 0.048545883 |
| 0.061 | 0.011354200 | 0.019281666 | 0.030602916 | 0.046008189 |
| 0.060 | 0.010705521 | 0.018204056 | 0.028931423 | 0.043555440 |
| 0.059 | 0.010082603 | 0.017167392 | 0.027320449 | 0.041186850 |
| 0.058 | 0.009484942 | 0.016170989 | 0.025769173 | 0.038901578 |
| 0.057 | 0.008912027 | 0.015214145 | 0.024276742 | 0.036698732 |
| 0.056 | 0.008363340 | 0.014296148 | 0.022842275 | 0.034577364 |
| 0.055 | 0.007838362 | 0.013416267 | 0.021464862 | 0.032536473 |
| 0.054 | 0.007336566 | 0.012573763 | 0.020143567 | 0.030575010 |
| 0.053 | 0.006857420 | 0.011767882 | 0.018877423 | 0.028691870 |
| 0.052 | 0.006400390 | 0.010997858 | 0.017665439 | 0.026885901 |
| 0.051 | 0.005964937 | 0.010262913 | 0.016506599 | 0.025155901 |

TABLE 1a:   CLUSTERING PROBABILITY   P(n;N,p)

n = 5

| P | N = 10 | N = 11 | N = 12 | N = 13 |
|---|--------|--------|--------|--------|
| 0.050 | 0.00555C519 | 0.009562260 | 0.015399860 | 0.023500620 |
| 0.049 | 0.005156591 | 0.008895099 | 0.014344156 | 0.021918761 |
| 0.048 | 0.004782604 | 0.008260621 | 0.013338397 | 0.020408981 |
| 0.047 | 0.004428008 | 0.007658010 | 0.012381472 | 0.018969894 |
| 0.046 | 0.004092250 | 0.007086437 | 0.011472247 | 0.017600068 |
| 0.045 | 0.003774775 | 0.006545069 | 0.010609570 | 0.016298031 |
| 0.044 | 0.003475028 | 0.006033063 | 0.009792266 | 0.015062270 |
| 0.043 | 0.003192452 | 0.005549571 | 0.009019144 | 0.013891235 |
| 0.042 | 0.002926490 | 0.005093739 | 0.008288998 | 0.012783337 |
| 0.041 | 0.002676585 | 0.004664708 | 0.007600602 | 0.011736951 |
| 0.040 | 0.002442181 | 0.004261612 | 0.006952717 | 0.010750422 |
| 0.039 | 0.002222720 | 0.003883586 | 0.006344053 | 0.009822059 |
| 0.038 | 0.002017650 | 0.003529757 | 0.005773463 | 0.008950144 |
| 0.037 | 0.001826416 | 0.003199256 | 0.005239553 | 0.008132932 |
| 0.036 | 0.001648469 | 0.002891207 | 0.004741079 | 0.007368650 |
| 0.035 | 0.001483260 | 0.002604739 | 0.004276749 | 0.006655503 |
| 0.034 | 0.001330245 | 0.002338977 | 0.003845265 | 0.005991677 |
| 0.033 | 0.001188883 | 0.002093052 | 0.003445322 | 0.005375338 |
| 0.032 | 0.001058637 | 0.001866095 | 0.003075617 | 0.004804637 |
| 0.031 | 0.000938975 | 0.001657242 | 0.002734840 | 0.004277711 |
| 0.030 | 0.000829371 | 0.001465633 | 0.002421685 | 0.003792688 |
| 0.029 | 0.000729305 | 0.001290413 | 0.002134847 | 0.003347688 |
| 0.028 | 0.000638262 | 0.001130736 | 0.001873026 | 0.002940828 |
| 0.027 | 0.000555736 | 0.000985762 | 0.001634927 | 0.002570221 |
| 0.026 | 0.000481226 | 0.000854661 | 0.001419264 | 0.002233986 |
| 0.025 | 0.000414243 | 0.000736614 | 0.001224760 | 0.001930245 |
| 0.024 | 0.000354304 | 0.000630812 | 0.001050152 | 0.001657130 |
| 0.023 | 0.000300936 | 0.000536460 | 0.000894189 | 0.001412784 |
| 0.022 | 0.000253678 | 0.000452776 | 0.000755639 | 0.001195368 |
| 0.021 | 0.000212078 | 0.000378994 | 0.000633288 | 0.001003064 |
| 0.020 | 0.000175695 | 0.000314365 | 0.000525945 | 0.000834077 |
| 0.019 | 0.000144103 | 0.000258156 | 0.000432439 | 0.000686639 |
| 0.018 | 0.000116886 | 0.000209656 | 0.000351630 | 0.000559019 |
| 0.017 | 0.000093644 | 0.000168174 | 0.000282405 | 0.000449520 |
| 0.016 | 0.000073990 | 0.000133041 | 0.000223683 | 0.000356487 |
| 0.015 | 0.000057552 | 0.000103611 | 0.000174417 | 0.000278313 |
| 0.014 | 0.000043976 | 0.000079266 | 0.000133598 | 0.000213442 |
| 0.013 | 0.000032921 | 0.000059413 | 0.000100259 | 0.000160374 |
| 0.012 | 0.000024067 | 0.000043487 | 0.000073473 | 0.000117672 |
| 0.011 | 0.000017110 | 0.000030954 | 0.000052363 | 0.000083965 |
| 0.010 | 0.000011767 | 0.000021314 | 0.000036100 | 0.000057957 |
| 0.009 | 0.000007774 | 0.000014098 | 0.000023906 | 0.000038428 |
| 0.008 | 0.000004887 | 0.000008873 | 0.000015064 | 0.000024244 |
| 0.007 | 0.000002884 | 0.000005243 | 0.000008913 | 0.000014361 |
| 0.006 | 0.000001567 | 0.000002853 | 0.000004856 | 0.000007833 |
| 0.005 | 0.000000761 | 0.000001387 | 0.000002363 | 0.000003817 |
| 0.004 | 0.000000314 | 0.000000573 | 0.000000977 | 0.000001580 |
| 0.003 | 0.000000100 | 0.000000183 | 0.000000312 | 0.000000505 |
| 0.002 | 0.000000020 | 0.000000036 | 0.000000062 | 0.000000101 |
| 0.001 | 0.000000001 | 0.000000002 | 0.000000004 | 0.000000006 |

## TABLE 1a:   CLUSTERING PROBABILITY   P(n;N,p)

n = 5

| P | N =14 | N =15 | N =16 | N =17 |
|---|---|---|---|---|
| 0.250 | 0.997004002 | 0.999770666 | 0.999994406 | 1 |
| 0.249 | 0.996736949 | 0.999739016 | 0.999993230 | 1.000000000 |
| 0.248 | 0.996450619 | 0.999703565 | 0.999991825 | 1.000000000 |
| 0.247 | 0.996143982 | 0.999663929 | 0.999990150 | 1.000000000 |
| 0.246 | 0.995815977 | 0.999619695 | 0.999988157 | 1.000000000 |
| 0.245 | 0.995465515 | 0.999570419 | 0.999985791 | 0.999999999 |
| 0.244 | 0.995091476 | 0.999515624 | 0.999982988 | 0.999999998 |
| 0.243 | 0.994692709 | 0.999454800 | 0.999979675 | 0.999999995 |
| 0.242 | 0.994268039 | 0.999387400 | 0.999975766 | 0.999999990 |
| 0.241 | 0.993816258 | 0.999312842 | 0.999971167 | 0.999999980 |
| 0.240 | 0.993336136 | 0.999230507 | 0.999965766 | 0.999999963 |
| 0.239 | 0.992826413 | 0.999139734 | 0.999959439 | 0.999999937 |
| 0.238 | 0.992285807 | 0.999039828 | 0.999952045 | 0.999999896 |
| 0.237 | 0.991713009 | 0.998930047 | 0.999943427 | 0.999999834 |
| 0.236 | 0.991106690 | 0.998809612 | 0.999933406 | 0.999999743 |
| 0.235 | 0.990465496 | 0.998677700 | 0.999921784 | 0.999999613 |
| 0.234 | 0.989788054 | 0.998533445 | 0.999908339 | 0.999999430 |
| 0.233 | 0.989072971 | 0.998375937 | 0.999892825 | 0.999999179 |
| 0.232 | 0.988318835 | 0.998204221 | 0.999874972 | 0.999998839 |
| 0.231 | 0.987524219 | 0.998017299 | 0.999854478 | 0.999998384 |
| 0.230 | 0.986687677 | 0.997814126 | 0.999831013 | 0.999997784 |
| 0.229 | 0.985807752 | 0.997593613 | 0.999804216 | 0.999997003 |
| 0.228 | 0.984882972 | 0.997354624 | 0.999773691 | 0.999995995 |
| 0.227 | 0.983911855 | 0.997095979 | 0.999739006 | 0.999994708 |
| 0.226 | 0.982892909 | 0.996816451 | 0.999699691 | 0.999993080 |
| 0.225 | 0.981824633 | 0.996514769 | 0.999655239 | 0.999991038 |
| 0.224 | 0.980705520 | 0.996189615 | 0.999605097 | 0.999988496 |
| 0.223 | 0.979534060 | 0.995839627 | 0.999548672 | 0.999985355 |
| 0.222 | 0.978308735 | 0.995463400 | 0.999485324 | 0.999981501 |
| 0.221 | 0.977028031 | 0.995059485 | 0.999414366 | 0.999976802 |
| 0.220 | 0.975690431 | 0.994626388 | 0.999335062 | 0.999971108 |
| 0.219 | 0.974294420 | 0.994162577 | 0.999246624 | 0.999964247 |
| 0.218 | 0.972838488 | 0.993666474 | 0.999148214 | 0.999956024 |
| 0.217 | 0.971321131 | 0.993136466 | 0.999038938 | 0.999946220 |
| 0.216 | 0.969740849 | 0.992570898 | 0.998917847 | 0.999934587 |
| 0.215 | 0.968096156 | 0.991968080 | 0.998783934 | 0.999920850 |
| 0.214 | 0.966385573 | 0.991326283 | 0.998636136 | 0.999904697 |
| 0.213 | 0.964607636 | 0.990643746 | 0.998473329 | 0.999885785 |
| 0.212 | 0.962760896 | 0.989918674 | 0.998294329 | 0.999863733 |
| 0.211 | 0.960843919 | 0.989149242 | 0.998097891 | 0.999838118 |
| 0.210 | 0.958855291 | 0.988333596 | 0.997882706 | 0.999808474 |
| 0.209 | 0.956793617 | 0.987469852 | 0.997647406 | 0.999774292 |
| 0.208 | 0.954657524 | 0.986556104 | 0.997390555 | 0.999735011 |
| 0.207 | 0.952445666 | 0.985590421 | 0.997110659 | 0.999690021 |
| 0.206 | 0.950156719 | 0.984570851 | 0.996806156 | 0.999638656 |
| 0.205 | 0.947789389 | 0.983495426 | 0.996475422 | 0.999580193 |
| 0.204 | 0.945342410 | 0.982362158 | 0.996116771 | 0.999513850 |
| 0.203 | 0.942814549 | 0.981169047 | 0.995728451 | 0.999438781 |
| 0.202 | 0.940204605 | 0.979914084 | 0.995308651 | 0.999354076 |
| 0.201 | 0.937511412 | 0.978595246 | 0.994855498 | 0.999258755 |

TABLE 1a: CLUSTERING PROBABILITY P(n;N,p)

n = 5

| P | N = 14 | N = 15 | N = 16 | N = 17 |
|---|--------|--------|--------|--------|
| 0.200 | 0.934733840 | 0.977210510 | 0.994367055 | 0.999151768 |
| 0.199 | 0.931870799 | 0.975757846 | 0.993841330 | 0.999031994 |
| 0.198 | 0.928921236 | 0.974235224 | 0.993276270 | 0.998898233 |
| 0.197 | 0.925884144 | 0.972640617 | 0.992669767 | 0.998749210 |
| 0.196 | 0.922758554 | 0.970972003 | 0.992019658 | 0.998583571 |
| 0.195 | 0.919543545 | 0.969227369 | 0.991323727 | 0.998399879 |
| 0.194 | 0.916238244 | 0.967404713 | 0.990579707 | 0.998196616 |
| 0.193 | 0.912841821 | 0.965502044 | 0.989785282 | 0.997972177 |
| 0.192 | 0.909353500 | 0.963517392 | 0.988938091 | 0.997724874 |
| 0.191 | 0.905772553 | 0.961448806 | 0.988035731 | 0.997452933 |
| 0.190 | 0.902098305 | 0.959294358 | 0.987075755 | 0.997154492 |
| 0.189 | 0.898330135 | 0.957052144 | 0.986055680 | 0.996827600 |
| 0.188 | 0.894467478 | 0.954720293 | 0.984972991 | 0.996470221 |
| 0.187 | 0.890509821 | 0.952296964 | 0.983825139 | 0.996080229 |
| 0.186 | 0.886456712 | 0.949780352 | 0.982609549 | 0.995655414 |
| 0.185 | 0.882307757 | 0.947168690 | 0.981323620 | 0.995193477 |
| 0.184 | 0.878062621 | 0.944460254 | 0.979964734 | 0.994692033 |
| 0.183 | 0.873721029 | 0.941653361 | 0.978530254 | 0.994148615 |
| 0.182 | 0.869282769 | 0.938746379 | 0.977017532 | 0.993560671 |
| 0.181 | 0.864747689 | 0.935737725 | 0.975423912 | 0.992925571 |
| 0.180 | 0.860115704 | 0.932625868 | 0.973746734 | 0.992240602 |
| 0.179 | 0.855386790 | 0.929409337 | 0.971983337 | 0.991502978 |
| 0.178 | 0.850560990 | 0.926086714 | 0.970131068 | 0.990709840 |
| 0.177 | 0.845638411 | 0.922656649 | 0.968187281 | 0.989858257 |
| 0.176 | 0.840619229 | 0.919117853 | 0.966149345 | 0.988945231 |
| 0.175 | 0.835503683 | 0.915469104 | 0.964014647 | 0.987967702 |
| 0.174 | 0.830292083 | 0.911709251 | 0.961780598 | 0.986922550 |
| 0.173 | 0.824984805 | 0.907837215 | 0.959444638 | 0.985806601 |
| 0.172 | 0.819582294 | 0.903851993 | 0.957004238 | 0.984616629 |
| 0.171 | 0.814085061 | 0.899752656 | 0.954456909 | 0.983349364 |
| 0.170 | 0.808493689 | 0.895538358 | 0.951800204 | 0.982001494 |
| 0.169 | 0.802808829 | 0.891208334 | 0.949031723 | 0.980569674 |
| 0.168 | 0.797031200 | 0.886761900 | 0.946149120 | 0.979050528 |
| 0.167 | 0.791161591 | 0.882198462 | 0.943150105 | 0.977440656 |
| 0.166 | 0.785200859 | 0.877517512 | 0.940032449 | 0.975736642 |
| 0.165 | 0.779149931 | 0.872718632 | 0.936793994 | 0.973935057 |
| 0.164 | 0.773009802 | 0.867801495 | 0.933432650 | 0.972032468 |
| 0.163 | 0.766781537 | 0.862765869 | 0.929946406 | 0.970025444 |
| 0.162 | 0.760466267 | 0.857611615 | 0.926333329 | 0.967910561 |
| 0.161 | 0.754065193 | 0.852338692 | 0.922591575 | 0.965684413 |
| 0.160 | 0.747579582 | 0.846947155 | 0.918719388 | 0.963343613 |
| 0.159 | 0.741010768 | 0.841437158 | 0.914715108 | 0.960884808 |
| 0.158 | 0.734360153 | 0.835808957 | 0.910577174 | 0.958304679 |
| 0.157 | 0.727629202 | 0.830062907 | 0.906304127 | 0.955599952 |
| 0.156 | 0.720819448 | 0.824199466 | 0.901894615 | 0.952767406 |
| 0.155 | 0.713932487 | 0.818219194 | 0.897347399 | 0.949803878 |
| 0.154 | 0.706969977 | 0.812122755 | 0.892661354 | 0.946706272 |
| 0.153 | 0.699933640 | 0.805910916 | 0.887835475 | 0.943471569 |
| 0.152 | 0.692825261 | 0.799584549 | 0.882868876 | 0.940096827 |
| 0.151 | 0.685646682 | 0.793144629 | 0.877760801 | 0.936579199 |

NEFF and NAUS

TABLE 1a: CLUSTERING PROBABILITY P(n;N,p)

n = 5

| P | N=14 | N=15 | N=16 | N=17 |
|---|---|---|---|---|
| 0.150 | 0.678399809 | 0.786592239 | 0.872510621 | 0.932915931 |
| 0.149 | 0.671086602 | 0.779928563 | 0.867117838 | 0.929104375 |
| 0.148 | 0.663709080 | 0.773154892 | 0.861582091 | 0.925141994 |
| 0.147 | 0.656269320 | 0.766272619 | 0.855903156 | 0.921026372 |
| 0.146 | 0.648769449 | 0.759283244 | 0.850080948 | 0.916755215 |
| 0.145 | 0.641211651 | 0.752188368 | 0.844115527 | 0.912326367 |
| 0.144 | 0.633598161 | 0.744989695 | 0.838007096 | 0.907737807 |
| 0.143 | 0.625931263 | 0.737689032 | 0.831756004 | 0.902987664 |
| 0.142 | 0.618213290 | 0.730288288 | 0.825362750 | 0.898074218 |
| 0.141 | 0.610446624 | 0.722789469 | 0.818827983 | 0.892995909 |
| 0.140 | 0.602633691 | 0.715194683 | 0.812152501 | 0.887751343 |
| 0.139 | 0.594776960 | 0.707506136 | 0.805337257 | 0.882339296 |
| 0.138 | 0.586878943 | 0.699726128 | 0.798383355 | 0.876758720 |
| 0.137 | 0.578942193 | 0.691857054 | 0.791292054 | 0.871008752 |
| 0.136 | 0.570969302 | 0.683901406 | 0.784064767 | 0.865088712 |
| 0.135 | 0.562962895 | 0.675861763 | 0.776703060 | 0.858998116 |
| 0.134 | 0.554925637 | 0.667740796 | 0.769208655 | 0.852736672 |
| 0.133 | 0.546860220 | 0.659541263 | 0.761583427 | 0.846304292 |
| 0.132 | 0.538769372 | 0.651266008 | 0.753829404 | 0.839701091 |
| 0.131 | 0.530655846 | 0.642917958 | 0.745948770 | 0.832927389 |
| 0.130 | 0.522522424 | 0.634500121 | 0.737943855 | 0.825983720 |
| 0.129 | 0.514371912 | 0.626015584 | 0.729817145 | 0.818870830 |
| 0.128 | 0.506207137 | 0.617467508 | 0.721571270 | 0.811589682 |
| 0.127 | 0.498030948 | 0.608859132 | 0.713209013 | 0.804141454 |
| 0.126 | 0.489846211 | 0.600193760 | 0.704733296 | 0.796527547 |
| 0.125 | 0.481655810 | 0.591474769 | 0.696147190 | 0.788749581 |
| 0.124 | 0.473462640 | 0.582705596 | 0.687453903 | 0.780809395 |
| 0.123 | 0.465269610 | 0.573889745 | 0.678656783 | 0.772709055 |
| 0.122 | 0.457079635 | 0.565030775 | 0.669759312 | 0.764450845 |
| 0.121 | 0.448895641 | 0.556132301 | 0.660765105 | 0.756037272 |
| 0.120 | 0.440720555 | 0.547197993 | 0.651677908 | 0.747471063 |
| 0.119 | 0.432557309 | 0.538231567 | 0.642501590 | 0.738755165 |
| 0.118 | 0.424408833 | 0.529236785 | 0.633240142 | 0.729892745 |
| 0.117 | 0.416278055 | 0.520217453 | 0.623897675 | 0.720887182 |
| 0.116 | 0.408167899 | 0.511177414 | 0.614478411 | 0.711742071 |
| 0.115 | 0.400081282 | 0.502120545 | 0.604986686 | 0.702461217 |
| 0.114 | 0.392021112 | 0.493050757 | 0.595426936 | 0.693048633 |
| 0.113 | 0.383990283 | 0.483971986 | 0.585803701 | 0.683508535 |
| 0.112 | 0.375991677 | 0.474888193 | 0.576121617 | 0.673845339 |
| 0.111 | 0.368028159 | 0.465803358 | 0.566385409 | 0.664063656 |
| 0.110 | 0.360102575 | 0.456721477 | 0.556599888 | 0.654168288 |
| 0.109 | 0.352217750 | 0.447646558 | 0.546769945 | 0.644164222 |
| 0.108 | 0.344376486 | 0.438582619 | 0.536900548 | 0.634056623 |
| 0.107 | 0.336581558 | 0.429533681 | 0.526996731 | 0.623850832 |
| 0.106 | 0.328835713 | 0.420503763 | 0.517063594 | 0.613552356 |
| 0.105 | 0.321141670 | 0.411496884 | 0.507106294 | 0.603166863 |
| 0.104 | 0.313502111 | 0.402517052 | 0.497130040 | 0.592700176 |
| 0.103 | 0.305919687 | 0.393568265 | 0.487140085 | 0.582158261 |
| 0.102 | 0.298397010 | 0.384654504 | 0.477141723 | 0.571547226 |
| 0.101 | 0.290936653 | 0.375779731 | 0.467140282 | 0.560873308 |

TABLE 1a:   CLUSTERING PROBABILITY   P(n;N,p)

n = 5

| P | N =14 | N =15 | N =16 | N =17 |
|---|---|---|---|---|
| 0.100 | 0.283541146 | 0.366947883 | 0.457141115 | 0.550142867 |
| 0.099 | 0.276212977 | 0.358162869 | 0.447149596 | 0.539362378 |
| 0.098 | 0.268954587 | 0.349428567 | 0.437171112 | 0.528538420 |
| 0.097 | 0.261768369 | 0.340748816 | 0.427211060 | 0.517677668 |
| 0.096 | 0.254656667 | 0.332127418 | 0.417274836 | 0.506786886 |
| 0.095 | 0.247621770 | 0.323568130 | 0.407367828 | 0.495872914 |
| 0.094 | 0.240665916 | 0.315074659 | 0.397495416 | 0.484942659 |
| 0.093 | 0.233791285 | 0.306650662 | 0.387662957 | 0.474003089 |
| 0.092 | 0.226999998 | 0.298299741 | 0.377875783 | 0.463061217 |
| 0.091 | 0.220294118 | 0.290025434 | 0.368139194 | 0.452124095 |
| 0.090 | 0.213675644 | 0.281831221 | 0.358458451 | 0.441198804 |
| 0.089 | 0.207146512 | 0.273720510 | 0.348838767 | 0.430292441 |
| 0.088 | 0.200708593 | 0.265696641 | 0.339285303 | 0.419412107 |
| 0.087 | 0.194363689 | 0.257762878 | 0.329803161 | 0.408564904 |
| 0.086 | 0.188113536 | 0.249922408 | 0.320397378 | 0.397757917 |
| 0.085 | 0.181959795 | 0.242178333 | 0.311072915 | 0.386998203 |
| 0.084 | 0.175904059 | 0.234533675 | 0.301834659 | 0.376292788 |
| 0.083 | 0.169947844 | 0.226991364 | 0.292687408 | 0.365648647 |
| 0.082 | 0.164092593 | 0.219554238 | 0.283635868 | 0.355072700 |
| 0.081 | 0.158339671 | 0.212225042 | 0.274684649 | 0.344571797 |
| 0.080 | 0.152690365 | 0.205006422 | 0.265838257 | 0.334152710 |
| 0.079 | 0.147145883 | 0.197900922 | 0.257101086 | 0.323822121 |
| 0.078 | 0.141707352 | 0.190910982 | 0.248477415 | 0.313586612 |
| 0.077 | 0.136375819 | 0.184038938 | 0.239971401 | 0.303452656 |
| 0.076 | 0.131152245 | 0.177287011 | 0.231587074 | 0.293426602 |
| 0.075 | 0.126037510 | 0.170657315 | 0.223328331 | 0.283514672 |
| 0.074 | 0.121032406 | 0.164151845 | 0.215198929 | 0.273722944 |
| 0.073 | 0.116137643 | 0.157772481 | 0.207202483 | 0.264057346 |
| 0.072 | 0.111353841 | 0.151520983 | 0.199342458 | 0.254523647 |
| 0.071 | 0.106681534 | 0.145398989 | 0.191622168 | 0.245127445 |
| 0.070 | 0.102121168 | 0.139408014 | 0.184044766 | 0.235874160 |
| 0.069 | 0.097673098 | 0.133549444 | 0.176613244 | 0.226769023 |
| 0.068 | 0.093337594 | 0.127824542 | 0.169330424 | 0.217817068 |
| 0.067 | 0.089114833 | 0.122234438 | 0.162198961 | 0.209023126 |
| 0.066 | 0.085004901 | 0.116780134 | 0.155221332 | 0.200391813 |
| 0.065 | 0.081007798 | 0.111462496 | 0.148399835 | 0.191927524 |
| 0.064 | 0.077123428 | 0.106282261 | 0.141736588 | 0.183634428 |
| 0.063 | 0.073351607 | 0.101240027 | 0.135233520 | 0.175516454 |
| 0.062 | 0.069692061 | 0.096336260 | 0.128892375 | 0.167577293 |
| 0.061 | 0.066144424 | 0.091571286 | 0.122714704 | 0.159820384 |
| 0.060 | 0.062708238 | 0.086945297 | 0.116701883 | 0.152248912 |
| 0.059 | 0.059382958 | 0.082458345 | 0.110855013 | 0.144865801 |
| 0.058 | 0.056167946 | 0.078110345 | 0.105175118 | 0.137673709 |
| 0.057 | 0.053062476 | 0.073901074 | 0.099662940 | 0.130675023 |
| 0.056 | 0.050065732 | 0.069830170 | 0.094319040 | 0.123871855 |
| 0.055 | 0.047176811 | 0.065897133 | 0.089143779 | 0.117266037 |
| 0.054 | 0.044394719 | 0.062101325 | 0.084137310 | 0.110859119 |
| 0.053 | 0.041718380 | 0.058441972 | 0.079299587 | 0.104652363 |
| 0.052 | 0.039146629 | 0.054918161 | 0.074630355 | 0.098646745 |
| 0.051 | 0.036678216 | 0.051528846 | 0.070129158 | 0.092842948 |

TABLE 1a:   CLUSTERING PROBABILITY   $P(n;N,p)$

$\underline{n = 5}$

| P | N =14 | N =15 | N =16 | N =17 |
|---|---|---|---|---|
| 0.050 | 0.034311809 | 0.048272844 | 0.065795335 | 0.087241363 |
| 0.049 | 0.032045993 | 0.045148840 | 0.061628021 | 0.081842088 |
| 0.048 | 0.029879272 | 0.042155386 | 0.057626148 | 0.076644926 |
| 0.047 | 0.027810071 | 0.039290904 | 0.053788449 | 0.071649387 |
| 0.046 | 0.025836737 | 0.036553689 | 0.050113456 | 0.066854686 |
| 0.045 | 0.023957542 | 0.033941907 | 0.046599502 | 0.062259746 |
| 0.044 | 0.022170684 | 0.031453601 | 0.043244728 | 0.057863196 |
| 0.043 | 0.020474288 | 0.029086695 | 0.040047080 | 0.053663379 |
| 0.042 | 0.018866411 | 0.026838989 | 0.037004315 | 0.049658352 |
| 0.041 | 0.017345042 | 0.024708172 | 0.034114003 | 0.045845885 |
| 0.040 | 0.015908106 | 0.022691818 | 0.031373534 | 0.042223472 |
| 0.039 | 0.014553464 | 0.020787392 | 0.028780118 | 0.038788331 |
| 0.038 | 0.013278920 | 0.018992255 | 0.026330792 | 0.035537411 |
| 0.037 | 0.012082221 | 0.017303664 | 0.024022423 | 0.032467398 |
| 0.036 | 0.010961062 | 0.015718781 | 0.021851717 | 0.029574719 |
| 0.035 | 0.009913085 | 0.014234676 | 0.019815221 | 0.026855551 |
| 0.034 | 0.008935889 | 0.012848328 | 0.017909334 | 0.024305832 |
| 0.033 | 0.008027029 | 0.011556636 | 0.016130306 | 0.021921261 |
| 0.032 | 0.007184021 | 0.010356420 | 0.014474255 | 0.019697317 |
| 0.031 | 0.006404345 | 0.009244429 | 0.012937165 | 0.017629263 |
| 0.030 | 0.005685454 | 0.008217347 | 0.011514903 | 0.015712158 |
| 0.029 | 0.005024769 | 0.007271796 | 0.010203222 | 0.013940868 |
| 0.028 | 0.004419695 | 0.006404349 | 0.008997770 | 0.012310082 |
| 0.027 | 0.003867616 | 0.005611529 | 0.007894105 | 0.010814321 |
| 0.026 | 0.003365906 | 0.004889823 | 0.006887700 | 0.009447952 |
| 0.025 | 0.002911932 | 0.004235687 | 0.005973955 | 0.008205207 |
| 0.024 | 0.002503059 | 0.003645553 | 0.005148210 | 0.007080193 |
| 0.023 | 0.002136657 | 0.003115841 | 0.004405759 | 0.006066915 |
| 0.022 | 0.001810107 | 0.002642962 | 0.003741855 | 0.005159290 |
| 0.021 | 0.001520804 | 0.002223334 | 0.003151732 | 0.004351165 |
| 0.020 | 0.001266167 | 0.001853387 | 0.002630614 | 0.003636338 |
| 0.019 | 0.001043647 | 0.001529573 | 0.002173733 | 0.003008582 |
| 0.018 | 0.000850726 | 0.001248381 | 0.001776340 | 0.002461660 |
| 0.017 | 0.000684935 | 0.001006342 | 0.001433725 | 0.001989354 |
| 0.016 | 0.000543851 | 0.000800045 | 0.001141233 | 0.001585488 |
| 0.015 | 0.000425114 | 0.000626146 | 0.000894281 | 0.001243950 |
| 0.014 | 0.000326426 | 0.000481383 | 0.000688376 | 0.000958724 |
| 0.013 | 0.000245568 | 0.000362587 | 0.000519138 | 0.000723914 |
| 0.012 | 0.000180403 | 0.000266695 | 0.000382314 | 0.000533777 |
| 0.011 | 0.000128885 | 0.000190768 | 0.000273806 | 0.000382752 |
| 0.010 | 0.000089071 | 0.000131999 | 0.000189688 | 0.000265489 |
| 0.009 | 0.000059130 | 0.000087734 | 0.000126232 | 0.000176892 |
| 0.008 | 0.000037350 | 0.000055486 | 0.000079930 | 0.000112144 |
| 0.007 | 0.000022152 | 0.000032948 | 0.000047521 | 0.000066754 |
| 0.006 | 0.000012098 | 0.000018016 | 0.000026015 | 0.000036589 |
| 0.005 | 0.000005903 | 0.000008801 | 0.000012724 | 0.000017918 |
| 0.004 | 0.000002446 | 0.000003652 | 0.000005286 | 0.000007452 |
| 0.003 | 0.000000783 | 0.000001170 | 0.000001696 | 0.000002394 |
| 0.002 | 0.000000156 | 0.000000234 | 0.000000340 | 0.000000480 |
| 0.001 | 0.000000010 | 0.000000015 | 0.000000022 | 0.000000030 |

TABLE 1a: CLUSTERING PROBABILITY P(n;N,p)

n = 5

| P | N =18 | P | N =18 | N =19 |
|---|-------|---|-------|-------|
| 0.200 | 0.99934102 | 0.150 | 0.96975804 | 0.988724462 |
| 0.199 | 0.99991955 | 0.149 | 0.96745609 | 0.987577998 |
| 0.198 | 0.99990198 | 0.148 | 0.96502692 | 0.986339882 |
| 0.197 | 0.99988432 | 0.147 | 0.96246233 | 0.985005094 |
| 0.196 | 0.99986208 | 0.146 | 0.95975851 | 0.983568519 |
| 0.195 | 0.99983026 | 0.145 | 0.95691614 | 0.982024950 |
| 0.194 | 0.99980130 | 0.144 | 0.95391695 | 0.980369104 |
| 0.193 | 0.99976504 | 0.143 | 0.95070561 | 0.978595631 |
| 0.192 | 0.99972869 | 0.142 | 0.94746827 | 0.976699126 |
| 0.191 | 0.99967267 | 0.141 | 0.94400626 | 0.974674140 |
| 0.190 | 0.99961606 | 0.140 | 0.94038340 | 0.972515197 |
| 0.189 | 0.99955263 | 0.139 | 0.93658867 | 0.970216805 |
| 0.188 | 0.99947168 | 0.138 | 0.93262563 | 0.967773471 |
| 0.187 | 0.99939762 | 0.137 | 0.92848211 | 0.965179714 |
| 0.186 | 0.99929894 | 0.136 | 0.92417331 | 0.962430083 |
| 0.185 | 0.99918824 | 0.135 | 0.91967781 | 0.959519170 |
| 0.184 | 0.99906574 | 0.134 | 0.91499875 | 0.956441628 |
| 0.183 | 0.99892627 | 0.133 | 0.91013342 | 0.953192185 |
| 0.182 | 0.99877252 | 0.132 | 0.90507918 | 0.949765662 |
| 0.181 | 0.99859876 | 0.131 | 0.89983372 | 0.946156991 |
| 0.180 | 0.99840458 | 0.130 | 0.89439484 | 0.942361226 |
| 0.179 | 0.99818813 | 0.129 | 0.88876064 | 0.938373567 |
| 0.178 | 0.99794743 | 0.128 | 0.88292941 | 0.934189373 |
| 0.177 | 0.99768038 | 0.127 | 0.87689971 | 0.929804178 |
| 0.176 | 0.99738478 | 0.126 | 0.87067032 | 0.925213711 |
| 0.175 | 0.99705830 | 0.125 | 0.86424030 | 0.920413910 |
| 0.174 | 0.99669850 | 0.124 | 0.85760897 | 0.915400940 |
| 0.173 | 0.99630281 | 0.123 | 0.85077593 | 0.910171207 |
| 0.172 | 0.99586855 | 0.122 | 0.84374104 | 0.904721377 |
| 0.171 | 0.99539292 | 0.121 | 0.83650447 | 0.899048388 |
| 0.170 | 0.99487300 | 0.120 | 0.82906666 | 0.893149468 |
| 0.169 | 0.99430575 | 0.119 | 0.82142836 | 0.887022148 |
| 0.168 | 0.99368802 | 0.118 | 0.81359061 | 0.880664278 |
| 0.167 | 0.99301656 | 0.117 | 0.80555476 | 0.874074035 |
| 0.166 | 0.99228798 | 0.116 | 0.79732247 | 0.867249942 |
| 0.165 | 0.99149881 | 0.115 | 0.78889569 | 0.860190879 |
| 0.164 | 0.99064547 | 0.114 | 0.78027671 | 0.852896091 |
| 0.163 | 0.98972428 | 0.113 | 0.77146811 | 0.845365200 |
| 0.162 | 0.98873145 | 0.112 | 0.76247280 | 0.837598216 |
| 0.161 | 0.98766314 | 0.111 | 0.75329398 | 0.829595547 |
| 0.160 | 0.98651539 | 0.110 | 0.74393517 | 0.821358002 |
| 0.159 | 0.98528418 | 0.109 | 0.73440021 | 0.812886802 |
| 0.158 | 0.98396542 | 0.108 | 0.72469324 | 0.804183586 |
| 0.157 | 0.98255494 | 0.107 | 0.71481870 | 0.795250415 |
| 0.156 | 0.98104854 | 0.106 | 0.70478134 | 0.786089773 |
| 0.155 | 0.97944195 | 0.105 | 0.69458619 | 0.776704576 |
| 0.154 | 0.97773088 | 0.104 | 0.68423857 | 0.767098170 |
| 0.153 | 0.97591094 | 0.103 | 0.67374412 | 0.757274331 |
| 0.152 | 0.97397792 | 0.102 | 0.66310871 | 0.747237265 |
| 0.151 | 0.97192732 | 0.101 | 0.65233851 | 0.736991611 |

TABLE 1a:　CLUSTERING PROBABILITY　$P(n;N,p)$

$n = 5$

| P | N =18 | N =19 | P | N =18 | N =19 |
|---|---|---|---|---|---|
| 0.100 | 0.641439957 | 0.726542429 | 0.050 | 0.112900614 | 0.142967332 |
| 0.099 | 0.630419713 | 0.715895206 | 0.049 | 0.106086325 | 0.134571243 |
| 0.098 | 0.619284712 | 0.705055845 | 0.048 | 0.099510229 | 0.126445733 |
| 0.097 | 0.608042116 | 0.694030659 | 0.047 | 0.093173257 | 0.118593987 |
| 0.096 | 0.596699312 | 0.682826366 | 0.046 | 0.087075931 | 0.111018642 |
| 0.095 | 0.585263900 | 0.671450079 | 0.045 | 0.081218355 | 0.103721778 |
| 0.094 | 0.573743686 | 0.659909299 | 0.044 | 0.075600223 | 0.096704909 |
| 0.093 | 0.562146664 | 0.648211899 | 0.043 | 0.070220812 | 0.089968982 |
| 0.092 | 0.550481009 | 0.636366118 | 0.042 | 0.065078986 | 0.083514377 |
| 0.091 | 0.538755060 | 0.624380543 | 0.041 | 0.060173201 | 0.077340897 |
| 0.090 | 0.526977307 | 0.612264101 | 0.040 | 0.055501503 | 0.071447779 |
| 0.089 | 0.515156381 | 0.600026037 | 0.039 | 0.051061532 | 0.065833687 |
| 0.088 | 0.503301035 | 0.587675905 | 0.038 | 0.046850533 | 0.060496718 |
| 0.087 | 0.491420130 | 0.575223548 | 0.037 | 0.042865356 | 0.055434409 |
| 0.086 | 0.479522625 | 0.562679077 | 0.036 | 0.039102463 | 0.050643740 |
| 0.085 | 0.467617553 | 0.550052860 | 0.035 | 0.035557942 | 0.046121143 |
| 0.084 | 0.455714014 | 0.537355496 | 0.034 | 0.032227509 | 0.041862511 |
| 0.083 | 0.443821154 | 0.524597798 | 0.033 | 0.029106523 | 0.037863213 |
| 0.082 | 0.431948149 | 0.511790770 | 0.032 | 0.026189993 | 0.034118098 |
| 0.081 | 0.420104193 | 0.498945589 | 0.031 | 0.023472595 | 0.030621518 |
| 0.080 | 0.408298474 | 0.486073578 | 0.030 | 0.020948684 | 0.027367342 |
| 0.079 | 0.396540167 | 0.473186187 | 0.029 | 0.018612305 | 0.024348970 |
| 0.078 | 0.384838408 | 0.460294969 | 0.028 | 0.016457216 | 0.021559360 |
| 0.077 | 0.373202284 | 0.447411554 | 0.027 | 0.014476899 | 0.018991041 |
| 0.076 | 0.361640814 | 0.434547627 | 0.026 | 0.012664580 | 0.016636145 |
| 0.075 | 0.350162932 | 0.421714903 | 0.025 | 0.011013252 | 0.014486427 |
| 0.074 | 0.338777469 | 0.408925102 | 0.024 | 0.009515693 | 0.012533295 |
| 0.073 | 0.327493140 | 0.396189924 | 0.023 | 0.008164489 | 0.010767837 |
| 0.072 | 0.316318525 | 0.383521024 | 0.022 | 0.006952058 | 0.009180857 |
| 0.071 | 0.305262054 | 0.370929986 | 0.021 | 0.005870676 | 0.007762906 |
| 0.070 | 0.294331991 | 0.358428299 | 0.020 | 0.004912504 | 0.006504317 |
| 0.069 | 0.283536414 | 0.346027332 | 0.019 | 0.004069618 | 0.005395248 |
| 0.068 | 0.272883207 | 0.333738306 | 0.018 | 0.003334037 | 0.004425716 |
| 0.067 | 0.262380040 | 0.321572272 | 0.017 | 0.002697754 | 0.003585649 |
| 0.066 | 0.252034354 | 0.309540086 | 0.016 | 0.002152774 | 0.002864924 |
| 0.065 | 0.241853347 | 0.297652383 | 0.015 | 0.001691147 | 0.002253422 |
| 0.064 | 0.231843962 | 0.285919552 | 0.014 | 0.001305008 | 0.001741077 |
| 0.063 | 0.222012868 | 0.274351719 | 0.013 | 0.000986610 | 0.001317928 |
| 0.062 | 0.212366455 | 0.262958713 | 0.012 | 0.000728376 | 0.000974184 |
| 0.061 | 0.202910812 | 0.251750054 | 0.011 | 0.000522935 | 0.000700277 |
| 0.060 | 0.193651720 | 0.240734924 | 0.010 | 0.000363171 | 0.000486932 |
| 0.059 | 0.184594643 | 0.229922147 | 0.009 | 0.000242272 | 0.000325232 |
| 0.058 | 0.175744707 | 0.219320172 | 0.008 | 0.000153782 | 0.000206693 |
| 0.057 | 0.167106701 | 0.208937049 | 0.007 | 0.000091651 | 0.000123335 |
| 0.056 | 0.158685060 | 0.198780409 | 0.006 | 0.000050297 | 0.000067767 |
| 0.055 | 0.150483855 | 0.188857453 | 0.005 | 0.000024660 | 0.000033266 |
| 0.054 | 0.142506791 | 0.179174926 | 0.004 | 0.000010269 | 0.000013869 |
| 0.053 | 0.134571192 | 0.169739106 | 0.003 | 0.000003303 | 0.000004467 |
| 0.052 | 0.127237997 | 0.160555788 | 0.002 | 0.000000663 | 0.000000898 |
| 0.051 | 0.119951754 | 0.151630268 | 0.001 | 0.000000042 | 0.000000057 |

TABLE 1a: CLUSTERING PROBABILITY P(n;N,p)

n = 6

| P | N ≡12 | N ≡13 | N ≡14 | N ≡15 |
|---|---|---|---|---|
| 0.400 | 0.989877314 | 0.999245348 | 0.999976875 | 0.999999803 |
| 0.399 | 0.989355615 | 0.999183727 | 0.999974142 | 0.999999772 |
| 0.398 | 0.988813903 | 0.999117891 | 0.999971121 | 0.999999735 |
| 0.397 | 0.988251708 | 0.999047611 | 0.999967785 | 0.999999693 |
| 0.396 | 0.987668558 | 0.998972647 | 0.999964106 | 0.999999646 |
| 0.395 | 0.987063981 | 0.998892753 | 0.999960052 | 0.999999591 |
| 0.394 | 0.986437504 | 0.998807672 | 0.999955591 | 0.999999528 |
| 0.393 | 0.985788654 | 0.998717138 | 0.999950687 | 0.999999457 |
| 0.392 | 0.985116956 | 0.998620876 | 0.999945301 | 0.999999376 |
| 0.391 | 0.984421940 | 0.998518601 | 0.999939392 | 0.999999283 |
| 0.390 | 0.983703132 | 0.998410018 | 0.999932915 | 0.999999178 |
| 0.389 | 0.982960061 | 0.998294823 | 0.999925822 | 0.999999058 |
| 0.388 | 0.982192257 | 0.998172701 | 0.999918064 | 0.999998923 |
| 0.387 | 0.981399251 | 0.998043328 | 0.999909585 | 0.999998769 |
| 0.386 | 0.980580577 | 0.997906369 | 0.999900326 | 0.999998595 |
| 0.385 | 0.979735770 | 0.997761479 | 0.999890226 | 0.999998399 |
| 0.384 | 0.978864368 | 0.997608303 | 0.999879218 | 0.999998177 |
| 0.383 | 0.977965911 | 0.997446477 | 0.999867231 | 0.999997926 |
| 0.382 | 0.977039942 | 0.997275623 | 0.999854190 | 0.999997644 |
| 0.381 | 0.976086009 | 0.997095356 | 0.999840013 | 0.999997326 |
| 0.380 | 0.975103661 | 0.996905279 | 0.999824615 | 0.999996968 |
| 0.379 | 0.974092454 | 0.996704985 | 0.999807905 | 0.999996566 |
| 0.378 | 0.973051945 | 0.996494058 | 0.999789786 | 0.999996115 |
| 0.377 | 0.971981698 | 0.996272067 | 0.999770156 | 0.999995609 |
| 0.376 | 0.970881281 | 0.996038577 | 0.999748905 | 0.999995041 |
| 0.375 | 0.969750267 | 0.995793138 | 0.999725917 | 0.999994406 |
| 0.374 | 0.968588233 | 0.995535291 | 0.999701072 | 0.999993696 |
| 0.373 | 0.967394764 | 0.995264567 | 0.999674238 | 0.999992903 |
| 0.372 | 0.966169450 | 0.994980489 | 0.999645279 | 0.999992017 |
| 0.371 | 0.964911886 | 0.994682565 | 0.999614051 | 0.999991028 |
| 0.370 | 0.963621677 | 0.994370299 | 0.999580400 | 0.999989926 |
| 0.369 | 0.962298430 | 0.994043181 | 0.999544165 | 0.999988699 |
| 0.368 | 0.960941763 | 0.993700694 | 0.999505177 | 0.999987334 |
| 0.367 | 0.959551300 | 0.993342310 | 0.999463256 | 0.999985816 |
| 0.366 | 0.958126671 | 0.992967493 | 0.999418213 | 0.999984129 |
| 0.365 | 0.956667517 | 0.992575697 | 0.999369851 | 0.999982256 |
| 0.364 | 0.955173484 | 0.992166369 | 0.999317961 | 0.999980179 |
| 0.363 | 0.953644227 | 0.991738947 | 0.999262323 | 0.999977877 |
| 0.362 | 0.952079412 | 0.991292859 | 0.999202708 | 0.999975327 |
| 0.361 | 0.950478711 | 0.990827528 | 0.999138875 | 0.999972505 |
| 0.360 | 0.948841805 | 0.990342367 | 0.999070570 | 0.999969383 |
| 0.359 | 0.947168387 | 0.989836783 | 0.998997529 | 0.999965934 |
| 0.358 | 0.945458157 | 0.989310176 | 0.998919474 | 0.999962124 |
| 0.357 | 0.943710825 | 0.988761939 | 0.998836116 | 0.999957920 |
| 0.356 | 0.941926112 | 0.988191459 | 0.998747152 | 0.999953283 |
| 0.355 | 0.940103747 | 0.987598119 | 0.998652265 | 0.999948173 |
| 0.354 | 0.938243472 | 0.986981293 | 0.998551124 | 0.999942545 |
| 0.353 | 0.936345036 | 0.986340352 | 0.998443385 | 0.999936350 |
| 0.352 | 0.934408202 | 0.985674665 | 0.998328690 | 0.999929537 |
| 0.351 | 0.932432742 | 0.984983592 | 0.998206664 | 0.999922048 |

TABLE 1a:　CLUSTERING PROBABILITY　$P(n;N,p)$

$n = 6$

| P | N =12 | N =13 | N =14 | N =15 |
|---|---|---|---|---|
| 0.350 | 0.930418440 | 0.984266493 | 0.998076920 | 0.999913820 |
| 0.349 | 0.928365089 | 0.983522724 | 0.997939051 | 0.999904788 |
| 0.348 | 0.926272495 | 0.982751638 | 0.997792640 | 0.999894879 |
| 0.347 | 0.924140475 | 0.981952587 | 0.997637249 | 0.999884014 |
| 0.346 | 0.921966858 | 0.981124920 | 0.997472426 | 0.999872107 |
| 0.345 | 0.919757482 | 0.980267987 | 0.997297704 | 0.999859068 |
| 0.344 | 0.917506202 | 0.979381136 | 0.997112597 | 0.999844796 |
| 0.343 | 0.915214878 | 0.978463715 | 0.996916604 | 0.999829183 |
| 0.342 | 0.912883388 | 0.977515074 | 0.996709204 | 0.999812115 |
| 0.341 | 0.910511619 | 0.976534564 | 0.996489864 | 0.999793466 |
| 0.340 | 0.908099470 | 0.975521537 | 0.996258029 | 0.999773100 |
| 0.339 | 0.905646852 | 0.974475349 | 0.996013130 | 0.999750872 |
| 0.338 | 0.903153691 | 0.973395357 | 0.995754581 | 0.999726624 |
| 0.337 | 0.900619922 | 0.972280924 | 0.995481777 | 0.999700188 |
| 0.336 | 0.898045493 | 0.971131416 | 0.995194097 | 0.999671381 |
| 0.335 | 0.895430366 | 0.969946204 | 0.994890904 | 0.999640007 |
| 0.334 | 0.892774514 | 0.968724665 | 0.994571544 | 0.999605854 |
| 0.333 | 0.890077922 | 0.967466182 | 0.994235347 | 0.999568697 |
| 0.332 | 0.887340590 | 0.966170144 | 0.993881628 | 0.999528291 |
| 0.331 | 0.884562526 | 0.964835948 | 0.993509685 | 0.999484373 |
| 0.330 | 0.881743756 | 0.963462999 | 0.993118802 | 0.999436664 |
| 0.329 | 0.878884313 | 0.962050711 | 0.992708250 | 0.999384865 |
| 0.328 | 0.875984247 | 0.960598504 | 0.992277284 | 0.999328655 |
| 0.327 | 0.873043618 | 0.959105811 | 0.991825147 | 0.999267698 |
| 0.326 | 0.870062497 | 0.957572073 | 0.991351069 | 0.999201632 |
| 0.325 | 0.867040972 | 0.955996742 | 0.990854269 | 0.999130080 |
| 0.324 | 0.863979139 | 0.954379283 | 0.990333954 | 0.999052638 |
| 0.323 | 0.860877108 | 0.952719169 | 0.989789319 | 0.998968885 |
| 0.322 | 0.857735002 | 0.951015887 | 0.989219551 | 0.998878376 |
| 0.321 | 0.854552954 | 0.949268937 | 0.988623826 | 0.998780645 |
| 0.320 | 0.851331112 | 0.947477830 | 0.988001314 | 0.998675202 |
| 0.319 | 0.848069633 | 0.945642093 | 0.987351173 | 0.998561536 |
| 0.318 | 0.844768689 | 0.943761264 | 0.986672557 | 0.998439112 |
| 0.317 | 0.841428463 | 0.941834898 | 0.985964613 | 0.998307373 |
| 0.316 | 0.838049148 | 0.939862562 | 0.985226482 | 0.998165739 |
| 0.315 | 0.834630951 | 0.937843840 | 0.984457300 | 0.998013607 |
| 0.314 | 0.831174090 | 0.935778330 | 0.983656199 | 0.997850348 |
| 0.313 | 0.827678795 | 0.933665646 | 0.982822308 | 0.997675315 |
| 0.312 | 0.824145308 | 0.931505419 | 0.981954754 | 0.997487834 |
| 0.311 | 0.820573880 | 0.929297295 | 0.981052662 | 0.997287210 |
| 0.310 | 0.816964776 | 0.927040938 | 0.980115155 | 0.997072724 |
| 0.309 | 0.813318271 | 0.924736027 | 0.979141359 | 0.996843635 |
| 0.308 | 0.809634651 | 0.922382261 | 0.978130398 | 0.996599180 |
| 0.307 | 0.805914214 | 0.919979355 | 0.977081400 | 0.996338573 |
| 0.306 | 0.802157269 | 0.917527041 | 0.975993493 | 0.996061007 |
| 0.305 | 0.798364134 | 0.915025070 | 0.974865812 | 0.995765652 |
| 0.304 | 0.794535140 | 0.912473212 | 0.973697493 | 0.995451660 |
| 0.303 | 0.790670626 | 0.909871254 | 0.972487679 | 0.995118159 |
| 0.302 | 0.786770944 | 0.907219002 | 0.971235519 | 0.994764259 |
| 0.301 | 0.782836455 | 0.904516282 | 0.969940167 | 0.994389049 |

TABLE 1a:   CLUSTERING PROBABILITY   P(n;N,p)

n = 6

| P | N=12 | N=13 | N=14 | N=15 |
|------|------------|------------|------------|------------|
| 0.300 | 0.778867529 | 0.901762938 | 0.968600788 | 0.993991601 |
| 0.299 | 0.774864549 | 0.898958832 | 0.967216553 | 0.993570966 |
| 0.298 | 0.770827906 | 0.896103849 | 0.965786643 | 0.993126179 |
| 0.297 | 0.766757999 | 0.893197889 | 0.964310248 | 0.992656258 |
| 0.296 | 0.762655241 | 0.890240874 | 0.962786573 | 0.992160203 |
| 0.295 | 0.758520051 | 0.887232746 | 0.961214830 | 0.991637000 |
| 0.294 | 0.754352859 | 0.884173465 | 0.959594247 | 0.991085619 |
| 0.293 | 0.750154102 | 0.881063013 | 0.957924063 | 0.990505018 |
| 0.292 | 0.745924229 | 0.877901390 | 0.956203533 | 0.989894138 |
| 0.291 | 0.741663696 | 0.874688616 | 0.954431926 | 0.989251913 |
| 0.290 | 0.737372968 | 0.871424732 | 0.952608527 | 0.988577262 |
| 0.289 | 0.733052518 | 0.868109799 | 0.950732638 | 0.987869094 |
| 0.288 | 0.728702829 | 0.864743897 | 0.948803577 | 0.987126311 |
| 0.287 | 0.724324391 | 0.861327127 | 0.946820681 | 0.986347804 |
| 0.286 | 0.719917703 | 0.857859609 | 0.944783305 | 0.985532458 |
| 0.285 | 0.715483270 | 0.854341483 | 0.942690823 | 0.984679153 |
| 0.284 | 0.711021607 | 0.850772910 | 0.940542630 | 0.983786762 |
| 0.283 | 0.706533235 | 0.847154070 | 0.938338139 | 0.982854154 |
| 0.282 | 0.702018683 | 0.843485163 | 0.936076788 | 0.981880198 |
| 0.281 | 0.697478487 | 0.839766408 | 0.933758033 | 0.980863759 |
| 0.280 | 0.692913191 | 0.835998046 | 0.931381356 | 0.979803701 |
| 0.279 | 0.688323345 | 0.832180333 | 0.928946259 | 0.978698890 |
| 0.278 | 0.683709505 | 0.828313550 | 0.926452268 | 0.977548195 |
| 0.277 | 0.679072235 | 0.824397993 | 0.923898935 | 0.976350486 |
| 0.276 | 0.674412105 | 0.820433980 | 0.921285834 | 0.975104638 |
| 0.275 | 0.669729690 | 0.816421846 | 0.918612566 | 0.973809533 |
| 0.274 | 0.665025572 | 0.812361945 | 0.915878756 | 0.972464058 |
| 0.273 | 0.660300339 | 0.808254653 | 0.913084056 | 0.971067109 |
| 0.272 | 0.655554586 | 0.804100359 | 0.910228143 | 0.969617591 |
| 0.271 | 0.650788909 | 0.799899477 | 0.907310723 | 0.968114421 |
| 0.270 | 0.646003914 | 0.795652433 | 0.904331527 | 0.966556525 |
| 0.269 | 0.641200209 | 0.791359675 | 0.901290314 | 0.964942846 |
| 0.268 | 0.636378409 | 0.787021668 | 0.898186871 | 0.963272337 |
| 0.267 | 0.631539133 | 0.782638895 | 0.895021014 | 0.961543970 |
| 0.266 | 0.626683004 | 0.778211854 | 0.891792586 | 0.959756731 |
| 0.265 | 0.621810649 | 0.773741063 | 0.888501459 | 0.957909626 |
| 0.264 | 0.616922701 | 0.769227056 | 0.885147534 | 0.956001680 |
| 0.263 | 0.612019795 | 0.764670384 | 0.881730742 | 0.954031937 |
| 0.262 | 0.607102572 | 0.760071615 | 0.878251042 | 0.951999464 |
| 0.261 | 0.602171674 | 0.755431331 | 0.874708424 | 0.949903350 |
| 0.260 | 0.597227748 | 0.750750132 | 0.871102905 | 0.947742706 |
| 0.259 | 0.592271444 | 0.746028634 | 0.867434535 | 0.945516672 |
| 0.258 | 0.587303415 | 0.741267467 | 0.863703391 | 0.943224410 |
| 0.257 | 0.582324317 | 0.736467277 | 0.859909582 | 0.940865111 |
| 0.256 | 0.577334808 | 0.731628726 | 0.856053245 | 0.938437994 |
| 0.255 | 0.572335550 | 0.726752488 | 0.852134551 | 0.935942306 |
| 0.254 | 0.567327205 | 0.721839253 | 0.848153695 | 0.933377326 |
| 0.253 | 0.562310439 | 0.716889725 | 0.844110908 | 0.930742363 |
| 0.252 | 0.557285920 | 0.711904622 | 0.840006447 | 0.928036757 |
| 0.251 | 0.552254315 | 0.706684674 | 0.835840600 | 0.925259883 |

TABLE 1a:   CLUSTERING PROBABILITY   P(n;N,p)

n = 6

| P | N = 12 | N = 13 | N = 14 | N = 15 |
|---|---|---|---|---|
| 0.250 | 0.54721€296 | 0.701830626 | 0.831613686 | 0.922411148 |
| 0.249 | 0.542172535 | 0.696743233 | 0.827326052 | 0.919489994 |
| 0.248 | 0.537123704 | 0.691623267 | 0.822978077 | 0.916495899 |
| 0.247 | 0.532070477 | 0.686471506 | 0.818570166 | 0.913428377 |
| 0.246 | 0.527013530 | 0.681288746 | 0.814102757 | 0.910286979 |
| 0.245 | 0.521953537 | 0.676075791 | 0.809576315 | 0.907071293 |
| 0.244 | 0.516891174 | 0.670833456 | 0.804991335 | 0.903780946 |
| 0.243 | 0.511827117 | 0.665562567 | 0.800348341 | 0.900415604 |
| 0.242 | 0.506762041 | 0.660263963 | 0.795647883 | 0.896974973 |
| 0.241 | 0.501696624 | 0.654938489 | 0.790890542 | 0.893458798 |
| 0.240 | 0.496631539 | 0.649587003 | 0.786076927 | 0.889866864 |
| 0.239 | 0.491567463 | 0.644210370 | 0.781207673 | 0.886199000 |
| 0.238 | 0.486505068 | 0.638809467 | 0.776283444 | 0.882455074 |
| 0.237 | 0.481445028 | 0.633385177 | 0.771304930 | 0.878634997 |
| 0.236 | 0.476388016 | 0.627938392 | 0.766272847 | 0.874738722 |
| 0.235 | 0.471334701 | 0.622470012 | 0.761187941 | 0.870766245 |
| 0.234 | 0.466285754 | 0.616980946 | 0.756050979 | 0.866717605 |
| 0.233 | 0.461241842 | 0.611472107 | 0.750862757 | 0.862592885 |
| 0.232 | 0.456203630 | 0.605944419 | 0.745624096 | 0.858392209 |
| 0.231 | 0.451171783 | 0.600398809 | 0.740335840 | 0.854115748 |
| 0.230 | 0.446146962 | 0.594836212 | 0.734998858 | 0.849763714 |
| 0.229 | 0.441129827 | 0.589257568 | 0.729614045 | 0.845336363 |
| 0.228 | 0.436121033 | 0.583663823 | 0.724182316 | 0.840833998 |
| 0.227 | 0.431121235 | 0.578055927 | 0.718704612 | 0.836256962 |
| 0.226 | 0.426131085 | 0.572434836 | 0.713181894 | 0.831605644 |
| 0.225 | 0.421151229 | 0.566801510 | 0.707615147 | 0.826880477 |
| 0.224 | 0.416182312 | 0.561156911 | 0.702005375 | 0.822081937 |
| 0.223 | 0.411224976 | 0.555502007 | 0.696353606 | 0.817210544 |
| 0.222 | 0.406279858 | 0.549837767 | 0.690660887 | 0.812266861 |
| 0.221 | 0.401347591 | 0.544165166 | 0.684928283 | 0.807251495 |
| 0.220 | 0.396428807 | 0.538485178 | 0.679156882 | 0.802165095 |
| 0.219 | 0.391524130 | 0.532798780 | 0.673347786 | 0.797008354 |
| 0.218 | 0.386634181 | 0.527106952 | 0.667502121 | 0.791782005 |
| 0.217 | 0.381759578 | 0.521410674 | 0.661621024 | 0.786486827 |
| 0.216 | 0.376900932 | 0.515710927 | 0.655705655 | 0.781123636 |
| 0.215 | 0.372058852 | 0.510008692 | 0.649757187 | 0.775693294 |
| 0.214 | 0.367233940 | 0.504304953 | 0.643776809 | 0.770196699 |
| 0.213 | 0.362426792 | 0.498600690 | 0.637765726 | 0.764634792 |
| 0.212 | 0.357638002 | 0.492896886 | 0.631725156 | 0.759008554 |
| 0.211 | 0.352868155 | 0.487194519 | 0.625656334 | 0.753319004 |
| 0.210 | 0.348117832 | 0.481494570 | 0.619560504 | 0.747567201 |
| 0.209 | 0.343387610 | 0.475798016 | 0.613438927 | 0.741754239 |
| 0.208 | 0.338678056 | 0.470105832 | 0.607292873 | 0.735881253 |
| 0.207 | 0.333989735 | 0.464418990 | 0.601123623 | 0.729949412 |
| 0.206 | 0.329323203 | 0.458738463 | 0.594932472 | 0.723959923 |
| 0.205 | 0.324679011 | 0.453065216 | 0.588720722 | 0.717914026 |
| 0.204 | 0.320057703 | 0.447400214 | 0.582489685 | 0.711812997 |
| 0.203 | 0.315459817 | 0.441744417 | 0.576240681 | 0.705658147 |
| 0.202 | 0.310885883 | 0.436098781 | 0.569975041 | 0.699450816 |
| 0.201 | 0.306336424 | 0.430464256 | 0.563694101 | 0.693192379 |

TABLE 1a:  CLUSTERING PROBABILITY  P(n;N,p)

n = 6

| P | N = 12 | N = 13 | N = 14 | N = 15 |
|---|---|---|---|---|
| 0.200 | 0.301811958 | 0.424841791 | 0.557399203 | 0.686884242 |
| 0.199 | 0.297312993 | 0.419232326 | 0.551091696 | 0.680527840 |
| 0.198 | 0.292840032 | 0.413636798 | 0.544772935 | 0.674124639 |
| 0.197 | 0.288393569 | 0.408056137 | 0.538444280 | 0.667676132 |
| 0.196 | 0.283974091 | 0.402491266 | 0.532107093 | 0.661183841 |
| 0.195 | 0.279582078 | 0.396943105 | 0.525762740 | 0.654649314 |
| 0.194 | 0.275217999 | 0.391412562 | 0.519412592 | 0.648074124 |
| 0.193 | 0.270882319 | 0.385900543 | 0.513058018 | 0.641459869 |
| 0.192 | 0.266575492 | 0.380407943 | 0.506700392 | 0.634808171 |
| 0.191 | 0.262297965 | 0.374935651 | 0.500341086 | 0.628120674 |
| 0.190 | 0.258050177 | 0.369484547 | 0.493981475 | 0.621399045 |
| 0.189 | 0.253832557 | 0.364055504 | 0.487622929 | 0.614644971 |
| 0.188 | 0.249645527 | 0.358649385 | 0.481266821 | 0.607860156 |
| 0.187 | 0.245489499 | 0.353267043 | 0.474914520 | 0.601046326 |
| 0.186 | 0.241364877 | 0.347909325 | 0.468567391 | 0.594205223 |
| 0.185 | 0.237272056 | 0.342577066 | 0.462226798 | 0.587338603 |
| 0.184 | 0.233211421 | 0.337271092 | 0.455894101 | 0.580448241 |
| 0.183 | 0.229183350 | 0.331992217 | 0.449570654 | 0.573535924 |
| 0.182 | 0.225188209 | 0.326741247 | 0.443257807 | 0.566603450 |
| 0.181 | 0.221226357 | 0.321518978 | 0.436956902 | 0.559652633 |
| 0.180 | 0.217298142 | 0.316326191 | 0.430669277 | 0.552685294 |
| 0.179 | 0.213403904 | 0.311163659 | 0.424396263 | 0.545703265 |
| 0.178 | 0.209543971 | 0.306032143 | 0.418139180 | 0.538708387 |
| 0.177 | 0.205718665 | 0.300932392 | 0.411899344 | 0.531702506 |
| 0.176 | 0.201928294 | 0.295865143 | 0.405678058 | 0.524687477 |
| 0.175 | 0.198173159 | 0.290831119 | 0.399476618 | 0.517665157 |
| 0.174 | 0.194453551 | 0.285831035 | 0.393296309 | 0.510637410 |
| 0.173 | 0.190769750 | 0.280865588 | 0.387138406 | 0.503606099 |
| 0.172 | 0.187122026 | 0.275935465 | 0.381004171 | 0.496573092 |
| 0.171 | 0.183510639 | 0.271041339 | 0.374894856 | 0.489540254 |
| 0.170 | 0.179935839 | 0.266183871 | 0.368811699 | 0.482509453 |
| 0.169 | 0.176397867 | 0.261363705 | 0.362755925 | 0.475482552 |
| 0.168 | 0.172896951 | 0.256581475 | 0.356728747 | 0.468461412 |
| 0.167 | 0.169433310 | 0.251837799 | 0.350731363 | 0.461447889 |
| 0.166 | 0.166007154 | 0.247133279 | 0.344764955 | 0.454443837 |
| 0.165 | 0.162618680 | 0.242468507 | 0.338830691 | 0.447451100 |
| 0.164 | 0.159268077 | 0.237844056 | 0.332929724 | 0.440471515 |
| 0.163 | 0.155955522 | 0.233260486 | 0.327063190 | 0.433506912 |
| 0.162 | 0.152681181 | 0.228718342 | 0.321232208 | 0.426559111 |
| 0.161 | 0.149445210 | 0.224218155 | 0.315437880 | 0.419629921 |
| 0.160 | 0.146247755 | 0.219760437 | 0.309681290 | 0.412721137 |
| 0.159 | 0.143088951 | 0.215345688 | 0.303963504 | 0.405834546 |
| 0.158 | 0.139968922 | 0.210974391 | 0.298285570 | 0.398971916 |
| 0.157 | 0.136887782 | 0.206647013 | 0.292648517 | 0.392135003 |
| 0.156 | 0.133845633 | 0.202364005 | 0.287053353 | 0.385325547 |
| 0.155 | 0.130842567 | 0.198125803 | 0.281501066 | 0.378545270 |
| 0.154 | 0.127878666 | 0.193932825 | 0.275992627 | 0.371795877 |
| 0.153 | 0.124954001 | 0.189785474 | 0.270528982 | 0.365079052 |
| 0.152 | 0.122068631 | 0.185684136 | 0.265111058 | 0.358396462 |
| 0.151 | 0.119222606 | 0.181629180 | 0.259739760 | 0.351749752 |

TABLE 1a:   CLUSTERING PROBABILITY  P(n;N,p)

n = 6

| P | N = 12 | N = 13 | N = 14 | N = 15 |
|---|---|---|---|---|
| 0.150 | 0.116415964 | 0.177620959 | 0.254415971 | 0.345140543 |
| 0.149 | 0.113648733 | 0.173659809 | 0.249140552 | 0.338570438 |
| 0.148 | 0.110920930 | 0.169746050 | 0.243914341 | 0.332041013 |
| 0.147 | 0.108232563 | 0.165879982 | 0.238738153 | 0.325553819 |
| 0.146 | 0.105583626 | 0.162061892 | 0.233612780 | 0.319110384 |
| 0.145 | 0.102974106 | 0.158292047 | 0.228538989 | 0.312712208 |
| 0.144 | 0.100403976 | 0.154570698 | 0.223517526 | 0.306360767 |
| 0.143 | 0.097873202 | 0.150898078 | 0.218549109 | 0.300057504 |
| 0.142 | 0.095381737 | 0.147274404 | 0.213634435 | 0.293803839 |
| 0.141 | 0.092929523 | 0.143699875 | 0.208774173 | 0.287601158 |
| 0.140 | 0.090516496 | 0.140174673 | 0.203968970 | 0.281450820 |
| 0.139 | 0.088142576 | 0.136698961 | 0.199219446 | 0.275354152 |
| 0.138 | 0.085807676 | 0.133272886 | 0.194526195 | 0.269312450 |
| 0.137 | 0.083511699 | 0.129896578 | 0.189889787 | 0.263326977 |
| 0.136 | 0.081254535 | 0.126570149 | 0.185310764 | 0.257398963 |
| 0.135 | 0.079036068 | 0.123293693 | 0.180789643 | 0.251529605 |
| 0.134 | 0.076856168 | 0.120067288 | 0.176326916 | 0.245720068 |
| 0.133 | 0.074714699 | 0.116890994 | 0.171923047 | 0.239971477 |
| 0.132 | 0.072611511 | 0.113764853 | 0.167578472 | 0.234284927 |
| 0.131 | 0.070546447 | 0.110688890 | 0.163293604 | 0.228661475 |
| 0.130 | 0.068519340 | 0.107663114 | 0.159068826 | 0.223102141 |
| 0.129 | 0.066530012 | 0.104687515 | 0.154904496 | 0.217607909 |
| 0.128 | 0.064578277 | 0.101762066 | 0.150800943 | 0.212179727 |
| 0.127 | 0.062663940 | 0.098886725 | 0.146758471 | 0.206818501 |
| 0.126 | 0.060786795 | 0.096061431 | 0.142777355 | 0.201525105 |
| 0.125 | 0.058946627 | 0.093286106 | 0.138857844 | 0.196300369 |
| 0.124 | 0.057143214 | 0.090560656 | 0.135000160 | 0.191145086 |
| 0.123 | 0.055376323 | 0.087884970 | 0.131204496 | 0.186060011 |
| 0.122 | 0.053645712 | 0.085258921 | 0.127471020 | 0.181045858 |
| 0.121 | 0.051951132 | 0.082682364 | 0.123799871 | 0.176103302 |
| 0.120 | 0.050292325 | 0.080155139 | 0.120191162 | 0.171232976 |
| 0.119 | 0.048669024 | 0.077677070 | 0.116644577 | 0.166435474 |
| 0.118 | 0.047080953 | 0.075247963 | 0.113161375 | 0.161711351 |
| 0.117 | 0.045527830 | 0.072867609 | 0.109740386 | 0.157061117 |
| 0.116 | 0.044009362 | 0.070535785 | 0.106382016 | 0.152485245 |
| 0.115 | 0.042525252 | 0.068252249 | 0.103086240 | 0.147984165 |
| 0.114 | 0.041075191 | 0.066016746 | 0.099853010 | 0.143558267 |
| 0.113 | 0.039658866 | 0.063829005 | 0.096682250 | 0.139207898 |
| 0.112 | 0.038275955 | 0.061688740 | 0.093573856 | 0.134933365 |
| 0.111 | 0.036926129 | 0.059595650 | 0.090527700 | 0.130734934 |
| 0.110 | 0.035609052 | 0.057549419 | 0.087543627 | 0.126612829 |
| 0.109 | 0.034324381 | 0.055549716 | 0.084621455 | 0.122567232 |
| 0.108 | 0.033071766 | 0.053596198 | 0.081760980 | 0.118598285 |
| 0.107 | 0.031850852 | 0.051688506 | 0.078961967 | 0.114706088 |
| 0.106 | 0.030661276 | 0.049826267 | 0.076224161 | 0.110890702 |
| 0.105 | 0.029502669 | 0.048009096 | 0.073547278 | 0.107152144 |
| 0.104 | 0.028374657 | 0.046236594 | 0.070931012 | 0.103490392 |
| 0.103 | 0.027276860 | 0.044508348 | 0.068375031 | 0.099905385 |
| 0.102 | 0.026208892 | 0.042823934 | 0.065878980 | 0.096397020 |
| 0.101 | 0.025170361 | 0.041182913 | 0.063442480 | 0.092965153 |

TABLE 1a:   CLUSTERING PROBABILITY   P(n;N,p)

n = 6

| P | N =12 | N =13 | N =14 | N =15 |
|---|---|---|---|---|
| 0.100 | 0.024160872 | 0.039584838 | 0.061065128 | 0.085609603 |
| 0.099 | 0.023180022 | 0.038029245 | 0.058746498 | 0.083330148 |
| 0.098 | 0.022227407 | 0.036515662 | 0.056486142 | 0.083126527 |
| 0.097 | 0.021302614 | 0.035043604 | 0.054283588 | 0.079998441 |
| 0.096 | 0.020405230 | 0.033612577 | 0.052138345 | 0.076945552 |
| 0.095 | 0.019534836 | 0.032222073 | 0.050049897 | 0.073967486 |
| 0.094 | 0.018691007 | 0.030871577 | 0.048017709 | 0.071063830 |
| 0.093 | 0.017873319 | 0.029560562 | 0.046041226 | 0.068234135 |
| 0.092 | 0.017081341 | 0.028288492 | 0.044119871 | 0.065477914 |
| 0.091 | 0.016314640 | 0.027054821 | 0.042253048 | 0.062794647 |
| 0.090 | 0.015572779 | 0.025858997 | 0.040440141 | 0.060183777 |
| 0.089 | 0.014855320 | 0.024700454 | 0.038680517 | 0.057644713 |
| 0.088 | 0.014161822 | 0.023578623 | 0.036973524 | 0.055176830 |
| 0.087 | 0.013491841 | 0.022492924 | 0.035318491 | 0.052779470 |
| 0.086 | 0.012844930 | 0.021442772 | 0.033714732 | 0.050451942 |
| 0.085 | 0.012220644 | 0.020427572 | 0.032161543 | 0.048193524 |
| 0.084 | 0.011618532 | 0.019446724 | 0.030658204 | 0.046003461 |
| 0.083 | 0.011038145 | 0.018499623 | 0.029203981 | 0.043880970 |
| 0.082 | 0.010479032 | 0.017585655 | 0.027798122 | 0.041825237 |
| 0.081 | 0.009940739 | 0.016704203 | 0.026439865 | 0.039835419 |
| 0.080 | 0.009422815 | 0.015854645 | 0.025128432 | 0.037910648 |
| 0.079 | 0.008924806 | 0.015036352 | 0.023863031 | 0.036050024 |
| 0.078 | 0.008446259 | 0.014248694 | 0.022642860 | 0.034252626 |
| 0.077 | 0.007986722 | 0.013491034 | 0.021467103 | 0.032517503 |
| 0.076 | 0.007545743 | 0.012762735 | 0.020334936 | 0.030843685 |
| 0.075 | 0.007122868 | 0.012063154 | 0.019245522 | 0.029230175 |
| 0.074 | 0.006717649 | 0.011391648 | 0.018198016 | 0.027675954 |
| 0.073 | 0.006329634 | 0.010747570 | 0.017191563 | 0.026179983 |
| 0.072 | 0.005958376 | 0.010130273 | 0.016225300 | 0.024741203 |
| 0.071 | 0.005603429 | 0.009539108 | 0.015298358 | 0.023358535 |
| 0.070 | 0.005264348 | 0.008973425 | 0.014409858 | 0.022030881 |
| 0.069 | 0.004940691 | 0.008432575 | 0.013558918 | 0.020757128 |
| 0.068 | 0.004632017 | 0.007915908 | 0.012744651 | 0.019536147 |
| 0.067 | 0.004337891 | 0.007422776 | 0.011966162 | 0.018366793 |
| 0.066 | 0.004057876 | 0.006952531 | 0.011222555 | 0.017247908 |
| 0.065 | 0.003791544 | 0.006504526 | 0.010512931 | 0.016178322 |
| 0.064 | 0.003538466 | 0.006078119 | 0.009836388 | 0.015156854 |
| 0.063 | 0.003298217 | 0.005672666 | 0.009192022 | 0.014182312 |
| 0.062 | 0.003070379 | 0.005287531 | 0.008578930 | 0.013253496 |
| 0.061 | 0.002854536 | 0.004922078 | 0.007996208 | 0.012369199 |
| 0.060 | 0.002650275 | 0.004575675 | 0.007442952 | 0.011528206 |
| 0.059 | 0.002457190 | 0.004247696 | 0.006918262 | 0.010729300 |
| 0.058 | 0.002274879 | 0.003937519 | 0.006421239 | 0.009971257 |
| 0.057 | 0.002102945 | 0.003644527 | 0.005950986 | 0.009252853 |
| 0.056 | 0.001940997 | 0.003368108 | 0.005506613 | 0.008572862 |
| 0.055 | 0.001788648 | 0.003107659 | 0.005087233 | 0.007930058 |
| 0.054 | 0.001645518 | 0.002862579 | 0.004691969 | 0.007323216 |
| 0.053 | 0.001511233 | 0.002632278 | 0.004319934 | 0.006751114 |
| 0.052 | 0.001385424 | 0.002416172 | 0.003970272 | 0.006212536 |
| 0.051 | 0.001267729 | 0.002213684 | 0.003642120 | 0.005706268 |

TABLE 1a:   CLUSTERING PROBABILITY   P(n;N,p)

n = 6

| P | N =12 | N =13 | N =14 | N =15 |
|---|---|---|---|---|
| C.050 | 0.001157793 | 0.002024247 | 0.003334628 | 0.005231105 |
| 0.049 | 0.001055268 | 0.001847300 | 0.003046953 | 0.004785850 |
| 0.048 | 0.000959810 | 0.001682293 | 0.002778265 | 0.004369314 |
| 0.047 | 0.000871086 | 0.001528686 | 0.002527744 | 0.003980319 |
| 0.046 | 0.000788768 | 0.001385947 | 0.002294581 | 0.003617699 |
| C.045 | 0.000712536 | 0.001253556 | 0.002077980 | 0.003280301 |
| 0.044 | 0.000642078 | 0.001131001 | 0.001877159 | 0.002966987 |
| 0.043 | 0.000577089 | 0.001017784 | 0.001691348 | 0.002676634 |
| 0.042 | 0.000517272 | 0.000913416 | 0.001519792 | 0.002408135 |
| 0.041 | 0.000462337 | 0.000817420 | 0.001361754 | 0.002160403 |
| C.040 | 0.000412005 | 0.000729331 | 0.001216508 | 0.001932367 |
| 0.039 | 0.000366004 | 0.000648696 | 0.001083348 | 0.001722981 |
| 0.038 | 0.000324067 | 0.000575076 | 0.000961584 | 0.001531216 |
| 0.037 | 0.000285941 | 0.000508043 | 0.000850543 | 0.001356069 |
| 0.036 | 0.000251378 | 0.000447181 | 0.000749572 | 0.001196558 |
| 0.035 | 0.000220139 | 0.000392090 | 0.000658034 | 0.001051727 |
| 0.034 | 0.000191995 | 0.000342381 | 0.000575313 | 0.000920645 |
| 0.033 | 0.000166725 | 0.000297680 | 0.000500813 | 0.000802409 |
| 0.032 | 0.000144116 | 0.000257626 | 0.000433956 | 0.000696142 |
| 0.031 | 0.000123964 | 0.000221872 | 0.000374187 | 0.000600997 |
| C.030 | 0.000106076 | 0.000190086 | 0.000320971 | 0.000516154 |
| 0.029 | 0.000090265 | 0.000161950 | 0.000273794 | 0.000440824 |
| C.028 | 0.000076354 | 0.000137158 | 0.000232162 | 0.000374250 |
| 0.027 | 0.000064176 | 0.000115421 | 0.000195606 | 0.000315704 |
| 0.026 | 0.000053570 | 0.000096464 | 0.000163677 | 0.000264492 |
| 0.025 | 0.000044387 | 0.000080025 | 0.000135948 | 0.000219950 |
| 0.024 | 0.000036485 | 0.000065857 | 0.000112015 | 0.000181447 |
| 0.023 | 0.000029730 | 0.000053728 | 0.000091495 | 0.000148388 |
| 0.022 | 0.000023997 | 0.000043420 | 0.000074030 | 0.000120207 |
| 0.021 | 0.000019170 | 0.000034742 | 0.000059282 | 0.000096375 |
| 0.020 | 0.000015141 | 0.000027462 | 0.000046935 | 0.000076395 |
| 0.019 | 0.000011810 | 0.000021446 | 0.000036697 | 0.000059803 |
| 0.018 | 0.000009085 | 0.000016517 | 0.000028297 | 0.000046169 |
| 0.C17 | 0.000006881 | 0.000012526 | 0.000021485 | 0.000035096 |
| 0.016 | 0.000005123 | 0.000009336 | 0.000016032 | 0.000026220 |
| 0.015 | 0.000003740 | 0.000006823 | 0.000011731 | 0.000019209 |
| 0.014 | 0.000002670 | 0.000004877 | 0.000008395 | 0.000013763 |
| 0.013 | 0.000001858 | 0.000003398 | 0.000005856 | 0.000009611 |
| 0.012 | 0.000001255 | 0.000002298 | 0.000003965 | 0.000006516 |
| 0.011 | 0.000000819 | 0.000001501 | 0.000002593 | 0.000004266 |
| 0.010 | 0.000000512 | 0.000000940 | 0.000001627 | 0.000002679 |
| 0.009 | 0.000000305 | 0.000000560 | 0.000000970 | 0.000001600 |
| 0.008 | 0.000000171 | 0.000000314 | 0.000000544 | 0.000000898 |
| 0.007 | 0.000000088 | 0.000000162 | 0.000000282 | 0.000000466 |
| 0.006 | 0.000000041 | 0.000000076 | 0.000000132 | 0.000000218 |
| 0.005 | 0.000000017 | 0.000000031 | 0.000000054 | 0.000000089 |
| 0.004 | 0.000000006 | 0.000000010 | 0.000000018 | 0.000000029 |
| 0.003 | 0.000000001 | 0.000000002 | 0.000000004 | 0.000000007 |
| 0.002 | 0.000000000 | 0.000000000 | 0.000000001 | 0.000000001 |
| 0.001 | 0.000000000 | 0.000000000 | 0.000000000 | 0.000000000 |

TABLE 1a:  CLUSTERING PROBABILITY  P(n;N,p)

n = 6

| P | N =16 | N =17 | N =18 |
|------|-------------|-------------|-------------|
| 0.300 | 0.999646158 | 0.999993327 | 0.999999961 |
| 0.299 | 0.999593912 | 0.999991591 | 0.999999945 |
| 0.298 | 0.999535654 | 0.999989477 | 0.999999923 |
| 0.297 | 0.999470878 | 0.999986918 | 0.999999895 |
| 0.296 | 0.999399049 | 0.999983837 | 0.999999856 |
| 0.295 | 0.999319604 | 0.999980149 | 0.999999805 |
| 0.294 | 0.999231954 | 0.999975755 | 0.999999739 |
| 0.293 | 0.999135479 | 0.999970547 | 0.999999652 |
| 0.292 | 0.999029529 | 0.999964401 | 0.999999541 |
| 0.291 | 0.998913426 | 0.999957180 | 0.999999398 |
| 0.290 | 0.998786462 | 0.999948734 | 0.999999215 |
| 0.289 | 0.998647898 | 0.999938893 | 0.999998985 |
| 0.288 | 0.998496965 | 0.999927471 | 0.999998694 |
| 0.287 | 0.998332863 | 0.999914264 | 0.999998331 |
| 0.286 | 0.998154762 | 0.999899046 | 0.999997879 |
| 0.285 | 0.997961802 | 0.999881571 | 0.999997319 |
| 0.284 | 0.997753090 | 0.999861570 | 0.999996630 |
| 0.283 | 0.997527704 | 0.999838750 | 0.999995784 |
| 0.282 | 0.997284693 | 0.999812793 | 0.999994753 |
| 0.281 | 0.997023073 | 0.999783355 | 0.999993500 |
| 0.280 | 0.996741834 | 0.999750063 | 0.999991985 |
| 0.279 | 0.996439932 | 0.999712516 | 0.999990161 |
| 0.278 | 0.996116298 | 0.999670280 | 0.999987972 |
| 0.277 | 0.995769832 | 0.999622894 | 0.999985357 |
| 0.276 | 0.995399409 | 0.999569859 | 0.999982243 |
| 0.275 | 0.995003873 | 0.999510646 | 0.999978551 |
| 0.274 | 0.994582044 | 0.999444688 | 0.999974188 |
| 0.273 | 0.994132716 | 0.999371382 | 0.999969049 |
| 0.272 | 0.993654659 | 0.999290089 | 0.999963017 |
| 0.271 | 0.993146616 | 0.999200130 | 0.999955959 |
| 0.270 | 0.992607312 | 0.999100787 | 0.999947728 |
| 0.269 | 0.992035445 | 0.998991301 | 0.999938157 |
| 0.268 | 0.991429696 | 0.998870874 | 0.999927062 |
| 0.267 | 0.990788724 | 0.998738664 | 0.999914237 |
| 0.266 | 0.990111173 | 0.998593788 | 0.999899455 |
| 0.265 | 0.989395666 | 0.998435318 | 0.999882463 |
| 0.264 | 0.988640813 | 0.998262285 | 0.999862985 |
| 0.263 | 0.987845210 | 0.998073675 | 0.999840715 |
| 0.262 | 0.987007439 | 0.997868432 | 0.999815319 |
| 0.261 | 0.986126071 | 0.997645452 | 0.999786430 |
| 0.260 | 0.985199668 | 0.997403591 | 0.999753648 |
| 0.259 | 0.984226784 | 0.997141661 | 0.999716540 |
| 0.258 | 0.983205967 | 0.996858427 | 0.999674631 |
| 0.257 | 0.982135759 | 0.996552616 | 0.999627411 |
| 0.256 | 0.981014698 | 0.996222908 | 0.999574326 |
| 0.255 | 0.979841325 | 0.995867944 | 0.999514781 |
| 0.254 | 0.978614179 | 0.995486323 | 0.999448133 |
| 0.253 | 0.977331802 | 0.995076604 | 0.999373694 |
| 0.252 | 0.975992738 | 0.994637305 | 0.999290727 |
| 0.251 | 0.974595540 | 0.994166910 | 0.999198445 |

TABLE 1a:   CLUSTERING PROBABILITY   P(n;N,p)

n = 6

| p | N =16 | N =17 | N =18 |
|---|---|---|---|
| 0.250 | 0.973138769 | 0.993663864 | 0.999096009 |
| 0.249 | 0.971620992 | 0.993126575 | 0.998982526 |
| 0.248 | 0.970040793 | 0.992553421 | 0.998857049 |
| 0.247 | 0.968396766 | 0.991942745 | 0.998718574 |
| 0.246 | 0.966687520 | 0.991292861 | 0.998566043 |
| 0.245 | 0.964911684 | 0.990602054 | 0.998398338 |
| 0.244 | 0.963067906 | 0.989868582 | 0.998214284 |
| 0.243 | 0.961154851 | 0.989090680 | 0.998012645 |
| 0.242 | 0.959171213 | 0.988266557 | 0.997792127 |
| 0.241 | 0.957115705 | 0.987394406 | 0.997551378 |
| 0.240 | 0.954987072 | 0.986472399 | 0.997288985 |
| 0.239 | 0.952784082 | 0.985498692 | 0.997003475 |
| 0.238 | 0.950505538 | 0.984471431 | 0.996693317 |
| 0.237 | 0.948150273 | 0.983388747 | 0.996356921 |
| 0.236 | 0.945717152 | 0.982248766 | 0.995992641 |
| 0.235 | 0.943205078 | 0.981049606 | 0.995598773 |
| 0.234 | 0.940612990 | 0.979789384 | 0.995173555 |
| 0.233 | 0.937939865 | 0.978466214 | 0.994715176 |
| 0.232 | 0.935184723 | 0.977078215 | 0.994221766 |
| 0.231 | 0.932346622 | 0.975623508 | 0.993691409 |
| 0.230 | 0.929424666 | 0.974100226 | 0.993122136 |
| 0.229 | 0.926418004 | 0.972506509 | 0.992511932 |
| 0.228 | 0.923325829 | 0.970840513 | 0.991858737 |
| 0.227 | 0.920147384 | 0.969100407 | 0.991160448 |
| 0.226 | 0.916881960 | 0.967284383 | 0.990414921 |
| 0.225 | 0.913528897 | 0.965390653 | 0.989619975 |
| 0.224 | 0.910087590 | 0.963417453 | 0.988773394 |
| 0.223 | 0.906557482 | 0.961363049 | 0.987872931 |
| 0.222 | 0.902938074 | 0.959225734 | 0.986916308 |
| 0.221 | 0.899228919 | 0.957003839 | 0.985901223 |
| 0.220 | 0.895429626 | 0.954695726 | 0.984825354 |
| 0.219 | 0.891539862 | 0.952299799 | 0.983686355 |
| 0.218 | 0.887559351 | 0.949814502 | 0.982481870 |
| 0.217 | 0.883487876 | 0.947238324 | 0.981209530 |
| 0.216 | 0.879325276 | 0.944569800 | 0.979866956 |
| 0.215 | 0.875071455 | 0.941807515 | 0.978451769 |
| 0.214 | 0.870726373 | 0.938950104 | 0.976961588 |
| 0.213 | 0.866290053 | 0.935996260 | 0.975394037 |
| 0.212 | 0.861762580 | 0.932944728 | 0.973746748 |
| 0.211 | 0.857144099 | 0.929794317 | 0.972017365 |
| 0.210 | 0.852434818 | 0.926543893 | 0.970203548 |
| 0.209 | 0.847635009 | 0.923192387 | 0.968302980 |
| 0.208 | 0.842745004 | 0.919738798 | 0.966313367 |
| 0.207 | 0.837765201 | 0.916182190 | 0.964232444 |
| 0.206 | 0.832696059 | 0.912521697 | 0.962057980 |
| 0.205 | 0.827538100 | 0.908756526 | 0.959787781 |
| 0.204 | 0.822291911 | 0.904885956 | 0.957419694 |
| 0.203 | 0.816958138 | 0.900909343 | 0.954951613 |
| 0.202 | 0.811537494 | 0.896826117 | 0.952381482 |
| 0.201 | 0.806030752 | 0.892635790 | 0.949707298 |

TABLE 1a:   CLUSTERING PROBABILITY   P(n;N,p)

n = 6

| p | N =16 | N =17 | N =18 |
|---|---|---|---|
| 0.200 | 0.800438747 | 0.883337950 | 0.946927116 |
| 0.199 | 0.794762378 | 0.883932268 | 0.944039054 |
| 0.198 | 0.789002602 | 0.879418498 | 0.941041295 |
| 0.197 | 0.783160440 | 0.874796475 | 0.937932092 |
| 0.196 | 0.777236971 | 0.870066122 | 0.934709771 |
| 0.195 | 0.771233336 | 0.865227445 | 0.931372735 |
| 0.194 | 0.765150732 | 0.860280536 | 0.927919469 |
| 0.193 | 0.758990418 | 0.855225576 | 0.924348540 |
| 0.192 | 0.752753706 | 0.850062830 | 0.920658605 |
| 0.191 | 0.746441968 | 0.844792656 | 0.916848409 |
| 0.190 | 0.740056632 | 0.839415496 | 0.912916792 |
| 0.189 | 0.733599178 | 0.833931883 | 0.908862691 |
| 0.188 | 0.727071142 | 0.828342438 | 0.904685143 |
| 0.187 | 0.720474111 | 0.822647872 | 0.900383285 |
| 0.186 | 0.713809725 | 0.816848984 | 0.895956361 |
| 0.185 | 0.707079674 | 0.810946661 | 0.891403721 |
| 0.184 | 0.700285697 | 0.804941880 | 0.886724826 |
| 0.183 | 0.693429581 | 0.798835706 | 0.881919247 |
| 0.182 | 0.686513160 | 0.792629290 | 0.876986669 |
| 0.181 | 0.679538311 | 0.786323871 | 0.871926892 |
| 0.180 | 0.672506959 | 0.779920774 | 0.866739834 |
| 0.179 | 0.665421068 | 0.773421410 | 0.861425529 |
| 0.178 | 0.658282643 | 0.766827275 | 0.855984134 |
| 0.177 | 0.651093732 | 0.760139946 | 0.850415922 |
| 0.176 | 0.643856417 | 0.753361085 | 0.844721292 |
| 0.175 | 0.636572818 | 0.746492434 | 0.838900763 |
| 0.174 | 0.629245090 | 0.739535816 | 0.832954978 |
| 0.173 | 0.621875421 | 0.732493131 | 0.826884700 |
| 0.172 | 0.614466030 | 0.725366357 | 0.820690821 |
| 0.171 | 0.607019167 | 0.718157546 | 0.814374350 |
| 0.170 | 0.599537107 | 0.710868826 | 0.807936425 |
| 0.169 | 0.592022156 | 0.703502396 | 0.801378303 |
| 0.168 | 0.584476640 | 0.696060524 | 0.794701365 |
| 0.167 | 0.576902911 | 0.688545548 | 0.787907112 |
| 0.166 | 0.569303342 | 0.680959870 | 0.780997169 |
| 0.165 | 0.561680322 | 0.673305959 | 0.773973277 |
| 0.164 | 0.554036261 | 0.665586343 | 0.766837297 |
| 0.163 | 0.546373583 | 0.657803612 | 0.759591208 |
| 0.162 | 0.538694726 | 0.649960412 | 0.752237104 |
| 0.161 | 0.531002140 | 0.642059445 | 0.744777191 |
| 0.160 | 0.523298285 | 0.634103465 | 0.737213789 |
| 0.159 | 0.515585629 | 0.626095277 | 0.729549327 |
| 0.158 | 0.507866646 | 0.618037733 | 0.721786341 |
| 0.157 | 0.500143814 | 0.609933731 | 0.713927474 |
| 0.156 | 0.492419617 | 0.601786211 | 0.705975468 |
| 0.155 | 0.484696535 | 0.593598151 | 0.697933170 |
| 0.154 | 0.476977050 | 0.585372570 | 0.689803521 |
| 0.153 | 0.469263641 | 0.577112518 | 0.681589557 |
| 0.152 | 0.461558779 | 0.568821079 | 0.673294405 |
| 0.151 | 0.453864933 | 0.560501364 | 0.664921281 |

TABLE 1a:  CLUSTERING PROBABILITY  P(n;N,p)

n = 6

| P | N =16 | N =17 | N =18 |
|-------|------------|------------|------------|
| 0.150 | 0.446184559 | 0.552156510 | 0.656473484 |
| 0.149 | 0.438520107 | 0.543789679 | 0.647954397 |
| 0.148 | 0.430874011 | 0.535404051 | 0.639367478 |
| 0.147 | 0.423248694 | 0.527002825 | 0.630716261 |
| 0.146 | 0.415646562 | 0.518589213 | 0.622004350 |
| 0.145 | 0.408070004 | 0.510166440 | 0.613235415 |
| 0.144 | 0.400521390 | 0.501737738 | 0.604413190 |
| 0.143 | 0.393003067 | 0.493306347 | 0.595541466 |
| 0.142 | 0.385517364 | 0.484875506 | 0.586624090 |
| 0.141 | 0.378066581 | 0.476448457 | 0.577664959 |
| 0.140 | 0.370652995 | 0.468028438 | 0.568668016 |
| 0.139 | 0.363278855 | 0.459618680 | 0.559637247 |
| 0.138 | 0.355946381 | 0.451222405 | 0.550576673 |
| 0.137 | 0.348657761 | 0.442842826 | 0.541490352 |
| 0.136 | 0.341415152 | 0.434483136 | 0.532382370 |
| 0.135 | 0.334220679 | 0.426146516 | 0.523256834 |
| 0.134 | 0.327076429 | 0.417836123 | 0.514117877 |
| 0.133 | 0.319984453 | 0.409555091 | 0.504969642 |
| 0.132 | 0.312946765 | 0.401306530 | 0.495816286 |
| 0.131 | 0.305965339 | 0.393093519 | 0.486661973 |
| 0.130 | 0.299042110 | 0.384919108 | 0.477510868 |
| 0.129 | 0.292178967 | 0.376786311 | 0.468367133 |
| 0.128 | 0.285377760 | 0.368698106 | 0.459234925 |
| 0.127 | 0.278640293 | 0.360657431 | 0.450118388 |
| 0.126 | 0.271968323 | 0.352667184 | 0.441021652 |
| 0.125 | 0.265363561 | 0.344730217 | 0.431948824 |
| 0.124 | 0.258827672 | 0.336849336 | 0.422903989 |
| 0.123 | 0.252362269 | 0.329027298 | 0.413891202 |
| 0.122 | 0.245968918 | 0.321266808 | 0.404914485 |
| 0.121 | 0.239649130 | 0.313570519 | 0.395977823 |
| 0.120 | 0.233404368 | 0.305941026 | 0.387085159 |
| 0.119 | 0.227236040 | 0.298380866 | 0.378240389 |
| 0.118 | 0.221145502 | 0.290892519 | 0.369447362 |
| 0.117 | 0.215134053 | 0.283478400 | 0.360709871 |
| 0.116 | 0.209202940 | 0.276140861 | 0.352031651 |
| 0.115 | 0.203353351 | 0.268882188 | 0.343416377 |
| 0.114 | 0.197586420 | 0.261704598 | 0.334867658 |
| 0.113 | 0.191903221 | 0.254610242 | 0.326389033 |
| 0.112 | 0.186304773 | 0.247601196 | 0.317983971 |
| 0.111 | 0.180792034 | 0.240679465 | 0.309655862 |
| 0.110 | 0.175365907 | 0.233846981 | 0.301408019 |
| 0.109 | 0.170027231 | 0.227105598 | 0.293243670 |
| 0.108 | 0.164776790 | 0.220457096 | 0.285165961 |
| 0.107 | 0.159615305 | 0.213903173 | 0.277177944 |
| 0.106 | 0.154543438 | 0.207445451 | 0.269282585 |
| 0.105 | 0.149561790 | 0.201085469 | 0.261482750 |
| 0.104 | 0.144670902 | 0.194824687 | 0.253781211 |
| 0.103 | 0.139871254 | 0.188664480 | 0.246180640 |
| 0.102 | 0.135163266 | 0.182606140 | 0.238683605 |
| 0.101 | 0.130547296 | 0.176650877 | 0.231292572 |

TABLE 1a:   CLUSTERING PROBABILITY   P(n;N,p)

n = 6

| p | N = 16 | N = 17 | N = 18 |
|---|---|---|---|
| 0.100 | 0.126023642 | 0.170799814 | 0.224009898 |
| 0.099 | 0.121592542 | 0.165053989 | 0.216837833 |
| 0.098 | 0.117254170 | 0.159414355 | 0.209778515 |
| 0.097 | 0.113008644 | 0.153881777 | 0.202833970 |
| 0.096 | 0.108856020 | 0.148457034 | 0.196006110 |
| 0.095 | 0.104796294 | 0.143140819 | 0.189296733 |
| 0.094 | 0.100829401 | 0.137933738 | 0.182707517 |
| 0.093 | 0.096955220 | 0.132836307 | 0.176240024 |
| 0.092 | 0.093173570 | 0.127848957 | 0.169895698 |
| 0.091 | 0.089484209 | 0.122972033 | 0.163675860 |
| 0.090 | 0.085886842 | 0.118205791 | 0.157581713 |
| 0.089 | 0.082381114 | 0.113550400 | 0.151614337 |
| 0.088 | 0.078966613 | 0.109005946 | 0.145774691 |
| 0.087 | 0.075642874 | 0.104572425 | 0.140063611 |
| 0.086 | 0.072409374 | 0.100249751 | 0.134481813 |
| 0.085 | 0.069265537 | 0.096037751 | 0.129029887 |
| 0.084 | 0.066210735 | 0.091936171 | 0.123708304 |
| 0.083 | 0.063244284 | 0.087944671 | 0.118517412 |
| 0.082 | 0.060365452 | 0.084062830 | 0.113457438 |
| 0.081 | 0.057573454 | 0.080290147 | 0.108528487 |
| 0.080 | 0.054867458 | 0.076626038 | 0.103730545 |
| 0.079 | 0.052246579 | 0.073069843 | 0.099063478 |
| 0.078 | 0.049709390 | 0.069620821 | 0.094527035 |
| 0.077 | 0.047256414 | 0.066278159 | 0.090120847 |
| 0.076 | 0.044885131 | 0.063040965 | 0.085844430 |
| 0.075 | 0.042594976 | 0.059908276 | 0.081697184 |
| 0.074 | 0.040384843 | 0.056879055 | 0.077678399 |
| 0.073 | 0.038253584 | 0.053952197 | 0.073787252 |
| 0.072 | 0.036200012 | 0.051126526 | 0.070022812 |
| 0.071 | 0.034222902 | 0.048400801 | 0.066384042 |
| 0.070 | 0.032320991 | 0.045773717 | 0.062869797 |
| 0.069 | 0.030492982 | 0.043243903 | 0.059478834 |
| 0.068 | 0.028737545 | 0.040809930 | 0.056209806 |
| 0.067 | 0.027053317 | 0.038470309 | 0.053061271 |
| 0.066 | 0.025438905 | 0.036223496 | 0.050031692 |
| 0.065 | 0.023892888 | 0.034067890 | 0.047119438 |
| 0.064 | 0.022413817 | 0.032001842 | 0.044322792 |
| 0.063 | 0.021000219 | 0.030023649 | 0.041639950 |
| 0.062 | 0.019650597 | 0.028131565 | 0.039069023 |
| 0.061 | 0.018363433 | 0.026323798 | 0.036608047 |
| 0.060 | 0.017137188 | 0.024598513 | 0.034254980 |
| 0.059 | 0.015970308 | 0.022953838 | 0.032007706 |
| 0.058 | 0.014861221 | 0.021387863 | 0.029864042 |
| 0.057 | 0.013808341 | 0.019898644 | 0.027821742 |
| 0.056 | 0.012810071 | 0.018484208 | 0.025878495 |
| 0.055 | 0.011864305 | 0.017142552 | 0.024031936 |
| 0.054 | 0.010970926 | 0.015871650 | 0.022279645 |
| 0.053 | 0.010126814 | 0.014669451 | 0.020619155 |
| 0.052 | 0.009330843 | 0.013533888 | 0.019047952 |
| 0.051 | 0.008581388 | 0.012462877 | 0.017563485 |

NEFF and NAUS


TABLE 1a:   CLUSTERING PROBABILITY   P(n;N,p)


                                                          n = 6

| p | N =16 | N =17 | N =18 |
|---|---|---|---|
| 0.050 | 0.007876321 | 0.011454318 | 0.016163164 |
| 0.049 | 0.007215518 | 0.010506107 | 0.014844370 |
| 0.048 | 0.006595860 | 0.009616128 | 0.013604454 |
| 0.047 | 0.006016233 | 0.008782265 | 0.012440749 |
| 0.046 | 0.005475031 | 0.008002398 | 0.011350566 |
| 0.045 | 0.004970662 | 0.007274415 | 0.010331206 |
| 0.044 | 0.004501542 | 0.006596205 | 0.009379961 |
| 0.043 | 0.004066106 | 0.005965668 | 0.008494118 |
| 0.042 | 0.003662803 | 0.005380719 | 0.007670968 |
| 0.041 | 0.003290102 | 0.004839234 | 0.006907807 |
| 0.040 | 0.002946492 | 0.004339312 | 0.006201940 |
| 0.039 | 0.002630486 | 0.003878771 | 0.005550690 |
| 0.038 | 0.002340620 | 0.003455655 | 0.004951400 |
| 0.037 | 0.002075458 | 0.003067988 | 0.004401438 |
| 0.036 | 0.001833592 | 0.002713823 | 0.003898202 |
| 0.035 | 0.001613644 | 0.002391248 | 0.003439126 |
| 0.034 | 0.001414269 | 0.002098387 | 0.003021680 |
| 0.033 | 0.001234154 | 0.001833408 | 0.002643382 |
| 0.032 | 0.001072024 | 0.001594516 | 0.002301796 |
| 0.031 | 0.000926640 | 0.001379968 | 0.001994538 |
| 0.030 | 0.000796800 | 0.001188064 | 0.001719284 |
| 0.029 | 0.000681343 | 0.001017157 | 0.001473769 |
| 0.028 | 0.000579151 | 0.000865655 | 0.001255794 |
| 0.027 | 0.000489147 | 0.000732018 | 0.001063228 |
| 0.026 | 0.000410297 | 0.000614766 | 0.000894014 |
| 0.025 | 0.000341614 | 0.000512478 | 0.000746171 |
| 0.024 | 0.000282156 | 0.000423794 | 0.000617798 |
| 0.023 | 0.000231026 | 0.000347419 | 0.000507075 |
| 0.022 | 0.000187378 | 0.000282121 | 0.000412269 |
| 0.021 | 0.000150410 | 0.000226735 | 0.000331733 |
| 0.020 | 0.000119371 | 0.000180162 | 0.000263911 |
| 0.019 | 0.000093558 | 0.000141373 | 0.000207340 |
| 0.018 | 0.000072315 | 0.000109405 | 0.000160648 |
| 0.017 | 0.000055037 | 0.000083366 | 0.000122560 |
| 0.016 | 0.000041167 | 0.000062431 | 0.000091893 |
| 0.015 | 0.000030196 | 0.000045847 | 0.000067564 |
| 0.014 | 0.000021660 | 0.000032926 | 0.000048580 |
| 0.013 | 0.000015144 | 0.000023049 | 0.000034048 |
| 0.012 | 0.000010279 | 0.000015663 | 0.000023165 |
| 0.011 | 0.000006738 | 0.000010279 | 0.000015220 |
| 0.010 | 0.000004237 | 0.000006471 | 0.000009594 |
| 0.009 | 0.000002534 | 0.000003875 | 0.000005751 |
| 0.008 | 0.000001424 | 0.000002180 | 0.000003239 |
| 0.007 | 0.000000740 | 0.000001134 | 0.000001687 |
| 0.006 | 0.000000347 | 0.000000532 | 0.000000792 |
| 0.005 | 0.000000141 | 0.000000217 | 0.000000323 |
| 0.004 | 0.000000047 | 0.000000072 | 0.000000107 |
| 0.003 | 0.000000011 | 0.000000017 | 0.000000026 |
| 0.002 | 0.000000001 | 0.000000002 | 0.000000003 |
| 0.001 | 0.000000000 | 0.000000000 | 0.000000000 |

TABLE 1a:   CLUSTERING PROBABILITY   $P(n;N,p)$

$n = 7$

| P | N = 14 | N = 15 | N = 16 | N = 17 |
|---|---|---|---|---|
| 0.400 | 0.982159473 | 0.997882612 | 0.999873464 | 0.999996801 |
| 0.399 | 0.981289661 | 0.997720916 | 0.999859514 | 0.999996320 |
| 0.398 | 0.980389493 | 0.997549161 | 0.999844220 | 0.999995774 |
| 0.397 | 0.979458402 | 0.997366875 | 0.999827471 | 0.999995154 |
| 0.396 | 0.978495830 | 0.997173575 | 0.999809151 | 0.999994451 |
| 0.395 | 0.977501220 | 0.996968764 | 0.999789134 | 0.999993655 |
| 0.394 | 0.976474027 | 0.996751932 | 0.999767288 | 0.999992756 |
| 0.393 | 0.975413705 | 0.996522557 | 0.999743470 | 0.999991740 |
| 0.392 | 0.974319721 | 0.996280103 | 0.999717530 | 0.999990595 |
| 0.391 | 0.973191546 | 0.996024021 | 0.999689310 | 0.999989305 |
| 0.390 | 0.972028658 | 0.995753752 | 0.999658640 | 0.999987855 |
| 0.389 | 0.970830545 | 0.995468722 | 0.999625342 | 0.999986226 |
| 0.388 | 0.969596702 | 0.995168345 | 0.999589226 | 0.999984399 |
| 0.387 | 0.968326633 | 0.994852024 | 0.999550092 | 0.999982351 |
| 0.386 | 0.967019851 | 0.994519151 | 0.999507731 | 0.999980060 |
| 0.385 | 0.965675879 | 0.994169106 | 0.999461918 | 0.999977498 |
| 0.384 | 0.964294248 | 0.993801257 | 0.999412420 | 0.999974637 |
| 0.383 | 0.962874503 | 0.993414962 | 0.999358990 | 0.999971447 |
| 0.382 | 0.961416195 | 0.993009569 | 0.999301368 | 0.999967892 |
| 0.381 | 0.959918890 | 0.992584415 | 0.999239283 | 0.999963937 |
| 0.380 | 0.958382162 | 0.992138829 | 0.999172448 | 0.999959540 |
| 0.379 | 0.956805598 | 0.991672130 | 0.999100562 | 0.999954657 |
| 0.378 | 0.955188798 | 0.991183626 | 0.999023312 | 0.999949241 |
| 0.377 | 0.953531371 | 0.990672621 | 0.998940368 | 0.999943240 |
| 0.376 | 0.951832943 | 0.990138408 | 0.998851386 | 0.999936596 |
| 0.375 | 0.950093148 | 0.989580273 | 0.998756007 | 0.999929249 |
| 0.374 | 0.948311636 | 0.988997498 | 0.998653855 | 0.999921132 |
| 0.373 | 0.946488070 | 0.988389354 | 0.998544538 | 0.999912174 |
| 0.372 | 0.944622126 | 0.987755110 | 0.998427649 | 0.999902297 |
| 0.371 | 0.942713493 | 0.987094030 | 0.998302763 | 0.999891418 |
| 0.370 | 0.940761876 | 0.986405371 | 0.998169437 | 0.999879445 |
| 0.369 | 0.938766993 | 0.985688388 | 0.998027213 | 0.999866283 |
| 0.368 | 0.936728576 | 0.984942333 | 0.997875614 | 0.999851825 |
| 0.367 | 0.934646373 | 0.984166453 | 0.997714145 | 0.999835960 |
| 0.366 | 0.932520146 | 0.983359997 | 0.997542292 | 0.999818566 |
| 0.365 | 0.930349671 | 0.982522210 | 0.997359527 | 0.999799512 |
| 0.364 | 0.928134741 | 0.981652337 | 0.997165298 | 0.999778660 |
| 0.363 | 0.925875164 | 0.980749623 | 0.996959039 | 0.999755859 |
| 0.362 | 0.923570762 | 0.979813315 | 0.996740162 | 0.999730950 |
| 0.361 | 0.921221375 | 0.978842661 | 0.996508063 | 0.999703761 |
| 0.360 | 0.918826855 | 0.977836910 | 0.996262119 | 0.999674110 |
| 0.359 | 0.916387074 | 0.976795318 | 0.996001686 | 0.999641799 |
| 0.358 | 0.913901916 | 0.975717140 | 0.995726105 | 0.999606622 |
| 0.357 | 0.911371285 | 0.974601639 | 0.995434695 | 0.999568356 |
| 0.356 | 0.908795097 | 0.973448082 | 0.995126759 | 0.999526764 |
| 0.355 | 0.906173286 | 0.972255742 | 0.994801580 | 0.999481594 |
| 0.354 | 0.903505803 | 0.971023901 | 0.994458426 | 0.999432579 |
| 0.353 | 0.900792613 | 0.969751845 | 0.994096544 | 0.999379436 |
| 0.352 | 0.898033699 | 0.968438871 | 0.993715166 | 0.999321862 |
| 0.351 | 0.895229059 | 0.967084284 | 0.993313505 | 0.999259539 |

NEFF and NAUS

TABLE 1a:  CLUSTERING PROBABILITY  P(n;N,p)

n = 7

| P | N =14 | N =15 | N =16 | N =17 |
|---|---|---|---|---|
| 0.350 | 0.892378708 | 0.965687401 | 0.992890758 | 0.999192128 |
| 0.349 | 0.889482677 | 0.964247546 | 0.992446108 | 0.999119274 |
| 0.348 | 0.886541012 | 0.962764057 | 0.991978719 | 0.999040597 |
| 0.347 | 0.883553776 | 0.961236284 | 0.991487743 | 0.998955700 |
| 0.346 | 0.880521049 | 0.959663588 | 0.990972314 | 0.998864163 |
| 0.345 | 0.877442926 | 0.958045345 | 0.990431557 | 0.998765544 |
| 0.344 | 0.874319518 | 0.956380945 | 0.989864579 | 0.998659377 |
| 0.343 | 0.871150952 | 0.954669791 | 0.989270478 | 0.998545173 |
| 0.342 | 0.867937370 | 0.952911305 | 0.988648338 | 0.998422419 |
| 0.341 | 0.864678932 | 0.951104921 | 0.987997236 | 0.998290576 |
| 0.340 | 0.861375811 | 0.949250092 | 0.987316235 | 0.998149082 |
| 0.339 | 0.858022197 | 0.947346289 | 0.986604392 | 0.997997345 |
| 0.338 | 0.854636295 | 0.945392998 | 0.985860755 | 0.997834750 |
| 0.337 | 0.851200326 | 0.943389727 | 0.985084366 | 0.997660655 |
| 0.336 | 0.847720526 | 0.941335998 | 0.984274260 | 0.997474388 |
| 0.335 | 0.844197144 | 0.939231358 | 0.983429468 | 0.997275252 |
| 0.334 | 0.840630448 | 0.937075370 | 0.982549018 | 0.997062523 |
| 0.333 | 0.837020716 | 0.934867618 | 0.981631936 | 0.996835448 |
| 0.332 | 0.833368245 | 0.932607708 | 0.980677244 | 0.996593247 |
| 0.331 | 0.829673343 | 0.930295266 | 0.979683966 | 0.996335115 |
| 0.330 | 0.825936334 | 0.927929941 | 0.978651128 | 0.996060217 |
| 0.329 | 0.822157557 | 0.925511401 | 0.977577756 | 0.995767695 |
| 0.328 | 0.818337364 | 0.923039339 | 0.976462880 | 0.995456663 |
| 0.327 | 0.814476119 | 0.920513469 | 0.975305536 | 0.995126211 |
| 0.326 | 0.810574203 | 0.917933529 | 0.974104763 | 0.994775403 |
| 0.325 | 0.806632007 | 0.915299279 | 0.972859609 | 0.994403281 |
| 0.324 | 0.802649939 | 0.912610503 | 0.971569129 | 0.994008862 |
| 0.323 | 0.798628416 | 0.909867008 | 0.970232387 | 0.993591140 |
| 0.322 | 0.794567871 | 0.907068625 | 0.968848459 | 0.993149091 |
| 0.321 | 0.790468749 | 0.904215209 | 0.967416429 | 0.992681665 |
| 0.320 | 0.786331505 | 0.901306639 | 0.965935396 | 0.992187797 |
| 0.319 | 0.782156609 | 0.898342817 | 0.964404472 | 0.991666399 |
| 0.318 | 0.777944543 | 0.895323672 | 0.962822782 | 0.991116368 |
| 0.317 | 0.773695799 | 0.892249155 | 0.961189470 | 0.990536584 |
| 0.316 | 0.769410881 | 0.889119242 | 0.959503692 | 0.989925909 |
| 0.315 | 0.765090305 | 0.885933933 | 0.957764626 | 0.989283194 |
| 0.314 | 0.760734599 | 0.882693253 | 0.955971465 | 0.988607274 |
| 0.313 | 0.756344300 | 0.879397253 | 0.954123424 | 0.987896973 |
| 0.312 | 0.751919956 | 0.876046006 | 0.952219737 | 0.987151104 |
| 0.311 | 0.747462126 | 0.872639610 | 0.950259661 | 0.986368472 |
| 0.310 | 0.742971380 | 0.869178188 | 0.948242474 | 0.985547871 |
| 0.309 | 0.738448296 | 0.865661889 | 0.946167477 | 0.984688090 |
| 0.308 | 0.733893464 | 0.862090882 | 0.944033995 | 0.983787915 |
| 0.307 | 0.729307481 | 0.858465363 | 0.941841379 | 0.982846123 |
| 0.306 | 0.724690955 | 0.854785553 | 0.939589005 | 0.981861494 |
| 0.305 | 0.720044503 | 0.851051694 | 0.937276273 | 0.980832802 |
| 0.304 | 0.715368750 | 0.847264054 | 0.934902614 | 0.979758826 |
| 0.303 | 0.710664329 | 0.843422923 | 0.932467483 | 0.978638344 |
| 0.302 | 0.705931883 | 0.839528616 | 0.929970366 | 0.977470139 |
| 0.301 | 0.701172062 | 0.835581470 | 0.927410775 | 0.976252998 |

TABLE 1a:   CLUSTERING PROBABILITY  P(n;N,p)

n = 7

| P | N =14 | N =15 | N =16 | N =17 |
|---|---|---|---|---|
| 0.300 | 0.696385524 | 0.831581846 | 0.924788256 | 0.974985716 |
| 0.299 | 0.691572934 | 0.827530126 | 0.922102380 | 0.973667094 |
| 0.298 | 0.686734964 | 0.823426716 | 0.919352752 | 0.972295944 |
| 0.297 | 0.681872295 | 0.819272044 | 0.916539006 | 0.970871090 |
| 0.296 | 0.676985613 | 0.815066560 | 0.913660810 | 0.969391365 |
| 0.295 | 0.672075611 | 0.810810736 | 0.910717862 | 0.967855621 |
| 0.294 | 0.667142988 | 0.806505065 | 0.907709893 | 0.966262722 |
| 0.293 | 0.662188449 | 0.802150062 | 0.904636665 | 0.964611551 |
| 0.292 | 0.657212706 | 0.797746261 | 0.901497977 | 0.962901007 |
| 0.291 | 0.652216475 | 0.793294218 | 0.898293658 | 0.961130012 |
| 0.290 | 0.647200478 | 0.788794510 | 0.895023570 | 0.959297507 |
| 0.289 | 0.642165441 | 0.784247733 | 0.891687613 | 0.957402458 |
| 0.288 | 0.637112096 | 0.779654502 | 0.888285717 | 0.955443854 |
| 0.287 | 0.632041179 | 0.775015451 | 0.884817847 | 0.953420709 |
| 0.286 | 0.626953431 | 0.770331235 | 0.881284003 | 0.951332065 |
| 0.285 | 0.621849595 | 0.765602525 | 0.877684220 | 0.949176991 |
| 0.284 | 0.616730420 | 0.760830012 | 0.874018566 | 0.946954586 |
| 0.283 | 0.611596656 | 0.756014403 | 0.870287145 | 0.944663981 |
| 0.282 | 0.606449059 | 0.751156425 | 0.866490094 | 0.942304336 |
| 0.281 | 0.601288387 | 0.746256819 | 0.862627585 | 0.939874846 |
| 0.280 | 0.596115399 | 0.741316344 | 0.858699825 | 0.937374740 |
| 0.279 | 0.590930858 | 0.736335774 | 0.854707056 | 0.934803282 |
| 0.278 | 0.585735530 | 0.731315902 | 0.850649551 | 0.932159772 |
| 0.277 | 0.580530181 | 0.726257532 | 0.846527621 | 0.929443547 |
| 0.276 | 0.575315579 | 0.721161486 | 0.842341609 | 0.926653983 |
| 0.275 | 0.570092496 | 0.716028599 | 0.838091892 | 0.923790495 |
| 0.274 | 0.564861702 | 0.710859720 | 0.833778881 | 0.920852537 |
| 0.273 | 0.559623970 | 0.705655712 | 0.829403019 | 0.917839606 |
| 0.272 | 0.554380073 | 0.700417453 | 0.824964783 | 0.914751237 |
| 0.271 | 0.549130784 | 0.695145830 | 0.820464683 | 0.911587011 |
| 0.270 | 0.543876876 | 0.689841745 | 0.815903261 | 0.908346549 |
| 0.269 | 0.538619124 | 0.684506111 | 0.811281091 | 0.905029519 |
| 0.268 | 0.533358301 | 0.679139852 | 0.806598778 | 0.901635630 |
| 0.267 | 0.528095180 | 0.673743905 | 0.801856960 | 0.898164638 |
| 0.266 | 0.522830532 | 0.668319214 | 0.797056305 | 0.894616342 |
| 0.265 | 0.517565130 | 0.662866736 | 0.792197511 | 0.890990589 |
| 0.264 | 0.512299743 | 0.657387437 | 0.787281306 | 0.887287270 |
| 0.263 | 0.507035139 | 0.651882291 | 0.782308449 | 0.883506325 |
| 0.262 | 0.501772085 | 0.646352281 | 0.777279726 | 0.879647737 |
| 0.261 | 0.496511346 | 0.640798400 | 0.772195953 | 0.875711540 |
| 0.260 | 0.491253685 | 0.635221646 | 0.767057973 | 0.871697812 |
| 0.259 | 0.485999861 | 0.629623026 | 0.761866657 | 0.867606679 |
| 0.258 | 0.480750633 | 0.624003554 | 0.756622904 | 0.863438316 |
| 0.257 | 0.475506755 | 0.618364249 | 0.751327638 | 0.859192943 |
| 0.256 | 0.470268979 | 0.612706138 | 0.745981808 | 0.854870829 |
| 0.255 | 0.465038053 | 0.607030251 | 0.740586390 | 0.850472290 |
| 0.254 | 0.459814723 | 0.601337625 | 0.735142383 | 0.845997688 |
| 0.253 | 0.454599728 | 0.595629301 | 0.729650812 | 0.841447435 |
| 0.252 | 0.449393806 | 0.589906324 | 0.724112722 | 0.836821986 |
| 0.251 | 0.444197691 | 0.584169742 | 0.718529184 | 0.832121846 |

TABLE 1a:　CLUSTERING PROBABILITY　P(n;N,p)

n = 7

| P | N ≡ 14 | N ≡ 15 | N ≡ 16 | N ≡ 17 |
|------|------------|------------|------------|------------|
| 0.250 | 0.439012110 | 0.578420607 | 0.712901289 | 0.827347564 |
| 0.249 | 0.433837788 | 0.572659974 | 0.707230150 | 0.822499737 |
| 0.248 | 0.428675443 | 0.566888900 | 0.701516899 | 0.817579007 |
| 0.247 | 0.423525790 | 0.561108442 | 0.695762691 | 0.812586061 |
| 0.246 | 0.418389537 | 0.555319662 | 0.689968697 | 0.807521630 |
| 0.245 | 0.413267388 | 0.549523620 | 0.684136108 | 0.802386492 |
| 0.244 | 0.408160040 | 0.543721379 | 0.678266134 | 0.797181465 |
| 0.243 | 0.403068185 | 0.537913999 | 0.672360000 | 0.791907415 |
| 0.242 | 0.397992509 | 0.532102542 | 0.666418947 | 0.786565246 |
| 0.241 | 0.392933692 | 0.526288069 | 0.660444234 | 0.781155907 |
| 0.240 | 0.387892407 | 0.520471640 | 0.654437133 | 0.775680387 |
| 0.239 | 0.382869319 | 0.514654311 | 0.648398931 | 0.770139717 |
| 0.238 | 0.377865090 | 0.508837139 | 0.642330927 | 0.764534966 |
| 0.237 | 0.372880371 | 0.503021177 | 0.636234435 | 0.758867244 |
| 0.236 | 0.367915808 | 0.497207475 | 0.630110778 | 0.753137699 |
| 0.235 | 0.362972039 | 0.491397081 | 0.623961293 | 0.747347515 |
| 0.234 | 0.358049695 | 0.485591039 | 0.617787326 | 0.741497914 |
| 0.233 | 0.353149399 | 0.479790386 | 0.611590231 | 0.735590154 |
| 0.232 | 0.348271766 | 0.473996159 | 0.605371373 | 0.729625527 |
| 0.231 | 0.343417402 | 0.468209388 | 0.599132126 | 0.723605359 |
| 0.230 | 0.338586908 | 0.462431096 | 0.592873868 | 0.717531010 |
| 0.229 | 0.333780873 | 0.456662304 | 0.586597986 | 0.711403872 |
| 0.228 | 0.328999879 | 0.450904024 | 0.580305873 | 0.705225366 |
| 0.227 | 0.324244499 | 0.445157263 | 0.573998926 | 0.698996945 |
| 0.226 | 0.319515298 | 0.439423020 | 0.567678547 | 0.692720090 |
| 0.225 | 0.314812832 | 0.433702289 | 0.561346141 | 0.686396310 |
| 0.224 | 0.310137647 | 0.427996055 | 0.555003116 | 0.680027142 |
| 0.223 | 0.305490280 | 0.422305294 | 0.548650882 | 0.673614147 |
| 0.222 | 0.300871258 | 0.416630977 | 0.542290852 | 0.667158912 |
| 0.221 | 0.296281100 | 0.410974063 | 0.535924437 | 0.660663046 |
| 0.220 | 0.291720314 | 0.405335504 | 0.529553050 | 0.654128181 |
| 0.219 | 0.287189399 | 0.399716242 | 0.523178102 | 0.647555972 |
| 0.218 | 0.282688844 | 0.394117211 | 0.516801003 | 0.640948091 |
| 0.217 | 0.278219128 | 0.388539333 | 0.510423161 | 0.634306232 |
| 0.216 | 0.273780719 | 0.382983521 | 0.504045980 | 0.627632105 |
| 0.215 | 0.269374076 | 0.377450677 | 0.497670862 | 0.620927436 |
| 0.214 | 0.264999646 | 0.371941692 | 0.491299204 | 0.614193969 |
| 0.213 | 0.260657867 | 0.366457447 | 0.484932397 | 0.607433460 |
| 0.212 | 0.256349166 | 0.360998810 | 0.478571828 | 0.600647678 |
| 0.211 | 0.252073960 | 0.355566639 | 0.472218876 | 0.593838406 |
| 0.210 | 0.247832653 | 0.350161777 | 0.465874914 | 0.587007437 |
| 0.209 | 0.243625641 | 0.344785059 | 0.459541307 | 0.580156571 |
| 0.208 | 0.239453307 | 0.339437303 | 0.453219413 | 0.573287619 |
| 0.207 | 0.235316023 | 0.334119318 | 0.446910579 | 0.566402400 |
| 0.206 | 0.231214152 | 0.328831897 | 0.440616143 | 0.559502736 |
| 0.205 | 0.227148043 | 0.323575822 | 0.434337434 | 0.552590456 |
| 0.204 | 0.223118036 | 0.318351860 | 0.428075769 | 0.545667391 |
| 0.203 | 0.219124458 | 0.313160764 | 0.421832454 | 0.538735377 |
| 0.202 | 0.215167626 | 0.308003274 | 0.415608783 | 0.531796249 |
| 0.201 | 0.211247845 | 0.302880115 | 0.409406038 | 0.524851843 |

TABLE 1a:   CLUSTERING PROBABILITY   P(n;N,p)

n = 7

| P | N =14 | N =15 | N =16 | N =17 |
|---|---|---|---|---|
| 0.200 | 0.207365408 | 0.297791997 | 0.403225487 | 0.517903995 |
| 0.199 | 0.203520597 | 0.292739616 | 0.397068385 | 0.510954536 |
| 0.198 | 0.199713683 | 0.287723654 | 0.390935972 | 0.504005297 |
| 0.197 | 0.195944924 | 0.282744777 | 0.384829474 | 0.497058104 |
| 0.196 | 0.192214568 | 0.277803634 | 0.378750102 | 0.490114775 |
| 0.195 | 0.188522849 | 0.272900862 | 0.372699049 | 0.483177124 |
| 0.194 | 0.184869992 | 0.268037079 | 0.366677496 | 0.476246958 |
| 0.193 | 0.181256208 | 0.263212889 | 0.360686602 | 0.469326072 |
| 0.192 | 0.177681697 | 0.258428880 | 0.354727513 | 0.462416255 |
| 0.191 | 0.174146649 | 0.253685624 | 0.348801355 | 0.455519283 |
| 0.190 | 0.170651239 | 0.248983674 | 0.342909237 | 0.448636919 |
| 0.189 | 0.167195633 | 0.244323571 | 0.337052248 | 0.441770917 |
| 0.188 | 0.163779983 | 0.239705836 | 0.331231459 | 0.434923013 |
| 0.187 | 0.160404431 | 0.235130975 | 0.325447921 | 0.428094932 |
| 0.186 | 0.157069106 | 0.230599475 | 0.319702666 | 0.421288380 |
| 0.185 | 0.153774127 | 0.226111810 | 0.313996705 | 0.414505047 |
| 0.184 | 0.150519598 | 0.221668433 | 0.308331028 | 0.407746608 |
| 0.183 | 0.147305614 | 0.217269782 | 0.302706605 | 0.401014715 |
| 0.182 | 0.144132258 | 0.212916277 | 0.297124383 | 0.394311004 |
| 0.181 | 0.140999601 | 0.208608321 | 0.291585289 | 0.387637090 |
| 0.180 | 0.137907701 | 0.204346300 | 0.286090228 | 0.380994565 |
| 0.179 | 0.134856607 | 0.200130582 | 0.280640081 | 0.374385002 |
| 0.178 | 0.131846355 | 0.195961518 | 0.275235707 | 0.367809948 |
| 0.177 | 0.128876970 | 0.191839440 | 0.269877942 | 0.361270928 |
| 0.176 | 0.125948463 | 0.187764665 | 0.264567601 | 0.354769443 |
| 0.175 | 0.123060838 | 0.183737490 | 0.259305471 | 0.348306967 |
| 0.174 | 0.120214085 | 0.179758195 | 0.254092320 | 0.341884951 |
| 0.173 | 0.117408182 | 0.175827043 | 0.248928888 | 0.335504817 |
| 0.172 | 0.114643099 | 0.171944279 | 0.243815893 | 0.329167960 |
| 0.171 | 0.111918791 | 0.168110130 | 0.238754029 | 0.322875749 |
| 0.170 | 0.109235205 | 0.164324806 | 0.233743964 | 0.316629521 |
| 0.169 | 0.106592275 | 0.160588498 | 0.228786342 | 0.310430586 |
| 0.168 | 0.103989926 | 0.156901381 | 0.223881782 | 0.304280224 |
| 0.167 | 0.101428070 | 0.153263612 | 0.219030877 | 0.298179686 |
| 0.166 | 0.098906611 | 0.149675330 | 0.214234196 | 0.292130188 |
| 0.165 | 0.096425440 | 0.146136655 | 0.209492282 | 0.286132919 |
| 0.164 | 0.093984439 | 0.142647693 | 0.204805652 | 0.280189034 |
| 0.163 | 0.091583478 | 0.139208530 | 0.200174799 | 0.274299654 |
| 0.162 | 0.089222420 | 0.135819235 | 0.195600189 | 0.268465870 |
| 0.161 | 0.086901114 | 0.132479861 | 0.191082263 | 0.262688739 |
| 0.160 | 0.084619401 | 0.129190442 | 0.186621435 | 0.256969282 |
| 0.159 | 0.082377112 | 0.125950996 | 0.182218094 | 0.251308489 |
| 0.158 | 0.080174069 | 0.122761524 | 0.177872605 | 0.245707313 |
| 0.157 | 0.078010082 | 0.119622010 | 0.173585303 | 0.240166674 |
| 0.156 | 0.075884954 | 0.116532421 | 0.169356500 | 0.234687456 |
| 0.155 | 0.073798477 | 0.113492708 | 0.165186483 | 0.229270507 |
| 0.154 | 0.071750435 | 0.110502805 | 0.161075510 | 0.223916641 |
| 0.153 | 0.069740601 | 0.107562628 | 0.157023816 | 0.218626634 |
| 0.152 | 0.067768742 | 0.104672081 | 0.153031609 | 0.213401227 |
| 0.151 | 0.065834615 | 0.101831047 | 0.149099071 | 0.208241126 |

### TABLE 1a:   CLUSTERING PROBABILITY   P(n;N,p)

n = 7

| P | N = 14 | N = 15 | N = 16 | N = 17 |
|---|---|---|---|---|
| 0.150 | 0.063937966 | 0.099039397 | 0.145226360 | 0.203146998 |
| 0.149 | 0.062078536 | 0.096296984 | 0.141413606 | 0.198119474 |
| 0.148 | 0.060256057 | 0.093603647 | 0.137660917 | 0.193159150 |
| 0.147 | 0.058470251 | 0.090959208 | 0.133968373 | 0.188266584 |
| 0.146 | 0.056720834 | 0.088363474 | 0.130336029 | 0.183442296 |
| 0.145 | 0.055007514 | 0.085816239 | 0.126763918 | 0.178686771 |
| 0.144 | 0.053329990 | 0.083317279 | 0.123252044 | 0.174000455 |
| 0.143 | 0.051687955 | 0.080866359 | 0.119800389 | 0.169383760 |
| 0.142 | 0.050081095 | 0.078463226 | 0.116408911 | 0.164837058 |
| 0.141 | 0.048509088 | 0.076107616 | 0.113077542 | 0.160360685 |
| 0.140 | 0.046971606 | 0.073799249 | 0.109806190 | 0.155954942 |
| 0.139 | 0.045468313 | 0.071537832 | 0.106594742 | 0.151620092 |
| 0.138 | 0.043998868 | 0.069323060 | 0.103443059 | 0.147356360 |
| 0.137 | 0.042562924 | 0.067154612 | 0.100350978 | 0.143163937 |
| 0.136 | 0.041160126 | 0.065032156 | 0.097318316 | 0.139042977 |
| 0.135 | 0.039790114 | 0.062955346 | 0.094344865 | 0.134993596 |
| 0.134 | 0.038452524 | 0.060923826 | 0.091430395 | 0.131015879 |
| 0.133 | 0.037146985 | 0.058937225 | 0.088574655 | 0.127109869 |
| 0.132 | 0.035873120 | 0.056995163 | 0.085777369 | 0.123275580 |
| 0.131 | 0.034630549 | 0.055097245 | 0.083038245 | 0.119512986 |
| 0.130 | 0.033418886 | 0.053243068 | 0.080356964 | 0.115822029 |
| 0.129 | 0.032237740 | 0.051432217 | 0.077733191 | 0.112202616 |
| 0.128 | 0.031086717 | 0.049664265 | 0.075166567 | 0.108654620 |
| 0.127 | 0.029965417 | 0.047938776 | 0.072656716 | 0.105177880 |
| 0.126 | 0.028873438 | 0.046255304 | 0.070203239 | 0.101772202 |
| 0.125 | 0.027810372 | 0.044613392 | 0.067805722 | 0.098437360 |
| 0.124 | 0.026775810 | 0.043012576 | 0.065463728 | 0.095173093 |
| 0.123 | 0.025769339 | 0.041452380 | 0.063176804 | 0.091979111 |
| 0.122 | 0.024790540 | 0.039932321 | 0.060944478 | 0.088855091 |
| 0.121 | 0.023838997 | 0.038451908 | 0.058766262 | 0.085800680 |
| 0.120 | 0.022914285 | 0.037010639 | 0.056641649 | 0.082815494 |
| 0.119 | 0.022015981 | 0.035608009 | 0.054570116 | 0.079899118 |
| 0.118 | 0.021143657 | 0.034243501 | 0.052551124 | 0.077051110 |
| 0.117 | 0.020296886 | 0.032916593 | 0.050584118 | 0.074270996 |
| 0.116 | 0.019475238 | 0.031626757 | 0.048668529 | 0.071558278 |
| 0.115 | 0.018678279 | 0.030373457 | 0.046803772 | 0.068912428 |
| 0.114 | 0.017905577 | 0.029156152 | 0.044989247 | 0.066332889 |
| 0.113 | 0.017156698 | 0.027974294 | 0.043224342 | 0.063819083 |
| 0.112 | 0.016431207 | 0.026827333 | 0.041508431 | 0.061370401 |
| 0.111 | 0.015728668 | 0.025714710 | 0.039840876 | 0.058986212 |
| 0.110 | 0.015048645 | 0.024635863 | 0.038221025 | 0.056665861 |
| 0.109 | 0.014390701 | 0.023590228 | 0.036648217 | 0.054408668 |
| 0.108 | 0.013754401 | 0.022577233 | 0.035121778 | 0.052213931 |
| 0.107 | 0.013139309 | 0.021596306 | 0.033641023 | 0.050080925 |
| 0.106 | 0.012544989 | 0.020646869 | 0.032205259 | 0.048008905 |
| 0.105 | 0.011971005 | 0.019728344 | 0.030813782 | 0.045997105 |
| 0.104 | 0.011416925 | 0.018840149 | 0.029465879 | 0.044044738 |
| 0.103 | 0.010882316 | 0.017981701 | 0.028160831 | 0.042151001 |
| 0.102 | 0.010366745 | 0.017152413 | 0.026897908 | 0.040315068 |
| 0.101 | 0.009869784 | 0.016351700 | 0.025676375 | 0.038536101 |

TABLE 1a:    CLUSTERING PROBABILITY   P(n;N,p)

n = 7

| P | N = 14 | N = 15 | N = 16 | N = 17 |
|---|---|---|---|---|
| 0.100 | 0.009391004 | 0.015578973 | 0.024495490 | 0.036813241 |
| 0.099 | 0.008929978 | 0.014833645 | 0.023354503 | 0.035145616 |
| 0.098 | 0.008486283 | 0.014115127 | 0.022252662 | 0.033532338 |
| 0.097 | 0.008059496 | 0.013422830 | 0.021189207 | 0.031972506 |
| 0.096 | 0.007649200 | 0.012756168 | 0.020163375 | 0.030465203 |
| 0.095 | 0.007254976 | 0.012114552 | 0.019174400 | 0.029009503 |
| 0.094 | 0.006876412 | 0.011497398 | 0.018221512 | 0.027604466 |
| 0.093 | 0.006513098 | 0.010904121 | 0.017303938 | 0.026249144 |
| 0.092 | 0.006164625 | 0.010334139 | 0.016420903 | 0.024942577 |
| 0.091 | 0.005830591 | 0.009786871 | 0.015571631 | 0.023683797 |
| 0.090 | 0.005510595 | 0.009261741 | 0.014755346 | 0.022471828 |
| 0.089 | 0.005204241 | 0.008758174 | 0.013971270 | 0.021305687 |
| 0.088 | 0.004911137 | 0.008275598 | 0.013218626 | 0.020184384 |
| 0.087 | 0.004630896 | 0.007813446 | 0.012496639 | 0.019106926 |
| 0.086 | 0.004363132 | 0.007371153 | 0.011804533 | 0.018072313 |
| 0.085 | 0.004107468 | 0.006948161 | 0.011141536 | 0.017079542 |
| 0.084 | 0.003863529 | 0.006543914 | 0.010506877 | 0.016127609 |
| 0.083 | 0.003630944 | 0.006157862 | 0.009899790 | 0.015215506 |
| 0.082 | 0.003409349 | 0.005789459 | 0.009319511 | 0.014342225 |
| 0.081 | 0.003198384 | 0.005438166 | 0.008765279 | 0.013506760 |
| 0.080 | 0.002997695 | 0.005103449 | 0.008236341 | 0.012708103 |
| 0.079 | 0.002806933 | 0.004784780 | 0.007731947 | 0.011945248 |
| 0.078 | 0.002625752 | 0.004481636 | 0.007251352 | 0.011217194 |
| 0.077 | 0.002453817 | 0.004193503 | 0.006793818 | 0.010522940 |
| 0.076 | 0.002290793 | 0.003919872 | 0.006358615 | 0.009861492 |
| 0.075 | 0.002136354 | 0.003660241 | 0.005945018 | 0.009231861 |
| 0.074 | 0.001990179 | 0.003414117 | 0.005552311 | 0.008633061 |
| 0.073 | 0.001851954 | 0.003181011 | 0.005179785 | 0.008064116 |
| 0.072 | 0.001721370 | 0.002960446 | 0.004826740 | 0.007524055 |
| 0.071 | 0.001598123 | 0.002751950 | 0.004492486 | 0.007011918 |
| 0.070 | 0.001481919 | 0.002555060 | 0.004176340 | 0.006526752 |
| 0.069 | 0.001372466 | 0.002369322 | 0.003877630 | 0.006067614 |
| 0.068 | 0.001269482 | 0.002194289 | 0.003595696 | 0.005633571 |
| 0.067 | 0.001172689 | 0.002029524 | 0.003329886 | 0.005223703 |
| 0.066 | 0.001081816 | 0.001874598 | 0.003079559 | 0.004837101 |
| 0.065 | 0.000996600 | 0.001729092 | 0.002844086 | 0.004472867 |
| 0.064 | 0.000916783 | 0.001592595 | 0.002622851 | 0.004130118 |
| 0.063 | 0.000842114 | 0.001464707 | 0.002415248 | 0.003807986 |
| 0.062 | 0.000772349 | 0.001345036 | 0.002220684 | 0.003505614 |
| 0.061 | 0.000707251 | 0.001233199 | 0.002038577 | 0.003222163 |
| 0.060 | 0.000646589 | 0.001128824 | 0.001868360 | 0.002956808 |
| 0.059 | 0.000590139 | 0.001031549 | 0.001709479 | 0.002708741 |
| 0.058 | 0.000537683 | 0.000941021 | 0.001561391 | 0.002477170 |
| 0.057 | 0.000489011 | 0.000856896 | 0.001423569 | 0.002261320 |
| 0.056 | 0.000443919 | 0.000778842 | 0.001295497 | 0.002060435 |
| 0.055 | 0.000402211 | 0.000706535 | 0.001176677 | 0.001873775 |
| 0.054 | 0.000363695 | 0.000639663 | 0.001066621 | 0.001700620 |
| 0.053 | 0.000328187 | 0.000577922 | 0.000964856 | 0.001540267 |
| 0.052 | 0.000295511 | 0.000521019 | 0.000870925 | 0.001392033 |
| 0.051 | 0.000265495 | 0.000468671 | 0.000784385 | 0.001255256 |

NEFF and NAUS

TABLE 1a:  CLUSTERING PROBABILITY  P(n;N,p)

n = 7

| P | N =14 | N =15 | N =16 | N =17 |
|---|---|---|---|---|
| 0.050 | 0.000237976 | 0.000420606 | 0.000704805 | 0.001129290 |
| 0.049 | 0.000212795 | 0.000376561 | 0.000631771 | 0.001013513 |
| 0.048 | 0.000189802 | 0.000336282 | 0.000564883 | 0.000907320 |
| 0.047 | 0.000168851 | 0.000299526 | 0.000503756 | 0.000810127 |
| 0.046 | 0.000149803 | 0.000266061 | 0.000448018 | 0.000721372 |
| 0.045 | 0.000132526 | 0.000235662 | 0.000397313 | 0.000640510 |
| 0.044 | 0.000116894 | 0.000208116 | 0.000351299 | 0.000567021 |
| 0.043 | 0.000102785 | 0.000183219 | 0.000309648 | 0.000500401 |
| 0.042 | 0.000090085 | 0.000160776 | 0.000272047 | 0.000440171 |
| 0.041 | 0.000078686 | 0.000140600 | 0.000238197 | 0.000385869 |
| 0.040 | 0.000068483 | 0.000122517 | 0.000207812 | 0.000337055 |
| 0.039 | 0.000059379 | 0.000106358 | 0.000180622 | 0.000293309 |
| 0.038 | 0.000051282 | 0.000091966 | 0.000156369 | 0.000254231 |
| 0.037 | 0.000044105 | 0.000079191 | 0.000134809 | 0.000219442 |
| 0.036 | 0.000037767 | 0.000067891 | 0.000115712 | 0.000188582 |
| 0.035 | 0.000032189 | 0.000057934 | 0.000098859 | 0.000161311 |
| 0.034 | 0.000027301 | 0.000049195 | 0.000084047 | 0.000137307 |
| 0.033 | 0.000023035 | 0.000041557 | 0.000071084 | 0.000116267 |
| 0.032 | 0.000019328 | 0.000034912 | 0.000059788 | 0.000097909 |
| 0.031 | 0.000016123 | 0.000029157 | 0.000049993 | 0.000081966 |
| 0.030 | 0.000013366 | 0.000024199 | 0.000041542 | 0.000068191 |
| 0.029 | 0.000011006 | 0.000019951 | 0.000034289 | 0.000056353 |
| 0.028 | 0.000008998 | 0.000016331 | 0.000028101 | 0.000046238 |
| 0.027 | 0.000007301 | 0.000013265 | 0.000022854 | 0.000037648 |
| 0.026 | 0.000005875 | 0.000010687 | 0.000018433 | 0.000030403 |
| 0.025 | 0.000004685 | 0.000008534 | 0.000014736 | 0.000024334 |
| 0.024 | 0.000003701 | 0.000006749 | 0.000011668 | 0.000019290 |
| 0.023 | 0.000002893 | 0.000005282 | 0.000009143 | 0.000015133 |
| 0.022 | 0.000002236 | 0.000004087 | 0.000007083 | 0.000011737 |
| 0.021 | 0.000001707 | 0.000003124 | 0.000005419 | 0.000008991 |
| 0.020 | 0.000001285 | 0.000002355 | 0.000004090 | 0.000006795 |
| 0.019 | 0.000000953 | 0.000001749 | 0.000003041 | 0.000005058 |
| 0.018 | 0.000000696 | 0.000001277 | 0.000002224 | 0.000003703 |
| 0.017 | 0.000000498 | 0.000000916 | 0.000001596 | 0.000002661 |
| 0.016 | 0.000000349 | 0.000000643 | 0.000001122 | 0.000001873 |
| 0.015 | 0.000000239 | 0.000000441 | 0.000000771 | 0.000001288 |
| 0.014 | 0.000000160 | 0.000000295 | 0.000000515 | 0.000000862 |
| 0.013 | 0.000000103 | 0.000000191 | 0.000000334 | 0.000000559 |
| 0.012 | 0.000000064 | 0.000000119 | 0.000000209 | 0.000000350 |
| 0.011 | 0.000000039 | 0.000000071 | 0.000000125 | 0.000000211 |
| 0.010 | 0.000000022 | 0.000000041 | 0.000000072 | 0.000000120 |
| 0.009 | 0.000000012 | 0.000000022 | 0.000000038 | 0.000000065 |
| 0.008 | 0.000000006 | 0.000000011 | 0.000000019 | 0.000000032 |
| 0.007 | 0.000000003 | 0.000000005 | 0.000000009 | 0.000000015 |
| 0.006 | 0.000000001 | 0.000000002 | 0.000000003 | 0.000000006 |
| 0.005 | 0.000000000 | 0.000000001 | 0.000000001 | 0.000000002 |
| 0.004 | 0.000000000 | 0.000000000 | 0.000000000 | 0.000000001 |
| 0.003 | 0.000000000 | 0.000000000 | 0.000000000 | 0.000000000 |

TABLE 1a:   CLUSTERING PROBABILITY  P(n;N,p)

n = 7

| P | N =18 | N =19 | N =20 | N =21 |
|------|------------|------------|------------|------------|
| 0.300 | 0.994714175 | 0.999561929 | 0.999986300 | 0.999999842 |
| 0.299 | 0.994287985 | 0.999495232 | 0.999982765 | 0.999999779 |
| 0.298 | 0.993833874 | 0.999420561 | 0.999978470 | 0.999999695 |
| 0.297 | 0.993350552 | 0.999337212 | 0.999973281 | 0.999999582 |
| 0.296 | 0.992836703 | 0.999244440 | 0.999967050 | 0.999999434 |
| 0.295 | 0.992290988 | 0.999141461 | 0.999959608 | 0.999999240 |
| 0.294 | 0.991712043 | 0.999027447 | 0.999950763 | 0.999998988 |
| 0.293 | 0.991098487 | 0.998901531 | 0.999940305 | 0.999998662 |
| 0.292 | 0.990448914 | 0.998762802 | 0.999927996 | 0.999998245 |
| 0.291 | 0.989761905 | 0.998610305 | 0.999913574 | 0.999997715 |
| 0.290 | 0.989036020 | 0.998443043 | 0.999896747 | 0.999997046 |
| 0.289 | 0.988269806 | 0.998259977 | 0.999877196 | 0.999996205 |
| 0.288 | 0.987461797 | 0.998060023 | 0.999854567 | 0.999995156 |
| 0.287 | 0.986610514 | 0.997842053 | 0.999828476 | 0.999993855 |
| 0.286 | 0.985714469 | 0.997604898 | 0.999798500 | 0.999992249 |
| 0.285 | 0.984772164 | 0.997347343 | 0.999764182 | 0.999990278 |
| 0.284 | 0.983782098 | 0.997068134 | 0.999725022 | 0.999987872 |
| 0.283 | 0.982742764 | 0.996765972 | 0.999680482 | 0.999984947 |
| 0.282 | 0.981652652 | 0.996439518 | 0.999629979 | 0.999981410 |
| 0.281 | 0.980510253 | 0.996087392 | 0.999572887 | 0.999977152 |
| 0.280 | 0.979314058 | 0.995708174 | 0.999508531 | 0.999972049 |
| 0.279 | 0.978062563 | 0.995300406 | 0.999436191 | 0.999965957 |
| 0.278 | 0.976754271 | 0.994862591 | 0.999355094 | 0.999958716 |
| 0.277 | 0.975387689 | 0.994393197 | 0.999264420 | 0.999950144 |
| 0.276 | 0.973961338 | 0.993890656 | 0.999163292 | 0.999940035 |
| 0.275 | 0.972473748 | 0.993353365 | 0.999050783 | 0.999928156 |
| 0.274 | 0.970923465 | 0.992779693 | 0.998925908 | 0.999914250 |
| 0.273 | 0.969309049 | 0.992167974 | 0.998787630 | 0.999898027 |
| 0.272 | 0.967629081 | 0.991516516 | 0.998634851 | 0.999879165 |
| 0.271 | 0.965882160 | 0.990823601 | 0.998466419 | 0.999857308 |
| 0.270 | 0.964066908 | 0.990087485 | 0.998281123 | 0.999832062 |
| 0.269 | 0.962181973 | 0.989306403 | 0.998077694 | 0.999802993 |
| 0.268 | 0.960226027 | 0.988478568 | 0.997854805 | 0.999769623 |
| 0.267 | 0.958197773 | 0.987602176 | 0.997611069 | 0.999731430 |
| 0.266 | 0.956095941 | 0.986675408 | 0.997345042 | 0.999687843 |
| 0.265 | 0.953919296 | 0.985696433 | 0.997055224 | 0.999638241 |
| 0.264 | 0.951666638 | 0.984663406 | 0.996740056 | 0.999581948 |
| 0.263 | 0.949336800 | 0.983574477 | 0.996397922 | 0.999518234 |
| 0.262 | 0.946928656 | 0.982427792 | 0.996027152 | 0.999446309 |
| 0.261 | 0.944441118 | 0.981221491 | 0.995626023 | 0.999365323 |
| 0.260 | 0.941873140 | 0.979953716 | 0.995192756 | 0.999274361 |
| 0.259 | 0.939223719 | 0.978622614 | 0.994725524 | 0.999172445 |
| 0.258 | 0.936491899 | 0.977226336 | 0.994222450 | 0.999058528 |
| 0.257 | 0.933676766 | 0.975763042 | 0.993681607 | 0.998931495 |
| 0.256 | 0.930777458 | 0.974230904 | 0.993101027 | 0.998790158 |
| 0.255 | 0.927793160 | 0.972628109 | 0.992478696 | 0.998633259 |
| 0.254 | 0.924723109 | 0.970952863 | 0.991812561 | 0.998459466 |
| 0.253 | 0.921566594 | 0.969203389 | 0.991100530 | 0.998267371 |
| 0.252 | 0.918322957 | 0.967377938 | 0.990340478 | 0.998055493 |
| 0.251 | 0.914991595 | 0.965474785 | 0.989530247 | 0.997822274 |

TABLE 1a:   CLUSTERING PROBABILITY   P(n;N,p)

n = 7

| P | N = 18 | N = 19 | N = 20 | N = 21 |
|---|---|---|---|---|
| 0.250 | 0.911571960 | 0.963492233 | 0.988667650 | 0.997566080 |
| 0.249 | 0.908063563 | 0.961428622 | 0.987750475 | 0.997285203 |
| 0.248 | 0.904465970 | 0.959282321 | 0.986776487 | 0.996977859 |
| 0.247 | 0.900778808 | 0.957051742 | 0.985743434 | 0.996642189 |
| 0.246 | 0.897001761 | 0.954735336 | 0.984649048 | 0.996276261 |
| 0.245 | 0.893134578 | 0.952331595 | 0.983491049 | 0.995878070 |
| 0.244 | 0.889177064 | 0.949839062 | 0.982267150 | 0.995445543 |
| 0.243 | 0.885129088 | 0.947256324 | 0.980975060 | 0.994976535 |
| 0.242 | 0.880990583 | 0.944582023 | 0.979612489 | 0.994468836 |
| 0.241 | 0.876761542 | 0.941814852 | 0.978177150 | 0.993920171 |
| 0.240 | 0.872442022 | 0.938953561 | 0.976666765 | 0.993328205 |
| 0.239 | 0.868032145 | 0.935996961 | 0.975079068 | 0.992690542 |
| 0.238 | 0.863532095 | 0.932943920 | 0.973411809 | 0.992004734 |
| 0.237 | 0.858942120 | 0.929793371 | 0.971662758 | 0.991268277 |
| 0.236 | 0.854262534 | 0.926544314 | 0.969829710 | 0.990478623 |
| 0.235 | 0.849493713 | 0.923195812 | 0.967910490 | 0.989633176 |
| 0.234 | 0.844636097 | 0.919747001 | 0.965902952 | 0.988729303 |
| 0.233 | 0.839690193 | 0.916197086 | 0.963804991 | 0.987764332 |
| 0.232 | 0.834656567 | 0.912545345 | 0.961614540 | 0.986735563 |
| 0.231 | 0.829535853 | 0.908791131 | 0.959329577 | 0.985640268 |
| 0.230 | 0.824328744 | 0.904933871 | 0.956948130 | 0.984475696 |
| 0.229 | 0.819036001 | 0.900973071 | 0.954468278 | 0.983239080 |
| 0.228 | 0.813658441 | 0.896908316 | 0.951888159 | 0.981927643 |
| 0.227 | 0.808196948 | 0.892739269 | 0.949205968 | 0.980538601 |
| 0.226 | 0.802652465 | 0.888465675 | 0.946419967 | 0.979069167 |
| 0.225 | 0.797025994 | 0.884087361 | 0.943528483 | 0.977516562 |
| 0.224 | 0.791318600 | 0.879604237 | 0.940529915 | 0.975878014 |
| 0.223 | 0.785531405 | 0.875016295 | 0.937422736 | 0.974150768 |
| 0.222 | 0.779665588 | 0.870323613 | 0.934205499 | 0.972332090 |
| 0.221 | 0.773722388 | 0.865526353 | 0.930876834 | 0.970419274 |
| 0.220 | 0.767703098 | 0.860624761 | 0.927435457 | 0.968409646 |
| 0.219 | 0.761609068 | 0.855619170 | 0.923880171 | 0.966300569 |
| 0.218 | 0.755441700 | 0.850509998 | 0.920209869 | 0.964089452 |
| 0.217 | 0.749202450 | 0.845297750 | 0.916423534 | 0.961773753 |
| 0.216 | 0.742892826 | 0.839983013 | 0.912520247 | 0.959350984 |
| 0.215 | 0.736514386 | 0.834566464 | 0.908499184 | 0.956818720 |
| 0.214 | 0.730068739 | 0.829048862 | 0.904359620 | 0.954174600 |
| 0.213 | 0.723557539 | 0.823431053 | 0.900100933 | 0.951416336 |
| 0.212 | 0.716982490 | 0.817713966 | 0.895722604 | 0.948541716 |
| 0.211 | 0.710345337 | 0.811898614 | 0.891224219 | 0.945548611 |
| 0.210 | 0.703647873 | 0.805986093 | 0.886605469 | 0.942434978 |
| 0.209 | 0.696891931 | 0.799977582 | 0.881866157 | 0.939198867 |
| 0.208 | 0.690079386 | 0.793874342 | 0.877006191 | 0.935838423 |
| 0.207 | 0.683212149 | 0.787677712 | 0.872025592 | 0.932351894 |
| 0.206 | 0.676292174 | 0.781389113 | 0.866924491 | 0.928737632 |
| 0.205 | 0.669321447 | 0.775010042 | 0.861703134 | 0.924994102 |
| 0.204 | 0.662301988 | 0.768542073 | 0.856361876 | 0.921119880 |
| 0.203 | 0.655235854 | 0.761986858 | 0.850901187 | 0.917113660 |
| 0.202 | 0.648125128 | 0.755346120 | 0.845321650 | 0.912974260 |
| 0.201 | 0.640971926 | 0.748621656 | 0.839623961 | 0.908700621 |

TABLE 1a:  CLUSTERING PROBABILITY P(n;N,p)

n = 7

| P | N =18 | N =19 | N =20 |
|---|---|---|---|
| 0.200 | 0.63377839 0 | 0.741815332 | 0.833808930 |
| 0.199 | 0.62654668 9 | 0.734929087 | 0.827877480 |
| 0.198 | 0.619279014 | 0.727964922 | 0.821830645 |
| 0.197 | 0.611977582 | 0.720924907 | 0.815669574 |
| 0.196 | 0.604644626 | 0.713811175 | 0.809395523 |
| 0.195 | 0.597282403 | 0.706625920 | 0.803009863 |
| 0.194 | 0.589893183 | 0.699371395 | 0.796514070 |
| 0.193 | 0.582479253 | 0.692049911 | 0.789909729 |
| 0.192 | 0.575042913 | 0.684663833 | 0.783198534 |
| 0.191 | 0.567586476 | 0.677215581 | 0.776382280 |
| 0.190 | 0.560112264 | 0.669707623 | 0.769462869 |
| 0.189 | 0.552622606 | 0.662142476 | 0.762442301 |
| 0.188 | 0.545119840 | 0.654522705 | 0.755322677 |
| 0.187 | 0.537606305 | 0.646850916 | 0.748106197 |
| 0.186 | 0.530084346 | 0.639129756 | 0.740795152 |
| 0.185 | 0.522556308 | 0.631361911 | 0.733391929 |
| 0.184 | 0.515024535 | 0.623550104 | 0.725899004 |
| 0.183 | 0.507491369 | 0.615697089 | 0.718318941 |
| 0.182 | 0.499959146 | 0.607805653 | 0.710654388 |
| 0.181 | 0.492430198 | 0.599878611 | 0.702908076 |
| 0.180 | 0.484906850 | 0.591918801 | 0.695082813 |
| 0.179 | 0.477391416 | 0.583929088 | 0.687181486 |
| 0.178 | 0.469886199 | 0.575912353 | 0.679207053 |
| 0.177 | 0.462393491 | 0.567871500 | 0.671162543 |
| 0.176 | 0.454915569 | 0.559809442 | 0.663051049 |
| 0.175 | 0.447454694 | 0.551729109 | 0.654875730 |
| 0.174 | 0.440013109 | 0.543633440 | 0.646639803 |
| 0.173 | 0.432593039 | 0.535525378 | 0.638346541 |
| 0.172 | 0.425196690 | 0.527407874 | 0.629999270 |
| 0.171 | 0.417826243 | 0.519283880 | 0.621601365 |
| 0.170 | 0.410483858 | 0.511156347 | 0.613156248 |
| 0.169 | 0.403171669 | 0.503028222 | 0.604667378 |
| 0.168 | 0.395891784 | 0.494902447 | 0.596138258 |
| 0.167 | 0.388646284 | 0.486781955 | 0.587572420 |
| 0.166 | 0.381437221 | 0.478669668 | 0.578973429 |
| 0.165 | 0.374266616 | 0.470568497 | 0.570344876 |
| 0.164 | 0.367136460 | 0.462481333 | 0.561690375 |
| 0.163 | 0.360048709 | 0.454411051 | 0.553013559 |
| 0.162 | 0.353005287 | 0.446360506 | 0.544318076 |
| 0.161 | 0.346008083 | 0.438332528 | 0.535607584 |
| 0.160 | 0.339058949 | 0.430329923 | 0.526885751 |
| 0.159 | 0.332159699 | 0.422355470 | 0.518156248 |
| 0.158 | 0.325312109 | 0.414411915 | 0.509422744 |
| 0.157 | 0.318517918 | 0.406501975 | 0.500688907 |
| 0.156 | 0.311778822 | 0.398628333 | 0.491958395 |
| 0.155 | 0.305096476 | 0.390793633 | 0.483234856 |
| 0.154 | 0.298472494 | 0.383000483 | 0.474521923 |
| 0.153 | 0.291908446 | 0.375251449 | 0.465823212 |
| 0.152 | 0.285405858 | 0.367549057 | 0.457142313 |
| 0.151 | 0.278966213 | 0.359895786 | 0.448482795 |

TABLE 1a:    CLUSTERING PROBABILITY   P(n;N,p)

n = 7

| p | N =18 | N =19 | N =20 |
|---|---|---|---|
| 0.150 | 0.272590945 | 0.352294073 | 0.439848195 |
| 0.149 | 0.266281447 | 0.344746303 | 0.431242018 |
| 0.148 | 0.260039061 | 0.337254814 | 0.422667734 |
| 0.147 | 0.253865083 | 0.329821893 | 0.414128774 |
| 0.146 | 0.247760763 | 0.322449774 | 0.405628525 |
| 0.145 | 0.241727299 | 0.315140637 | 0.397170332 |
| 0.144 | 0.235765842 | 0.307896605 | 0.388757488 |
| 0.143 | 0.229877495 | 0.300719746 | 0.380393236 |
| 0.142 | 0.224063308 | 0.293612070 | 0.372080766 |
| 0.141 | 0.218324283 | 0.286575525 | 0.363823209 |
| 0.140 | 0.212661371 | 0.279612001 | 0.355623636 |
| 0.139 | 0.207075472 | 0.272723323 | 0.347485057 |
| 0.138 | 0.201567436 | 0.265911257 | 0.339410416 |
| 0.137 | 0.196138060 | 0.259177501 | 0.331402590 |
| 0.136 | 0.190788090 | 0.252523691 | 0.323464386 |
| 0.135 | 0.185518223 | 0.245951397 | 0.315598539 |
| 0.134 | 0.180329101 | 0.239462121 | 0.307807709 |
| 0.133 | 0.175221316 | 0.233057299 | 0.300094480 |
| 0.132 | 0.170195408 | 0.226738298 | 0.292461359 |
| 0.131 | 0.165251867 | 0.220506418 | 0.284910772 |
| 0.130 | 0.160391128 | 0.214362890 | 0.277445062 |
| 0.129 | 0.155613579 | 0.208308874 | 0.270066490 |
| 0.128 | 0.150919553 | 0.202345461 | 0.262777233 |
| 0.127 | 0.146309335 | 0.196473673 | 0.255579378 |
| 0.126 | 0.141783157 | 0.190694459 | 0.248474928 |
| 0.125 | 0.137341202 | 0.185008701 | 0.241465794 |
| 0.124 | 0.132983602 | 0.179417207 | 0.234553800 |
| 0.123 | 0.128710439 | 0.173920717 | 0.227740676 |
| 0.122 | 0.124521748 | 0.168519898 | 0.221028063 |
| 0.121 | 0.120417511 | 0.163215349 | 0.214417505 |
| 0.120 | 0.116397665 | 0.158007595 | 0.207910458 |
| 0.119 | 0.112462099 | 0.152897095 | 0.201508280 |
| 0.118 | 0.108610651 | 0.147884236 | 0.195212237 |
| 0.117 | 0.104843116 | 0.142969335 | 0.189023498 |
| 0.116 | 0.101159242 | 0.138152640 | 0.182943140 |
| 0.115 | 0.097558729 | 0.133434331 | 0.176972141 |
| 0.114 | 0.094041235 | 0.128814520 | 0.171111389 |
| 0.113 | 0.090606373 | 0.124293251 | 0.165361673 |
| 0.112 | 0.087253712 | 0.119870502 | 0.159723688 |
| 0.111 | 0.083982778 | 0.115546183 | 0.154198035 |
| 0.110 | 0.080793057 | 0.111320140 | 0.148785222 |
| 0.109 | 0.077683993 | 0.107192154 | 0.143485661 |
| 0.108 | 0.074654988 | 0.103161944 | 0.138299673 |
| 0.107 | 0.071705408 | 0.099229163 | 0.133227486 |
| 0.106 | 0.068834579 | 0.095393406 | 0.128269236 |
| 0.105 | 0.066041789 | 0.091654204 | 0.123424969 |
| 0.104 | 0.063326292 | 0.088011031 | 0.118694641 |
| 0.103 | 0.060687304 | 0.084463300 | 0.114078119 |
| 0.102 | 0.058124008 | 0.081010371 | 0.109575185 |
| 0.101 | 0.055635555 | 0.077651543 | 0.105185530 |

TABLE 1a:  CLUSTERING PROBABILITY  P(n;N,p)

n = 7

| p | N =18 | N =19 | N =20 |
|---|---|---|---|
| 0.100 | 0.053221062 | 0.074386063 | 0.100908766 |
| 0.099 | 0.050879614 | 0.071213126 | 0.096744418 |
| 0.098 | 0.048610270 | 0.068131872 | 0.092691928 |
| 0.097 | 0.046412056 | 0.065141394 | 0.088750662 |
| 0.096 | 0.044283972 | 0.062240732 | 0.084919904 |
| 0.095 | 0.042224994 | 0.059428881 | 0.081198863 |
| 0.094 | 0.040234068 | 0.056704791 | 0.077586670 |
| 0.093 | 0.038310121 | 0.054067366 | 0.074082386 |
| 0.092 | 0.036452053 | 0.051515466 | 0.070685000 |
| 0.091 | 0.034658745 | 0.049047913 | 0.067393432 |
| 0.090 | 0.032929058 | 0.046663488 | 0.064206533 |
| 0.089 | 0.031261832 | 0.044360935 | 0.061123091 |
| 0.088 | 0.029655891 | 0.042138961 | 0.058141831 |
| 0.087 | 0.028110042 | 0.039996240 | 0.055261418 |
| 0.086 | 0.026623078 | 0.037931412 | 0.052480458 |
| 0.085 | 0.025193776 | 0.035943089 | 0.049797501 |
| 0.084 | 0.023820903 | 0.034029854 | 0.047211046 |
| 0.083 | 0.022503214 | 0.032190260 | 0.044719538 |
| 0.082 | 0.021239453 | 0.030422839 | 0.042321376 |
| 0.081 | 0.020028358 | 0.028726098 | 0.040014913 |
| 0.080 | 0.018868658 | 0.027098524 | 0.037798458 |
| 0.079 | 0.017759076 | 0.025538584 | 0.035670280 |
| 0.078 | 0.016698332 | 0.024044728 | 0.033628612 |
| 0.077 | 0.015685143 | 0.022615391 | 0.031671648 |
| 0.076 | 0.014718222 | 0.021248994 | 0.029797554 |
| 0.075 | 0.013796284 | 0.019943947 | 0.028004463 |
| 0.074 | 0.012913042 | 0.018698651 | 0.026290483 |
| 0.073 | 0.012032215 | 0.017511500 | 0.024653696 |
| 0.072 | 0.011287521 | 0.016380879 | 0.023092166 |
| 0.071 | 0.010532695 | 0.015305175 | 0.021603935 |
| 0.070 | 0.009816439 | 0.014282768 | 0.020187031 |
| 0.069 | 0.009137519 | 0.013312042 | 0.018839469 |
| 0.068 | 0.008494672 | 0.012391380 | 0.017559254 |
| 0.067 | 0.007886653 | 0.011519172 | 0.016344381 |
| 0.066 | 0.007312228 | 0.010693811 | 0.015192845 |
| 0.065 | 0.006770175 | 0.009913699 | 0.014102635 |
| 0.064 | 0.006259286 | 0.009177248 | 0.013071744 |
| 0.063 | 0.005778365 | 0.008482880 | 0.012098165 |
| 0.062 | 0.005326232 | 0.007829031 | 0.011179900 |
| 0.061 | 0.004901724 | 0.007214150 | 0.010314959 |
| 0.060 | 0.004503694 | 0.006636705 | 0.009501364 |
| 0.059 | 0.004131014 | 0.006095179 | 0.008737149 |
| 0.058 | 0.003782574 | 0.005588078 | 0.008020367 |
| 0.057 | 0.003457286 | 0.005113925 | 0.007349088 |
| 0.056 | 0.003154080 | 0.004671268 | 0.006721404 |
| 0.055 | 0.002871911 | 0.004258630 | 0.006135431 |
| 0.054 | 0.002609753 | 0.003874756 | 0.005589310 |
| 0.053 | 0.002366606 | 0.003518122 | 0.005081212 |
| 0.052 | 0.002141493 | 0.003187429 | 0.004609335 |
| 0.051 | 0.001933461 | 0.002881358 | 0.004171914 |

NEFF and NAUS

TABLE 1a:   CLUSTERING PROBABILITY   P(n;N,p)

n = 7

| p | N =18 | N =19 | N =20 |
|---|---|---|---|
| 0.050 | 0.001741582 | 0.002598620 | 0.003767213 |
| 0.049 | 0.001564956 | 0.002337959 | 0.003393535 |
| 0.048 | 0.001402706 | 0.002098150 | 0.003049223 |
| 0.047 | 0.001253984 | 0.001878002 | 0.002732654 |
| 0.046 | 0.001117969 | 0.001676358 | 0.002442252 |
| 0.045 | 0.000993865 | 0.001492097 | 0.002176480 |
| 0.044 | 0.000880908 | 0.001324132 | 0.001933846 |
| 0.043 | 0.000778357 | 0.001171414 | 0.001712904 |
| 0.042 | 0.000685504 | 0.001032931 | 0.001512254 |
| 0.041 | 0.000601667 | 0.000907707 | 0.001330544 |
| 0.040 | 0.000526191 | 0.000794806 | 0.001166470 |
| 0.039 | 0.000458452 | 0.000693327 | 0.001018777 |
| 0.038 | 0.000397853 | 0.000602412 | 0.000886261 |
| 0.037 | 0.000343826 | 0.000521236 | 0.000767766 |
| 0.036 | 0.000295831 | 0.000449018 | 0.000662191 |
| 0.035 | 0.000253354 | 0.000385010 | 0.000568482 |
| 0.034 | 0.000215912 | 0.000328507 | 0.000485637 |
| 0.033 | 0.000183048 | 0.000278839 | 0.000412709 |
| 0.032 | 0.000154330 | 0.000235374 | 0.000348796 |
| 0.031 | 0.000129354 | 0.000197520 | 0.000293052 |
| 0.030 | 0.000107744 | 0.000164718 | 0.000244678 |
| 0.029 | 0.000089145 | 0.000136448 | 0.000202927 |
| 0.028 | 0.000073232 | 0.000112223 | 0.000167100 |
| 0.027 | 0.000059699 | 0.000091594 | 0.000136545 |
| 0.026 | 0.000048266 | 0.000074142 | 0.000110660 |
| 0.025 | 0.000038678 | 0.000059483 | 0.000088887 |
| 0.024 | 0.000030697 | 0.000047266 | 0.000070714 |
| 0.023 | 0.000024110 | 0.000037167 | 0.000055671 |
| 0.022 | 0.000018722 | 0.000028896 | 0.000043333 |
| 0.021 | 0.000014359 | 0.000022188 | 0.000033313 |
| 0.020 | 0.000010864 | 0.000016806 | 0.000025263 |
| 0.019 | 0.000008096 | 0.000012540 | 0.000018872 |
| 0.018 | 0.000005934 | 0.000009202 | 0.000013865 |
| 0.017 | 0.000004270 | 0.000006628 | 0.000009999 |
| 0.016 | 0.000003009 | 0.000004676 | 0.000007062 |
| 0.015 | 0.000002071 | 0.000003222 | 0.000004872 |
| 0.014 | 0.000001388 | 0.000002162 | 0.000003273 |
| 0.013 | 0.000000902 | 0.000001407 | 0.000002132 |
| 0.012 | 0.000000566 | 0.000000883 | 0.000001340 |
| 0.011 | 0.000000340 | 0.000000532 | 0.000000808 |
| 0.010 | 0.000000195 | 0.000000305 | 0.000000463 |
| 0.009 | 0.000000105 | 0.000000164 | 0.000000250 |
| 0.008 | 0.000000052 | 0.000000082 | 0.000000125 |
| 0.007 | 0.000000024 | 0.000000037 | 0.000000057 |
| 0.006 | 0.000000010 | 0.000000015 | 0.000000023 |
| 0.005 | 0.000000003 | 0.000000005 | 0.000000008 |
| 0.004 | 0.000000001 | 0.000000001 | 0.000000002 |
| 0.003 | 0.000000000 | 0.000000000 | 0.000000000 |

TABLE 1a:   CLUSTERING PROBABILITY   P(n;N,p)

n = 8

| P | N≡16 | N≡17 | N≡18 | N≡19 |
|------|-------------|-------------|-------------|-------------|
| 0.450 | 0.999126351 | 0.999976094 | 0.999999708 | 0.999999998 |
| 0.449 | 0.999029105 | 0.999971941 | 0.999999636 | 0.999999998 |
| 0.448 | 0.998923548 | 0.999967181 | 0.999999549 | 0.999999997 |
| 0.447 | 0.998809192 | 0.999961744 | 0.999999443 | 0.999999997 |
| 0.446 | 0.998685533 | 0.999955550 | 0.999999316 | 0.999999995 |
| 0.445 | 0.998552052 | 0.999948514 | 0.999999163 | 0.999999994 |
| 0.444 | 0.998408219 | 0.999940545 | 0.999998980 | 0.999999992 |
| 0.443 | 0.998253489 | 0.999931543 | 0.999998761 | 0.999999990 |
| 0.442 | 0.998087304 | 0.999921400 | 0.999998501 | 0.999999988 |
| 0.441 | 0.997909094 | 0.999909998 | 0.999998194 | 0.999999984 |
| 0.440 | 0.997718275 | 0.999897214 | 0.999997830 | 0.999999980 |
| 0.439 | 0.997514252 | 0.999882912 | 0.999997403 | 0.999999975 |
| 0.438 | 0.997296420 | 0.999866949 | 0.999996902 | 0.999999968 |
| 0.437 | 0.997064162 | 0.999849171 | 0.999996316 | 0.999999960 |
| 0.436 | 0.996816848 | 0.999829412 | 0.999995632 | 0.999999950 |
| 0.435 | 0.996553843 | 0.999807497 | 0.999994837 | 0.999999938 |
| 0.434 | 0.996274497 | 0.999783240 | 0.999993915 | 0.999999923 |
| 0.433 | 0.995978155 | 0.999756441 | 0.999992848 | 0.999999905 |
| 0.432 | 0.995664152 | 0.999726890 | 0.999991617 | 0.999999884 |
| 0.431 | 0.995331814 | 0.999694365 | 0.999990201 | 0.999999857 |
| 0.430 | 0.994980461 | 0.999658628 | 0.999988575 | 0.999999826 |
| 0.429 | 0.994609407 | 0.999619431 | 0.999986713 | 0.999999788 |
| 0.428 | 0.994217957 | 0.999576510 | 0.999984585 | 0.999999742 |
| 0.427 | 0.993805414 | 0.999529589 | 0.999982159 | 0.999999688 |
| 0.426 | 0.993371072 | 0.999478377 | 0.999979399 | 0.999999623 |
| 0.425 | 0.992914223 | 0.999422568 | 0.999976266 | 0.999999547 |
| 0.424 | 0.992434156 | 0.999361841 | 0.999972716 | 0.999999456 |
| 0.423 | 0.991930154 | 0.999295861 | 0.999968702 | 0.999999348 |
| 0.422 | 0.991401501 | 0.999224276 | 0.999964174 | 0.999999222 |
| 0.421 | 0.990847476 | 0.999146718 | 0.999959073 | 0.999999073 |
| 0.420 | 0.990267359 | 0.999062806 | 0.999953340 | 0.999998899 |
| 0.419 | 0.989660429 | 0.998972139 | 0.999946907 | 0.999998694 |
| 0.418 | 0.989025963 | 0.998874303 | 0.999939702 | 0.999998456 |
| 0.417 | 0.988363242 | 0.998768864 | 0.999931646 | 0.999998179 |
| 0.416 | 0.987671546 | 0.998655376 | 0.999922655 | 0.999997856 |
| 0.415 | 0.986950158 | 0.998533372 | 0.999912636 | 0.999997482 |
| 0.414 | 0.986198364 | 0.998402370 | 0.999901490 | 0.999997049 |
| 0.413 | 0.985415454 | 0.998261871 | 0.999889111 | 0.999996549 |
| 0.412 | 0.984600719 | 0.998111360 | 0.999875383 | 0.999995972 |
| 0.411 | 0.983753459 | 0.997950303 | 0.999860183 | 0.999995308 |
| 0.410 | 0.982872975 | 0.997778151 | 0.999843379 | 0.999994546 |
| 0.409 | 0.981958577 | 0.997594337 | 0.999824828 | 0.999993673 |
| 0.408 | 0.981009580 | 0.997398279 | 0.999804379 | 0.999992673 |
| 0.407 | 0.980025307 | 0.997189377 | 0.999781869 | 0.999991531 |
| 0.406 | 0.979005089 | 0.996967014 | 0.999757125 | 0.999990228 |
| 0.405 | 0.977948263 | 0.996730559 | 0.999729963 | 0.999988746 |
| 0.404 | 0.976854179 | 0.996479364 | 0.999700186 | 0.999987061 |
| 0.403 | 0.975722193 | 0.996212765 | 0.999667585 | 0.999985150 |
| 0.402 | 0.974551672 | 0.995930082 | 0.999631938 | 0.999982985 |
| 0.401 | 0.973341995 | 0.995630622 | 0.999593011 | 0.999980537 |

TABLE 1a:  CLUSTERING PROBABILITY  P(n;N,p)

n = 8

| P | N =16 | N =17 | N =18 | N =19 |
|---|---|---|---|---|
| 0.400 | 0.972092552 | 0.995313676 | 0.999550554 | 0.999977772 |
| 0.399 | 0.970802743 | 0.994978521 | 0.999504304 | 0.999974655 |
| 0.398 | 0.969471981 | 0.994624420 | 0.999453983 | 0.999971146 |
| 0.397 | 0.968099694 | 0.994250622 | 0.999399296 | 0.999967202 |
| 0.396 | 0.966685320 | 0.993856364 | 0.999339934 | 0.999962775 |
| 0.395 | 0.965228313 | 0.993440872 | 0.999275572 | 0.999957812 |
| 0.394 | 0.963728140 | 0.993003357 | 0.999205865 | 0.999952258 |
| 0.393 | 0.962184283 | 0.992543020 | 0.999130455 | 0.999946050 |
| 0.392 | 0.960596240 | 0.992059054 | 0.999048963 | 0.999939121 |
| 0.391 | 0.958963523 | 0.991550638 | 0.998960993 | 0.999931397 |
| 0.390 | 0.957285662 | 0.991016946 | 0.998866130 | 0.999922798 |
| 0.389 | 0.955562201 | 0.990457139 | 0.998763941 | 0.999913238 |
| 0.388 | 0.953792702 | 0.989870373 | 0.998653972 | 0.999902623 |
| 0.387 | 0.951976744 | 0.989255798 | 0.998535750 | 0.999890851 |
| 0.386 | 0.950113924 | 0.988612554 | 0.998408783 | 0.999877813 |
| 0.385 | 0.948203854 | 0.987939779 | 0.998272556 | 0.999863390 |
| 0.384 | 0.946246168 | 0.987236605 | 0.998126536 | 0.999847454 |
| 0.383 | 0.944240515 | 0.986502159 | 0.997970168 | 0.999829867 |
| 0.382 | 0.942186565 | 0.985735569 | 0.997802875 | 0.999810482 |
| 0.381 | 0.940084004 | 0.984935957 | 0.997624061 | 0.999789139 |
| 0.380 | 0.937932539 | 0.984102445 | 0.997433106 | 0.999765667 |
| 0.379 | 0.935731897 | 0.983234156 | 0.997229370 | 0.999739884 |
| 0.378 | 0.933481822 | 0.982330212 | 0.997012191 | 0.999711592 |
| 0.377 | 0.931182079 | 0.981389738 | 0.996780886 | 0.999680584 |
| 0.376 | 0.928832455 | 0.980411862 | 0.996534749 | 0.999646633 |
| 0.375 | 0.926432752 | 0.979395713 | 0.996273053 | 0.999609503 |
| 0.374 | 0.923982795 | 0.978340429 | 0.995995051 | 0.999568937 |
| 0.373 | 0.921482431 | 0.977245148 | 0.995699974 | 0.999524665 |
| 0.372 | 0.918931523 | 0.976109019 | 0.995387030 | 0.999476399 |
| 0.371 | 0.916329958 | 0.974931196 | 0.995055409 | 0.999423832 |
| 0.370 | 0.913677641 | 0.973710843 | 0.994704281 | 0.999366639 |
| 0.369 | 0.910974499 | 0.972447131 | 0.994332792 | 0.999304477 |
| 0.368 | 0.908220479 | 0.971139245 | 0.993940073 | 0.999236980 |
| 0.367 | 0.905415548 | 0.969786378 | 0.993525234 | 0.999163763 |
| 0.366 | 0.902559694 | 0.968387736 | 0.993087366 | 0.999084420 |
| 0.365 | 0.899652926 | 0.966942539 | 0.992625542 | 0.998998521 |
| 0.364 | 0.896695271 | 0.965450020 | 0.992138819 | 0.998905613 |
| 0.363 | 0.893686780 | 0.963909428 | 0.991626236 | 0.998805220 |
| 0.362 | 0.890627523 | 0.962320028 | 0.991086818 | 0.998696842 |
| 0.361 | 0.887517588 | 0.960681101 | 0.990519573 | 0.998579953 |
| 0.360 | 0.884357086 | 0.958991945 | 0.989923496 | 0.998454000 |
| 0.359 | 0.881146147 | 0.957251878 | 0.989297569 | 0.998318407 |
| 0.358 | 0.877884922 | 0.955460236 | 0.988640762 | 0.998172569 |
| 0.357 | 0.874573579 | 0.953616375 | 0.987952032 | 0.998015852 |
| 0.356 | 0.871212310 | 0.951719673 | 0.987230329 | 0.997847598 |
| 0.355 | 0.867801323 | 0.949769528 | 0.986474591 | 0.997667116 |
| 0.354 | 0.864340846 | 0.947765362 | 0.985683751 | 0.997473691 |
| 0.353 | 0.860831129 | 0.945706618 | 0.984856732 | 0.997266576 |
| 0.352 | 0.857272437 | 0.943592763 | 0.983992456 | 0.997044995 |
| 0.351 | 0.853665056 | 0.941423290 | 0.983089838 | 0.996808144 |

TABLE 1a:   CLUSTERING PROBABILITY   P(n;N,p)

n = 8

| P | N =16 | N =17 | N =18 | N =19 |
|---|---|---|---|---|
| 0.350 | 0.850009292 | 0.939197714 | 0.982147791 | 0.996555188 |
| 0.349 | 0.846305467 | 0.936915577 | 0.981165228 | 0.996285263 |
| 0.348 | 0.842553922 | 0.934576446 | 0.980141059 | 0.995997475 |
| 0.347 | 0.838755017 | 0.932179917 | 0.979074198 | 0.995690902 |
| 0.346 | 0.834909129 | 0.929725608 | 0.977963563 | 0.995364592 |
| 0.345 | 0.831016652 | 0.927213169 | 0.976808073 | 0.995017564 |
| 0.344 | 0.827077998 | 0.924642274 | 0.975606654 | 0.994648810 |
| 0.343 | 0.823093597 | 0.922012628 | 0.974358242 | 0.994257292 |
| 0.342 | 0.819063893 | 0.919323961 | 0.973061777 | 0.993841947 |
| 0.341 | 0.814989351 | 0.916576035 | 0.971716214 | 0.993401685 |
| 0.340 | 0.810870447 | 0.913768638 | 0.970320515 | 0.992935389 |
| 0.339 | 0.806707677 | 0.910901589 | 0.968873659 | 0.992441920 |
| 0.338 | 0.802501550 | 0.907974736 | 0.967374638 | 0.991920114 |
| 0.337 | 0.798252594 | 0.904987956 | 0.965822458 | 0.991368785 |
| 0.336 | 0.793961347 | 0.901941155 | 0.964216146 | 0.990786725 |
| 0.335 | 0.789628367 | 0.898834270 | 0.962554746 | 0.990172708 |
| 0.334 | 0.785254222 | 0.895667267 | 0.960837320 | 0.989525488 |
| 0.333 | 0.780839498 | 0.892440141 | 0.959062954 | 0.988843804 |
| 0.332 | 0.776384793 | 0.889152918 | 0.957230756 | 0.988126379 |
| 0.331 | 0.771890718 | 0.885805653 | 0.955339857 | 0.987371922 |
| 0.330 | 0.767357899 | 0.882398431 | 0.953389414 | 0.986579131 |
| 0.329 | 0.762786973 | 0.878931368 | 0.951378610 | 0.985746692 |
| 0.328 | 0.758178592 | 0.875404606 | 0.949306654 | 0.984873287 |
| 0.327 | 0.753533419 | 0.871818320 | 0.947172785 | 0.983957586 |
| 0.326 | 0.748852128 | 0.868172712 | 0.944976271 | 0.982998259 |
| 0.325 | 0.744135407 | 0.864468013 | 0.942716409 | 0.981993969 |
| 0.324 | 0.739383953 | 0.860704486 | 0.940392529 | 0.980943382 |
| 0.323 | 0.734598476 | 0.856882418 | 0.938003991 | 0.979845162 |
| 0.322 | 0.729779696 | 0.853002129 | 0.935550192 | 0.978697977 |
| 0.321 | 0.724928342 | 0.849063962 | 0.933030558 | 0.977500500 |
| 0.320 | 0.720045155 | 0.845068293 | 0.930444552 | 0.976251409 |
| 0.319 | 0.715130885 | 0.841015522 | 0.927791674 | 0.974949392 |
| 0.318 | 0.710186291 | 0.836906078 | 0.925071455 | 0.973593148 |
| 0.317 | 0.705212142 | 0.832740416 | 0.922283468 | 0.972181385 |
| 0.316 | 0.700209214 | 0.828519018 | 0.919427320 | 0.970712830 |
| 0.315 | 0.695178292 | 0.824242393 | 0.916502655 | 0.969186221 |
| 0.314 | 0.690120171 | 0.819911073 | 0.913509158 | 0.967600319 |
| 0.313 | 0.685035652 | 0.815525619 | 0.910446550 | 0.965953900 |
| 0.312 | 0.679925541 | 0.811086615 | 0.907314591 | 0.964245764 |
| 0.311 | 0.674790656 | 0.806594670 | 0.904113082 | 0.962474735 |
| 0.310 | 0.669631818 | 0.802050415 | 0.900841861 | 0.960639660 |
| 0.309 | 0.664449855 | 0.797454509 | 0.897500809 | 0.958739415 |
| 0.308 | 0.659245603 | 0.792807631 | 0.894089843 | 0.956772901 |
| 0.307 | 0.654019900 | 0.788110483 | 0.890608923 | 0.954739053 |
| 0.306 | 0.648773593 | 0.783363791 | 0.887058048 | 0.952636835 |
| 0.305 | 0.643507531 | 0.778568300 | 0.883437257 | 0.950465244 |
| 0.304 | 0.638222570 | 0.773724779 | 0.879746630 | 0.948223315 |
| 0.303 | 0.632919570 | 0.768834017 | 0.875986288 | 0.945910115 |
| 0.302 | 0.627599393 | 0.763896822 | 0.872156389 | 0.943524751 |
| 0.301 | 0.622262907 | 0.758914022 | 0.868257134 | 0.941066368 |

NEFF and NAUS


TABLE 1a:   CLUSTERING PROBABILITY   $P(n;N,p)$


$$n = 8$$

| P | N ≡ 16 | N ≡ 17 | N ≡ 18 | N ≡ 19 |
|---|---|---|---|---|
| 0.300 | 0.61691C982 | 0.753886467 | 0.864288764 | 0.938534150 |
| 0.299 | 0.611544491 | 0.748815021 | 0.860251557 | 0.935927325 |
| C.298 | 0.606164311 | 0.743700570 | 0.856145833 | 0.933245162 |
| 0.297 | 0.600771320 | 0.738544015 | 0.851971951 | 0.930486972 |
| 0.296 | 0.595366398 | 0.733346277 | 0.847730308 | 0.927652114 |
| 0.295 | 0.589950428 | 0.728108290 | 0.843421341 | 0.924739991 |
| 0.294 | 0.584524293 | 0.722831006 | 0.839045523 | 0.921750052 |
| 0.293 | 0.579088877 | 0.717515392 | 0.834603367 | 0.918681794 |
| 0.292 | 0.573645067 | 0.712162429 | 0.830095423 | 0.915534763 |
| C.291 | 0.568193748 | 0.706773114 | 0.825522279 | 0.912308552 |
| C.290 | 0.562735806 | 0.701348456 | 0.820884556 | 0.909002807 |
| 0.289 | 0.557272126 | 0.695889477 | 0.816182916 | 0.905617221 |
| 0.288 | 0.551803593 | C.690397213 | 0.811418053 | 0.902151539 |
| 0.287 | 0.546331093 | 0.684872711 | 0.806590697 | 0.898605558 |
| 0.286 | 0.540855508 | 0.679317029 | 0.801701613 | 0.894979125 |
| 0.285 | 0.535377720 | 0.673731237 | 0.796751598 | 0.891272142 |
| 0.284 | 0.529898610 | 0.668116414 | 0.791741485 | 0.887484561 |
| 0.283 | 0.524419054 | 0.662473648 | 0.786672137 | 0.883616386 |
| 0.282 | 0.518939930 | 0.656804039 | 0.781544450 | 0.879667676 |
| 0.281 | 0.513462110 | 0.651108692 | 0.776359350 | 0.875638541 |
| C.280 | 0.507986464 | 0.645388724 | 0.771117796 | 0.871529146 |
| 0.279 | 0.502513859 | 0.639645254 | 0.765820773 | 0.867339705 |
| 0.278 | 0.497045159 | 0.633879413 | 0.760469299 | 0.863070490 |
| C.277 | 0.491581223 | 0.628092335 | 0.755064416 | 0.858721821 |
| 0.276 | 0.486122908 | 0.622285161 | 0.749607196 | 0.854294074 |
| 0.275 | 0.480671064 | 0.616459037 | 0.744098739 | 0.849787673 |
| 0.274 | 0.475226538 | 0.610615114 | 0.738540167 | 0.845203099 |
| 0.273 | 0.469790171 | 0.604754545 | 0.732932630 | 0.840540880 |
| 0.272 | 0.464362802 | 0.598878489 | 0.727277300 | 0.835801599 |
| 0.271 | 0.458945260 | 0.592988106 | 0.721575375 | 0.830985885 |
| C.270 | 0.453538371 | 0.587084559 | 0.715828074 | 0.826094422 |
| 0.269 | 0.448142955 | 0.581169015 | 0.710036637 | 0.821127941 |
| 0.268 | 0.442759825 | 0.575242638 | 0.704202326 | 0.816087221 |
| 0.267 | 0.437389789 | 0.569306596 | 0.698326422 | 0.810973092 |
| C.266 | 0.432033645 | 0.563362057 | 0.692410226 | 0.805786428 |
| 0.265 | 0.426692189 | 0.557410187 | 0.686455057 | 0.800528154 |
| C.264 | 0.421366205 | 0.551452153 | 0.680462250 | 0.795199238 |
| 0.263 | 0.416056472 | 0.545489120 | 0.674433159 | 0.789800694 |
| 0.262 | 0.410763762 | 0.539522251 | 0.668369152 | 0.784333581 |
| C.261 | 0.405488837 | 0.533552706 | 0.662271610 | 0.778799000 |
| 0.260 | 0.400232452 | 0.527581645 | 0.656141932 | 0.773198096 |
| 0.259 | 0.394995355 | 0.521610221 | 0.649981526 | 0.767532054 |
| 0.258 | 0.389778282 | 0.515639586 | 0.643791813 | 0.761802102 |
| 0.257 | 0.384581963 | 0.509670887 | 0.637574227 | 0.756009505 |
| C.256 | 0.379407119 | 0.503705266 | 0.631330210 | 0.750155567 |
| 0.255 | 0.374254460 | 0.497743859 | 0.625061215 | 0.744241630 |
| 0.254 | 0.369124689 | 0.491787797 | 0.618768702 | 0.738269072 |
| 0.253 | 0.364018496 | 0.485838207 | 0.612454138 | 0.732239306 |
| 0.252 | 0.358936565 | 0.479966206 | 0.606119001 | 0.726153779 |
| 0.251 | 0.353879567 | 0.473962905 | 0.599764769 | 0.720013970 |

TABLE 1a:   CLUSTERING PROBABILITY   P(n;N,p)

n = 8

| P | N =16 | N =17 | N =18 | N =19 |
|---|---|---|---|---|
| 0.250 | 0.348848165 | 0.468039410 | 0.593392930 | 0.713821391 |
| 0.249 | 0.343843011 | 0.462126817 | 0.587004973 | 0.707577583 |
| 0.248 | 0.338864745 | 0.456226213 | 0.580602391 | 0.701284117 |
| 0.247 | 0.333913998 | 0.450338677 | 0.574186680 | 0.694942591 |
| 0.246 | 0.328991389 | 0.444465281 | 0.567759338 | 0.688554631 |
| 0.245 | 0.324097528 | 0.438607084 | 0.561321862 | 0.682121886 |
| 0.244 | 0.319233012 | 0.432765138 | 0.554875751 | 0.675646033 |
| 0.243 | 0.314398426 | 0.426940482 | 0.548422501 | 0.669128767 |
| 0.242 | 0.309594346 | 0.421134148 | 0.541963609 | 0.662571807 |
| 0.241 | 0.304821334 | 0.415347154 | 0.535500568 | 0.655976893 |
| 0.240 | 0.300079941 | 0.409580507 | 0.529034867 | 0.649345783 |
| 0.239 | 0.295370707 | 0.403835204 | 0.522567991 | 0.642680250 |
| 0.238 | 0.290694159 | 0.398112227 | 0.516101423 | 0.635982088 |
| 0.237 | 0.286050813 | 0.392412549 | 0.509636637 | 0.629253103 |
| 0.236 | 0.281441170 | 0.386737127 | 0.503175103 | 0.622495115 |
| 0.235 | 0.276865721 | 0.381086907 | 0.496718282 | 0.615709955 |
| 0.234 | 0.272324946 | 0.375462822 | 0.490267629 | 0.608899469 |
| 0.233 | 0.267819307 | 0.369865788 | 0.483824589 | 0.602065509 |
| 0.232 | 0.263349260 | 0.364296712 | 0.477390599 | 0.595209936 |
| 0.231 | 0.258915243 | 0.358756482 | 0.470967087 | 0.588334619 |
| 0.230 | 0.254517684 | 0.353245974 | 0.464555467 | 0.581441433 |
| 0.229 | 0.250156996 | 0.347766048 | 0.458157145 | 0.574532256 |
| 0.228 | 0.245833581 | 0.342317551 | 0.451773513 | 0.567608972 |
| 0.227 | 0.241547828 | 0.336901311 | 0.445405953 | 0.560673465 |
| 0.226 | 0.237300110 | 0.331518144 | 0.439055832 | 0.553727619 |
| 0.225 | 0.233090790 | 0.326168848 | 0.432724503 | 0.546773320 |
| 0.224 | 0.228920216 | 0.320854205 | 0.426413305 | 0.539812451 |
| 0.223 | 0.224788723 | 0.315574981 | 0.420123562 | 0.532846891 |
| 0.222 | 0.220696634 | 0.310331927 | 0.413856584 | 0.525878517 |
| 0.221 | 0.216644256 | 0.305125773 | 0.407613662 | 0.518909201 |
| 0.220 | 0.212631884 | 0.299957237 | 0.401396072 | 0.511940805 |
| 0.219 | 0.208659801 | 0.294827017 | 0.395205074 | 0.504975188 |
| 0.218 | 0.204728274 | 0.289735795 | 0.389041909 | 0.498014198 |
| 0.217 | 0.200837559 | 0.284684233 | 0.382907798 | 0.491059672 |
| 0.216 | 0.196987895 | 0.279672978 | 0.376803946 | 0.484113439 |
| 0.215 | 0.193179511 | 0.274702659 | 0.370731538 | 0.477177314 |
| 0.214 | 0.189412622 | 0.269773886 | 0.364691740 | 0.470253100 |
| 0.213 | 0.185687427 | 0.264887251 | 0.358685696 | 0.463342585 |
| 0.212 | 0.182004115 | 0.260043327 | 0.352714531 | 0.456447543 |
| 0.211 | 0.178362859 | 0.255242672 | 0.346779349 | 0.449569730 |
| 0.210 | 0.174763819 | 0.250485821 | 0.340881231 | 0.442710888 |
| 0.209 | 0.171207143 | 0.245773294 | 0.335021240 | 0.435872738 |
| 0.208 | 0.167692964 | 0.241105590 | 0.329200413 | 0.429056984 |
| 0.207 | 0.164221404 | 0.236483190 | 0.323419765 | 0.422265309 |
| 0.206 | 0.160792568 | 0.231906558 | 0.317680292 | 0.415499377 |
| 0.205 | 0.157406551 | 0.227376136 | 0.311982962 | 0.408760827 |
| 0.204 | 0.154063434 | 0.222892350 | 0.306328721 | 0.402051280 |
| 0.203 | 0.150763284 | 0.218455604 | 0.300718494 | 0.395372330 |
| 0.202 | 0.147506157 | 0.214066286 | 0.295153179 | 0.388725549 |
| 0.201 | 0.144292093 | 0.209724763 | 0.289633650 | 0.382112483 |

NEFF and NAUS


              TABLE 1a:   CLUSTERING PROBABILITY   P(n;N,p)


                                                                        n = 8

| P | N = 16 | N = 17 | N = 18 | N = 19 |
|---|---|---|---|---|
| 0.200 | 0.141121123 | 0.205431383 | 0.284160757 | 0.375534654 |
| 0.199 | 0.137993261 | 0.201186477 | 0.278735326 | 0.368993557 |
| 0.198 | 0.134908512 | 0.196990354 | 0.273358156 | 0.362490658 |
| 0.197 | 0.131866866 | 0.192843305 | 0.268030022 | 0.356027399 |
| 0.196 | 0.128868302 | 0.188745603 | 0.262751675 | 0.349605192 |
| 0.195 | 0.125912785 | 0.184697502 | 0.257523838 | 0.343225418 |
| 0.194 | 0.123000269 | 0.180699235 | 0.252347208 | 0.336889432 |
| 0.193 | 0.120130695 | 0.176751018 | 0.247222459 | 0.330598558 |
| 0.192 | 0.117303994 | 0.172853047 | 0.242150237 | 0.324354087 |
| 0.191 | 0.114520082 | 0.169005499 | 0.237131161 | 0.318157281 |
| 0.190 | 0.111778865 | 0.165208535 | 0.232165825 | 0.312009370 |
| 0.189 | 0.109080238 | 0.161462292 | 0.227254797 | 0.305911553 |
| 0.188 | 0.106424082 | 0.157766894 | 0.222398618 | 0.299864994 |
| 0.187 | 0.103810268 | 0.154122444 | 0.217597803 | 0.293870826 |
| 0.186 | 0.101238657 | 0.150529025 | 0.212852839 | 0.287930147 |
| 0.185 | 0.098709097 | 0.146986704 | 0.208164188 | 0.282044023 |
| 0.184 | 0.096221427 | 0.143495529 | 0.203532286 | 0.276213484 |
| 0.183 | 0.093775472 | 0.140055531 | 0.198957540 | 0.270439528 |
| 0.182 | 0.091371049 | 0.136666722 | 0.194440333 | 0.264723117 |
| 0.181 | 0.089007965 | 0.133329096 | 0.189981020 | 0.259065177 |
| 0.180 | 0.086686015 | 0.130042631 | 0.185579929 | 0.253466600 |
| 0.179 | 0.084404984 | 0.126807284 | 0.181237365 | 0.247928243 |
| 0.178 | 0.082164649 | 0.123623000 | 0.176953603 | 0.242450926 |
| 0.177 | 0.079964774 | 0.120489703 | 0.172728892 | 0.237035434 |
| 0.176 | 0.077805117 | 0.117407300 | 0.168563458 | 0.231682518 |
| 0.175 | 0.075685424 | 0.114375685 | 0.164457499 | 0.226392888 |
| 0.174 | 0.073605433 | 0.111394730 | 0.160411185 | 0.221167223 |
| 0.173 | 0.071564873 | 0.108464296 | 0.156424664 | 0.216006163 |
| 0.172 | 0.069563463 | 0.105584225 | 0.152498057 | 0.210910313 |
| 0.171 | 0.067600916 | 0.102754343 | 0.148631459 | 0.205880241 |
| 0.170 | 0.065676934 | 0.099974462 | 0.144824940 | 0.200916478 |
| 0.169 | 0.063791212 | 0.097244377 | 0.141078545 | 0.196019520 |
| 0.168 | 0.061943438 | 0.094563869 | 0.137392296 | 0.191189828 |
| 0.167 | 0.060133289 | 0.091932703 | 0.133766188 | 0.186427823 |
| 0.166 | 0.058360438 | 0.089350630 | 0.130200192 | 0.181733895 |
| 0.165 | 0.056624550 | 0.086817386 | 0.126694257 | 0.177108393 |
| 0.164 | 0.054925281 | 0.084332695 | 0.123248305 | 0.172551634 |
| 0.163 | 0.053262281 | 0.081896264 | 0.119862237 | 0.168063898 |
| 0.162 | 0.051635195 | 0.079507789 | 0.116535930 | 0.163645430 |
| 0.161 | 0.050043660 | 0.077166951 | 0.113269238 | 0.159296439 |
| 0.160 | 0.048487306 | 0.074873418 | 0.110061992 | 0.155017100 |
| 0.159 | 0.046965760 | 0.072626847 | 0.106914001 | 0.150807553 |
| 0.158 | 0.045478639 | 0.070426882 | 0.103825052 | 0.146667902 |
| 0.157 | 0.044025559 | 0.068273152 | 0.100794910 | 0.142598221 |
| 0.156 | 0.042606127 | 0.066165279 | 0.097823320 | 0.138598545 |
| 0.155 | 0.041219948 | 0.064102870 | 0.094910004 | 0.134668879 |
| 0.154 | 0.039866620 | 0.062085522 | 0.092054665 | 0.130809193 |
| 0.153 | 0.038545737 | 0.060112821 | 0.089256985 | 0.127019427 |
| 0.152 | 0.037256889 | 0.058184343 | 0.086516628 | 0.123299485 |
| 0.151 | 0.035999661 | 0.056299654 | 0.083833235 | 0.119649241 |

TABLE 1a:   CLUSTERING PROBABILITY  P(n;N,p)

n = 8

| P | N = 16 | N = 17 | N = 18 | N = 19 |
|---|---|---|---|---|
| 0.150 | 0.034773637 | 0.054458308 | 0.081206432 | 0.116068538 |
| 0.149 | 0.033578393 | 0.052659851 | 0.078635824 | 0.112557187 |
| 0.148 | 0.032413506 | 0.050903822 | 0.076120999 | 0.109114967 |
| 0.147 | 0.031278545 | 0.049189747 | 0.073661526 | 0.105741631 |
| 0.146 | 0.030173082 | 0.047517146 | 0.071256959 | 0.102436898 |
| 0.145 | 0.029096681 | 0.045885530 | 0.068906833 | 0.099200459 |
| 0.144 | 0.028048906 | 0.044294403 | 0.066610667 | 0.096031979 |
| 0.143 | 0.027029318 | 0.042743262 | 0.064367966 | 0.092931093 |
| 0.142 | 0.026037479 | 0.041231595 | 0.062178218 | 0.089897407 |
| 0.141 | 0.025072944 | 0.039758885 | 0.060040896 | 0.086930502 |
| 0.140 | 0.024135270 | 0.038324607 | 0.057955459 | 0.084029934 |
| 0.139 | 0.023224013 | 0.036928232 | 0.055921353 | 0.081195230 |
| 0.138 | 0.022338727 | 0.035569224 | 0.053938009 | 0.078425895 |
| 0.137 | 0.021478964 | 0.034247041 | 0.052004846 | 0.075721408 |
| 0.136 | 0.020644277 | 0.032961138 | 0.050121270 | 0.073081223 |
| 0.135 | 0.019834219 | 0.031710965 | 0.048286677 | 0.070504774 |
| 0.134 | 0.019048341 | 0.030495966 | 0.046500448 | 0.067991471 |
| 0.133 | 0.018286196 | 0.029315584 | 0.044761957 | 0.065540701 |
| 0.132 | 0.017547336 | 0.028169255 | 0.043070565 | 0.063151832 |
| 0.131 | 0.016831314 | 0.027056415 | 0.041425624 | 0.060824210 |
| 0.130 | 0.016137685 | 0.025976495 | 0.039826478 | 0.058557163 |
| 0.129 | 0.015466002 | 0.024928926 | 0.038272460 | 0.056349997 |
| 0.128 | 0.014815822 | 0.023913134 | 0.036762896 | 0.054202004 |
| 0.127 | 0.014186702 | 0.022928546 | 0.035297104 | 0.052112455 |
| 0.126 | 0.013578200 | 0.021974585 | 0.033874395 | 0.050080605 |
| 0.125 | 0.012989877 | 0.021050675 | 0.032494074 | 0.048105696 |
| 0.124 | 0.012421297 | 0.020156239 | 0.031155438 | 0.046186949 |
| 0.123 | 0.011872023 | 0.019290698 | 0.029857781 | 0.044323576 |
| 0.122 | 0.011341623 | 0.018453476 | 0.028600390 | 0.042514773 |
| 0.121 | 0.010829666 | 0.017643994 | 0.027382549 | 0.040759722 |
| 0.120 | 0.010335724 | 0.016861675 | 0.026203535 | 0.039057594 |
| 0.119 | 0.009859374 | 0.016105945 | 0.025062626 | 0.037407548 |
| 0.118 | 0.009400193 | 0.015376228 | 0.023959093 | 0.035808734 |
| 0.117 | 0.008957763 | 0.014671952 | 0.022892208 | 0.034260290 |
| 0.116 | 0.008531669 | 0.013992546 | 0.021861238 | 0.032761345 |
| 0.115 | 0.008121499 | 0.013337440 | 0.020865451 | 0.031311022 |
| 0.114 | 0.007726847 | 0.012706069 | 0.019904113 | 0.029908434 |
| 0.113 | 0.007347307 | 0.012097870 | 0.018976489 | 0.028552687 |
| 0.112 | 0.006982482 | 0.011512283 | 0.018081847 | 0.027242883 |
| 0.111 | 0.006631975 | 0.010948750 | 0.017219451 | 0.025978117 |
| 0.110 | 0.006295395 | 0.010406718 | 0.016388571 | 0.024757481 |
| 0.109 | 0.005972355 | 0.009885640 | 0.015588475 | 0.023580062 |
| 0.108 | 0.005662474 | 0.009384969 | 0.014818435 | 0.022444944 |
| 0.107 | 0.005365375 | 0.008904167 | 0.014077725 | 0.021351210 |
| 0.106 | 0.005080684 | 0.008442696 | 0.013365621 | 0.020297940 |
| 0.105 | 0.004808035 | 0.008000028 | 0.012681405 | 0.019284214 |
| 0.104 | 0.004547066 | 0.007575637 | 0.012024360 | 0.018309111 |
| 0.103 | 0.004297419 | 0.007169003 | 0.011393775 | 0.017371713 |
| 0.102 | 0.004058742 | 0.006779613 | 0.010788944 | 0.016471102 |
| 0.101 | 0.003830689 | 0.006406959 | 0.010209164 | 0.015606360 |

TABLE 1a:   CLUSTERING PROBABILITY   P(n;N,p)

n = 8

| P | N =16 | N =17 | N =18 | N =19 |
|---|---|---|---|---|
| 0.100 | 0.003612920 | 0.006050539 | 0.009653741 | 0.014776576 |
| 0.099 | 0.003405098 | 0.005709857 | 0.009121985 | 0.013980839 |
| C.098 | 0.003206895 | 0.005384426 | 0.008613211 | 0.013218244 |
| 0.097 | 0.003017987 | 0.005073763 | 0.008126744 | 0.012487889 |
| 0.096 | 0.002838055 | 0.004777394 | 0.007661914 | 0.011788880 |
| 0.095 | 0.002666787 | 0.004494850 | 0.007218058 | 0.011120326 |
| 0.094 | 0.002503878 | 0.004225672 | 0.006794523 | 0.010481346 |
| 0.093 | 0.002349026 | 0.003969405 | 0.006390662 | 0.009871063 |
| 0.092 | 0.002201938 | 0.003725605 | 0.006005838 | 0.009288610 |
| C.091 | 0.002062326 | 0.003493834 | 0.005639422 | 0.008733129 |
| 0.090 | 0.001929907 | 0.003273663 | 0.005290794 | 0.008203767 |
| 0.C89 | 0.001804406 | 0.003064668 | 0.004959344 | 0.007699686 |
| 0.088 | 0.001685553 | 0.002866438 | 0.004644472 | 0.007220053 |
| 0.087 | 0.001573085 | 0.002678566 | 0.004345585 | 0.006764050 |
| 0.C86 | 0.001466744 | 0.002500656 | 0.004062104 | 0.006330866 |
| 0.085 | 0.001366279 | 0.002332317 | 0.003793458 | 0.005919704 |
| 0.084 | 0.001271447 | 0.002173172 | 0.003539087 | 0.005529779 |
| 0.083 | 0.001182007 | 0.002022846 | 0.003298442 | 0.005160317 |
| C.082 | 0.001097728 | 0.001880979 | 0.003070986 | 0.004810556 |
| C.081 | 0.001018384 | 0.001747214 | 0.002856190 | 0.004479750 |
| C.080 | 0.000943755 | 0.001621207 | 0.002653541 | 0.004167165 |
| 0.079 | 0.000873628 | 0.001502620 | 0.002462533 | 0.003872080 |
| 0.078 | 0.000807793 | 0.001391125 | 0.002282673 | 0.003593789 |
| 0.077 | 0.000746052 | 0.001286403 | 0.002113481 | 0.003331601 |
| 0.076 | 0.000688207 | 0.001188142 | 0.001954488 | 0.003084838 |
| 0.075 | 0.000634069 | 0.001096042 | 0.001805236 | 0.002852839 |
| 0.074 | 0.000583457 | 0.001009808 | 0.001665279 | 0.002634958 |
| 0.073 | 0.000536191 | 0.000929156 | 0.001534185 | 0.002430562 |
| 0.072 | 0.000492101 | 0.000853811 | 0.001411532 | 0.002239036 |
| 0.071 | 0.000451021 | 0.000783506 | 0.001296911 | 0.002059781 |
| 0.070 | 0.000412791 | 0.000717981 | 0.001189923 | 0.001892212 |
| 0.069 | 0.000377257 | 0.000656987 | 0.001090185 | 0.001735762 |
| 0.068 | 0.000344272 | 0.000600282 | 0.000997323 | 0.001589878 |
| 0.067 | 0.000313691 | 0.000547634 | 0.000910975 | 0.001454026 |
| 0.066 | 0.000285377 | 0.000498817 | 0.000830794 | 0.001327685 |
| 0.065 | 0.000259199 | 0.000453616 | 0.000756440 | 0.001210353 |
| 0.C64 | 0.000235029 | 0.000411822 | 0.000687589 | 0.001101544 |
| 0.063 | 0.000212746 | 0.000373235 | 0.000623928 | 0.001000786 |
| 0.062 | 0.000192234 | 0.000337662 | 0.000565154 | 0.000907626 |
| 0.061 | 0.000173382 | 0.000304919 | 0.000510976 | 0.000821625 |
| C.060 | 0.000156082 | 0.000274830 | 0.000461117 | 0.000742361 |
| 0.C59 | 0.000140234 | 0.000247225 | 0.000415307 | 0.000669429 |
| C.C58 | 0.000125740 | 0.000221943 | 0.000373291 | 0.000602439 |
| 0.057 | 0.000112509 | 0.000198830 | 0.000334823 | 0.000541016 |
| 0.056 | 0.000100452 | 0.000177737 | 0.000299667 | 0.000484800 |
| 0.055 | 0.000089485 | 0.000158526 | 0.000267601 | 0.000433450 |
| 0.054 | 0.000079531 | 0.000141062 | 0.000238409 | 0.000386636 |
| 0.053 | 0.000070514 | 0.000125219 | 0.000211890 | 0.000344044 |
| 0.C52 | 0.000062362 | 0.000110877 | 0.000187848 | 0.000305377 |
| 0.051 | 0.000055010 | 0.000097923 | 0.000166101 | 0.000270351 |

TABLE 1a:   CLUSTERING PROBABILITY   P(n;N,p)

n = 8

| P | N =16 | N =17 | N =18 | N =19 |
|---|---|---|---|---|
| 0.050 | 0.000048393 | 0.000086248 | 0.000146475 | 0.000238694 |
| 0.049 | 0.000042453 | 0.000075752 | 0.000128804 | 0.000210151 |
| 0.048 | 0.000037133 | 0.000066339 | 0.000112933 | 0.000184479 |
| 0.047 | 0.000032381 | 0.000057918 | 0.000098716 | 0.000161449 |
| 0.046 | 0.000028147 | 0.000050405 | 0.000086015 | 0.000140845 |
| 0.045 | 0.000024386 | 0.000043722 | 0.000074699 | 0.000122462 |
| 0.044 | 0.000021054 | 0.000037793 | 0.000064646 | 0.000106109 |
| 0.043 | 0.000018111 | 0.000032550 | 0.000055744 | 0.000091606 |
| 0.042 | 0.000015521 | 0.000027927 | 0.000047885 | 0.000078785 |
| 0.041 | 0.000013248 | 0.000023866 | 0.000040971 | 0.000067489 |
| 0.040 | 0.000011261 | 0.000020311 | 0.000034908 | 0.000057571 |
| 0.039 | 0.000009530 | 0.000017209 | 0.000029613 | 0.000048896 |
| 0.038 | 0.000008028 | 0.000014514 | 0.000025005 | 0.000041336 |
| 0.037 | 0.000006730 | 0.000012182 | 0.000021011 | 0.000034776 |
| 0.036 | 0.000005613 | 0.000010172 | 0.000017565 | 0.000029107 |
| 0.035 | 0.000004656 | 0.000008448 | 0.000014605 | 0.000024230 |
| 0.034 | 0.000003840 | 0.000006976 | 0.000012074 | 0.000020055 |
| 0.033 | 0.000003148 | 0.000005725 | 0.000009922 | 0.000016499 |
| 0.032 | 0.000002564 | 0.000004669 | 0.000008100 | 0.000013486 |
| 0.031 | 0.000002074 | 0.000003781 | 0.000006568 | 0.000010948 |
| 0.030 | 0.000001666 | 0.000003040 | 0.000005287 | 0.000008823 |
| 0.029 | 0.000001327 | 0.000002425 | 0.000004223 | 0.000007056 |
| 0.028 | 0.000001049 | 0.000001919 | 0.000003345 | 0.000005595 |
| 0.027 | 0.000000822 | 0.000001505 | 0.000002626 | 0.000004397 |
| 0.026 | 0.000000637 | 0.000001168 | 0.000002042 | 0.000003423 |
| 0.025 | 0.000000489 | 0.000000898 | 0.000001571 | 0.000002637 |
| 0.024 | 0.000000371 | 0.000000683 | 0.000001195 | 0.000002009 |
| 0.023 | 0.000000279 | 0.000000512 | 0.000000898 | 0.000001512 |
| 0.022 | 0.000000206 | 0.000000380 | 0.000000666 | 0.000001123 |
| 0.021 | 0.000000150 | 0.000000277 | 0.000000487 | 0.000000822 |
| 0.020 | 0.000000108 | 0.000000199 | 0.000000351 | 0.000000592 |
| 0.019 | 0.000000076 | 0.000000141 | 0.000000248 | 0.000000419 |
| 0.018 | 0.000000053 | 0.000000097 | 0.000000172 | 0.000000291 |
| 0.017 | 0.000000036 | 0.000000066 | 0.000000117 | 0.000000198 |
| 0.016 | 0.000000024 | 0.000000044 | 0.000000077 | 0.000000131 |
| 0.015 | 0.000000015 | 0.000000028 | 0.000000050 | 0.000000085 |
| 0.014 | 0.000000009 | 0.000000018 | 0.000000031 | 0.000000053 |
| 0.013 | 0.000000006 | 0.000000011 | 0.000000019 | 0.000000032 |
| 0.012 | 0.000000003 | 0.000000006 | 0.000000011 | 0.000000018 |
| 0.011 | 0.000000002 | 0.000000003 | 0.000000006 | 0.000000010 |
| 0.010 | 0.000000001 | 0.000000002 | 0.000000003 | 0.000000005 |
| 0.009 | 0.000000000 | 0.000000001 | 0.000000002 | 0.000000003 |
| 0.008 | 0.000000000 | 0.000000000 | 0.000000001 | 0.000000001 |
| 0.007 | 0.000000000 | 0.000000000 | 0.000000000 | 0.000000000 |

TABLE 1a:   CLUSTERING PROBABILITY   P(n;N,p)

n = 8

| P | N = 20 | N = 21 | N = 22 | N = 23 |
|---|---|---|---|---|
| 0.300 | 0.978671378 | 0.995052769 | 0.999464358 | 0.999975025 |
| 0.299 | 0.977378584 | 0.994607930 | 0.999380615 | 0.999968634 |
| 0.298 | 0.976026383 | 0.994130876 | 0.999286490 | 0.999960884 |
| 0.297 | 0.974613204 | 0.993619971 | 0.999181075 | 0.999951545 |
| 0.296 | 0.973137485 | 0.993073544 | 0.999063369 | 0.999940353 |
| 0.295 | 0.971597685 | 0.992489887 | 0.998932312 | 0.999927016 |
| 0.294 | 0.969992278 | 0.991867266 | 0.998786789 | 0.999911205 |
| 0.293 | 0.968319762 | 0.991203915 | 0.998625627 | 0.999892555 |
| 0.292 | 0.966578656 | 0.990498041 | 0.998447594 | 0.999870659 |
| 0.291 | 0.964767506 | 0.989747827 | 0.998251402 | 0.999845069 |
| 0.290 | 0.962884887 | 0.988951435 | 0.998035701 | 0.999815292 |
| 0.289 | 0.960929404 | 0.988107006 | 0.997799087 | 0.999780785 |
| 0.288 | 0.958899695 | 0.987212664 | 0.997540094 | 0.999740956 |
| 0.287 | 0.956794433 | 0.986266522 | 0.997257201 | 0.999695160 |
| 0.286 | 0.954612327 | 0.985266676 | 0.996948830 | 0.999642695 |
| 0.285 | 0.952352129 | 0.984211220 | 0.996613344 | 0.999582803 |
| 0.284 | 0.950012628 | 0.983098237 | 0.996249056 | 0.999514663 |
| 0.283 | 0.947592660 | 0.981925811 | 0.995854221 | 0.999437394 |
| 0.282 | 0.945091104 | 0.980692025 | 0.995427044 | 0.999350047 |
| 0.281 | 0.942506890 | 0.979394966 | 0.994965679 | 0.999251609 |
| 0.280 | 0.939838992 | 0.978032729 | 0.994468230 | 0.999140997 |
| 0.279 | 0.937086441 | 0.976603417 | 0.993932757 | 0.999017059 |
| 0.278 | 0.934248315 | 0.975105148 | 0.993357273 | 0.998878570 |
| 0.277 | 0.931323749 | 0.973536055 | 0.992739750 | 0.998724234 |
| 0.276 | 0.928311936 | 0.971894292 | 0.992078120 | 0.998552680 |
| 0.275 | 0.925212122 | 0.970178035 | 0.991370279 | 0.998362464 |
| 0.274 | 0.922023613 | 0.968385486 | 0.990614090 | 0.998152065 |
| 0.273 | 0.918745778 | 0.966514876 | 0.989807383 | 0.997919889 |
| 0.272 | 0.915378043 | 0.964564469 | 0.988947962 | 0.997664266 |
| 0.271 | 0.911919898 | 0.962532565 | 0.988033608 | 0.997383452 |
| 0.270 | 0.908370896 | 0.960417501 | 0.987062079 | 0.997075629 |
| 0.269 | 0.904730655 | 0.958217656 | 0.986031119 | 0.996738907 |
| 0.268 | 0.900998857 | 0.955931455 | 0.984938457 | 0.996371321 |
| 0.267 | 0.897175250 | 0.953557369 | 0.983781813 | 0.995970841 |
| 0.266 | 0.893259649 | 0.951093921 | 0.982558902 | 0.995535364 |
| 0.265 | 0.889251935 | 0.948539684 | 0.981267438 | 0.995062723 |
| 0.264 | 0.885152058 | 0.945893291 | 0.979905138 | 0.994550686 |
| 0.263 | 0.880960034 | 0.943153430 | 0.978469725 | 0.993996962 |
| 0.262 | 0.876675949 | 0.940318852 | 0.976958935 | 0.993399201 |
| 0.261 | 0.872299956 | 0.937388372 | 0.975370517 | 0.992754996 |
| 0.260 | 0.867832277 | 0.934360867 | 0.973702242 | 0.992061892 |
| 0.259 | 0.863273203 | 0.931235287 | 0.971951903 | 0.991317384 |
| 0.258 | 0.858623092 | 0.928010650 | 0.970117324 | 0.990518924 |
| 0.257 | 0.853882373 | 0.924686044 | 0.968196359 | 0.989663925 |
| 0.256 | 0.849051541 | 0.921260636 | 0.966186899 | 0.988749765 |
| 0.255 | 0.844131159 | 0.917733664 | 0.964086878 | 0.987773790 |
| 0.254 | 0.839121858 | 0.914104446 | 0.961894272 | 0.986733323 |
| 0.253 | 0.834024335 | 0.910372381 | 0.959607110 | 0.985625665 |
| 0.252 | 0.828839355 | 0.906536944 | 0.957223471 | 0.984448100 |
| 0.251 | 0.823567747 | 0.902597696 | 0.954741493 | 0.983197905 |

TABLE 1a:   CLUSTERING PROBABILITY   P(n;N,p)

n = 8

| P | N = 20 | N = 21 | N = 22 | N = 23 |
|---|---|---|---|---|
| 0.250 | 0.818210406 | 0.898554280 | 0.952159374 | 0.981872349 |
| 0.249 | 0.812768291 | 0.894406420 | 0.949475377 | 0.980468702 |
| 0.248 | 0.807242425 | 0.890153931 | 0.946687836 | 0.978984242 |
| 0.247 | 0.801633892 | 0.885796709 | 0.943795153 | 0.977416255 |
| 0.246 | 0.795943840 | 0.881334738 | 0.940795810 | 0.975762048 |
| 0.245 | 0.790173475 | 0.876768090 | 0.937688363 | 0.974018947 |
| 0.244 | 0.784324064 | 0.872096924 | 0.934471456 | 0.972184311 |
| 0.243 | 0.778396930 | 0.867321486 | 0.931143812 | 0.970255528 |
| 0.242 | 0.772393456 | 0.862442110 | 0.927704247 | 0.968230030 |
| 0.241 | 0.766315078 | 0.857459221 | 0.924151667 | 0.966105293 |
| 0.240 | 0.760163288 | 0.852373327 | 0.920485068 | 0.963878843 |
| 0.239 | 0.753939630 | 0.847185028 | 0.916703547 | 0.961548264 |
| 0.238 | 0.747645700 | 0.841895009 | 0.912806296 | 0.959111201 |
| 0.237 | 0.741283144 | 0.836504041 | 0.908792608 | 0.956565369 |
| 0.236 | 0.734853656 | 0.831012985 | 0.904661879 | 0.953908550 |
| 0.235 | 0.728358977 | 0.825422785 | 0.900413609 | 0.951138610 |
| 0.234 | 0.721800894 | 0.819734470 | 0.896047403 | 0.948253494 |
| 0.233 | 0.715181237 | 0.813949152 | 0.891562974 | 0.945251236 |
| 0.232 | 0.708501878 | 0.808068029 | 0.886960143 | 0.942129960 |
| 0.231 | 0.701764729 | 0.802092377 | 0.882238841 | 0.938887890 |
| 0.230 | 0.694971742 | 0.796023555 | 0.877399109 | 0.935523350 |
| 0.229 | 0.688124904 | 0.789863001 | 0.872441098 | 0.932034768 |
| 0.228 | 0.681226238 | 0.783612230 | 0.867365073 | 0.928420683 |
| 0.227 | 0.674277800 | 0.777272833 | 0.862171409 | 0.924679747 |
| 0.226 | 0.667281678 | 0.770846476 | 0.856860594 | 0.920810727 |
| 0.225 | 0.660239989 | 0.764334899 | 0.851433228 | 0.916812513 |
| 0.224 | 0.653154877 | 0.757739910 | 0.845890022 | 0.912684116 |
| 0.223 | 0.646028513 | 0.751063390 | 0.840231801 | 0.908424673 |
| 0.222 | 0.638863092 | 0.744307286 | 0.834459497 | 0.904033452 |
| 0.221 | 0.631660830 | 0.737473608 | 0.828574156 | 0.899509849 |
| 0.220 | 0.624423964 | 0.730564433 | 0.822576932 | 0.894853396 |
| 0.219 | 0.617154750 | 0.723581897 | 0.816469088 | 0.890063759 |
| 0.218 | 0.609855459 | 0.716528194 | 0.810251992 | 0.885140741 |
| 0.217 | 0.602528378 | 0.709405577 | 0.803927120 | 0.880084286 |
| 0.216 | 0.595175806 | 0.702216351 | 0.797496052 | 0.874894473 |
| 0.215 | 0.587800054 | 0.694962876 | 0.790960470 | 0.869571526 |
| 0.214 | 0.580403439 | 0.687647557 | 0.784322157 | 0.864115808 |
| 0.213 | 0.572988288 | 0.680272851 | 0.777582995 | 0.858527824 |
| 0.212 | 0.565556932 | 0.672841257 | 0.770744963 | 0.852808221 |
| 0.211 | 0.558111707 | 0.665355315 | 0.763810134 | 0.846957788 |
| 0.210 | 0.550654949 | 0.657817608 | 0.756780673 | 0.840977456 |
| 0.209 | 0.543188995 | 0.650230754 | 0.749658836 | 0.834868295 |
| 0.208 | 0.535716178 | 0.642597404 | 0.742446965 | 0.828631516 |
| 0.207 | 0.528238829 | 0.634920244 | 0.735147487 | 0.822268472 |
| 0.206 | 0.520759275 | 0.627201987 | 0.727762911 | 0.815780648 |
| 0.205 | 0.513279833 | 0.619445373 | 0.720295824 | 0.809169671 |
| 0.204 | 0.505802812 | 0.611653166 | 0.712748885 | 0.802437298 |
| 0.203 | 0.498330512 | 0.603828152 | 0.705124843 | 0.795585422 |
| 0.202 | 0.490865219 | 0.595973135 | 0.697426491 | 0.788616063 |
| 0.201 | 0.483409206 | 0.588090936 | 0.689656705 | 0.781531372 |

NEFF and NAUS

## TABLE 1a:   CLUSTERING PROBABILITY   P(n;N,p)

n = 8

| P | N = 20 | N = 21 | N = 22 | N = 23 |
|---|---|---|---|---|
| 0.200 | 0.475964731 | 0.580184386 | 0.681818421 | 0.774333625 |
| 0.199 | 0.468534034 | 0.572256332 | 0.673914635 | 0.767025217 |
| 0.198 | 0.461119337 | 0.564309624 | 0.665948396 | 0.759608668 |
| 0.197 | 0.453722844 | 0.556347122 | 0.657922811 | 0.752086611 |
| 0.196 | 0.446346734 | 0.548371686 | 0.649841033 | 0.744461793 |
| 0.195 | 0.438993166 | 0.540386178 | 0.641706261 | 0.736737070 |
| 0.194 | 0.431664273 | 0.532393456 | 0.633521739 | 0.728915406 |
| 0.193 | 0.424362164 | 0.524396376 | 0.625290746 | 0.720999866 |
| 0.192 | 0.417088920 | 0.516397785 | 0.617016600 | 0.712993612 |
| 0.191 | 0.409846594 | 0.508400522 | 0.608702648 | 0.704899903 |
| 0.190 | 0.402637209 | 0.500407412 | 0.600352264 | 0.696722087 |
| 0.189 | 0.395462759 | 0.492421267 | 0.591968851 | 0.688463596 |
| 0.188 | 0.388325203 | 0.484444883 | 0.583555826 | 0.680127946 |
| 0.187 | 0.381226471 | 0.476481036 | 0.575116628 | 0.671718728 |
| 0.186 | 0.374168456 | 0.468532482 | 0.566654708 | 0.663239608 |
| 0.185 | 0.367153017 | 0.460601951 | 0.558173526 | 0.654694316 |
| 0.184 | 0.360181975 | 0.452692151 | 0.549676547 | 0.646086648 |
| 0.183 | 0.353257117 | 0.444805760 | 0.541167243 | 0.637420456 |
| 0.182 | 0.346380189 | 0.436945427 | 0.532649079 | 0.628699648 |
| 0.181 | 0.339552899 | 0.429113769 | 0.524125521 | 0.619928177 |
| 0.180 | 0.332776915 | 0.421313371 | 0.515600023 | 0.611110043 |
| 0.179 | 0.326053866 | 0.413546779 | 0.507076031 | 0.602249281 |
| 0.178 | 0.319385335 | 0.405816504 | 0.498556973 | 0.593349965 |
| 0.177 | 0.312772868 | 0.398125019 | 0.490046262 | 0.584416191 |
| 0.176 | 0.306217964 | 0.390474753 | 0.481547288 | 0.575452085 |
| 0.175 | 0.299722081 | 0.382868094 | 0.473063416 | 0.566461787 |
| 0.174 | 0.293286631 | 0.375307386 | 0.464597986 | 0.557449455 |
| 0.173 | 0.286912982 | 0.367794927 | 0.456154305 | 0.548419253 |
| 0.172 | 0.280602457 | 0.360332966 | 0.447735646 | 0.539375350 |
| 0.171 | 0.274356332 | 0.352923707 | 0.439345248 | 0.530321916 |
| 0.170 | 0.268175837 | 0.345569301 | 0.430986307 | 0.521263112 |
| 0.169 | 0.262062157 | 0.338271848 | 0.422661979 | 0.512203093 |
| 0.168 | 0.256016428 | 0.331033395 | 0.414375376 | 0.503145996 |
| 0.167 | 0.250039739 | 0.323855937 | 0.406129559 | 0.494095939 |
| 0.166 | 0.244133130 | 0.316741413 | 0.397927542 | 0.485057019 |
| 0.165 | 0.238297596 | 0.309691704 | 0.389772284 | 0.476033300 |
| 0.164 | 0.232534082 | 0.302708639 | 0.381666692 | 0.467028816 |
| 0.163 | 0.226843484 | 0.295793983 | 0.373613612 | 0.458047564 |
| 0.162 | 0.221226651 | 0.288949446 | 0.365615834 | 0.449093500 |
| 0.161 | 0.215684382 | 0.282176679 | 0.357676084 | 0.440170532 |
| 0.160 | 0.210217429 | 0.275477269 | 0.349797025 | 0.431282521 |
| 0.159 | 0.204826493 | 0.268852746 | 0.341981255 | 0.422433275 |
| 0.158 | 0.199512228 | 0.262304576 | 0.334231305 | 0.413626544 |
| 0.157 | 0.194275240 | 0.255834163 | 0.326549637 | 0.404866019 |
| 0.156 | 0.189116084 | 0.249442849 | 0.318938640 | 0.396155325 |
| 0.155 | 0.184035269 | 0.243131912 | 0.311400634 | 0.387498021 |
| 0.154 | 0.179033255 | 0.236902569 | 0.303937863 | 0.378897594 |
| 0.153 | 0.174110453 | 0.230755970 | 0.296552498 | 0.370357459 |
| 0.152 | 0.169267228 | 0.224693204 | 0.289246632 | 0.361880952 |
| 0.151 | 0.164503895 | 0.218715293 | 0.282022282 | 0.353471330 |

TABLE 1a:   CLUSTERING PROBABILITY   P(n;N,p)

n = 8

| P | N = 20 | N = 21 | N = 22 | N = 23 |
|---|---|---|---|---|
| 0.150 | 0.159820726 | 0.212823196 | 0.274881385 | 0.345131770 |
| 0.149 | 0.155217941 | 0.207017810 | 0.267825799 | 0.336865359 |
| 0.148 | 0.150695717 | 0.201299963 | 0.260857304 | 0.328675101 |
| 0.147 | 0.146254184 | 0.195670424 | 0.253977596 | 0.320563908 |
| 0.146 | 0.141893425 | 0.190129893 | 0.247188292 | 0.312534600 |
| 0.145 | 0.137613481 | 0.184679009 | 0.240490923 | 0.304589903 |
| 0.144 | 0.133414344 | 0.179318346 | 0.233886941 | 0.296732449 |
| 0.143 | 0.129295966 | 0.174048417 | 0.227377714 | 0.288964769 |
| 0.142 | 0.125258253 | 0.168869668 | 0.220964525 | 0.281289298 |
| 0.141 | 0.121301068 | 0.163782486 | 0.214648575 | 0.273708367 |
| 0.140 | 0.117424233 | 0.158787192 | 0.208430980 | 0.266224207 |
| 0.139 | 0.113627527 | 0.153884050 | 0.202312775 | 0.258838946 |
| 0.138 | 0.109910688 | 0.149073258 | 0.196294907 | 0.251554607 |
| 0.137 | 0.106273414 | 0.144354956 | 0.190378243 | 0.244373107 |
| 0.136 | 0.102715364 | 0.139729225 | 0.184563565 | 0.237296258 |
| 0.135 | 0.099236157 | 0.135196084 | 0.178851574 | 0.230325766 |
| 0.134 | 0.095835374 | 0.130755495 | 0.173242885 | 0.223463229 |
| 0.133 | 0.092512560 | 0.126407363 | 0.167738035 | 0.216710138 |
| 0.132 | 0.089267221 | 0.122151535 | 0.162337476 | 0.210067876 |
| 0.131 | 0.086098830 | 0.117987802 | 0.157041582 | 0.203537720 |
| 0.130 | 0.083006824 | 0.113915900 | 0.151850645 | 0.197120837 |
| 0.129 | 0.079990605 | 0.109935511 | 0.146764879 | 0.190818288 |
| 0.128 | 0.077049545 | 0.106046265 | 0.141784418 | 0.184631027 |
| 0.127 | 0.074182981 | 0.102247738 | 0.136909320 | 0.178559901 |
| 0.126 | 0.071390219 | 0.098539456 | 0.132139567 | 0.172605651 |
| 0.125 | 0.068670538 | 0.094920894 | 0.127475062 | 0.166768912 |
| 0.124 | 0.066023184 | 0.091391481 | 0.122915639 | 0.161050217 |
| 0.123 | 0.063447376 | 0.087950596 | 0.118461056 | 0.155449992 |
| 0.122 | 0.060942307 | 0.084597571 | 0.114110999 | 0.149968563 |
| 0.121 | 0.058507143 | 0.081331695 | 0.109865085 | 0.144606152 |
| 0.120 | 0.056141025 | 0.078152213 | 0.105722861 | 0.139362883 |
| 0.119 | 0.053843068 | 0.075058325 | 0.101683807 | 0.134238780 |
| 0.118 | 0.051612368 | 0.072049193 | 0.097747338 | 0.129233770 |
| 0.117 | 0.049447995 | 0.069123937 | 0.093912802 | 0.124347685 |
| 0.116 | 0.047349001 | 0.066281639 | 0.090179486 | 0.119580260 |
| 0.115 | 0.045314415 | 0.063521343 | 0.086546617 | 0.114931141 |
| 0.114 | 0.043343251 | 0.060842061 | 0.083013359 | 0.110399882 |
| 0.113 | 0.041434502 | 0.058242765 | 0.079578822 | 0.105985947 |
| 0.112 | 0.039587147 | 0.055722399 | 0.076242057 | 0.101688715 |
| 0.111 | 0.037800148 | 0.053279874 | 0.073002064 | 0.097507480 |
| 0.110 | 0.036072452 | 0.050914071 | 0.069857789 | 0.093441453 |
| 0.109 | 0.034402996 | 0.048623842 | 0.066808127 | 0.089489766 |
| 0.108 | 0.032790700 | 0.046408013 | 0.063851927 | 0.085651470 |
| 0.107 | 0.031234477 | 0.044265386 | 0.060987989 | 0.081925545 |
| 0.106 | 0.029733228 | 0.042194737 | 0.058215071 | 0.078310894 |
| 0.105 | 0.028285846 | 0.040194821 | 0.055531886 | 0.074806352 |
| 0.104 | 0.026891216 | 0.038264372 | 0.052937110 | 0.071410683 |
| 0.103 | 0.025548216 | 0.036402104 | 0.050429378 | 0.068122589 |
| 0.102 | 0.024255718 | 0.034606716 | 0.048007290 | 0.064940707 |
| 0.101 | 0.023012591 | 0.032876888 | 0.045669412 | 0.061863612 |

NEFF and NAUS

TABLE 1a: CLUSTERING PROBABILITY P(n;N,p)

n = 8

| P | N =20 | N =21 | N =22 | N =23 |
|------|-------------|-------------|-------------|-------------|
| 0.100 | 0.021817700 | 0.031211287 | 0.043414277 | 0.058889826 |
| 0.099 | 0.020669907 | 0.029608567 | 0.041240390 | 0.056017812 |
| 0.098 | 0.019568072 | 0.028067370 | 0.039146226 | 0.053245985 |
| 0.097 | 0.018511056 | 0.026586329 | 0.037130237 | 0.050572706 |
| 0.096 | 0.017497721 | 0.025164068 | 0.035190850 | 0.047996295 |
| 0.095 | 0.016526930 | 0.023799204 | 0.033326470 | 0.045515024 |
| 0.094 | 0.015597548 | 0.022490350 | 0.031535484 | 0.043127127 |
| 0.093 | 0.014708446 | 0.021236113 | 0.029816263 | 0.040830800 |
| 0.092 | 0.013858496 | 0.020035099 | 0.028167160 | 0.038624205 |
| 0.091 | 0.013046580 | 0.018885913 | 0.026586518 | 0.036505469 |
| 0.090 | 0.012271583 | 0.017787159 | 0.025072669 | 0.034472694 |
| 0.089 | 0.011532399 | 0.016737447 | 0.023623935 | 0.032523953 |
| 0.088 | 0.010827931 | 0.015735385 | 0.022238631 | 0.030657298 |
| 0.087 | 0.010157089 | 0.014779589 | 0.020915070 | 0.028870759 |
| 0.086 | 0.009518795 | 0.013868681 | 0.019651561 | 0.027162350 |
| 0.085 | 0.008911982 | 0.013001289 | 0.018446413 | 0.025530069 |
| 0.084 | 0.008335593 | 0.012176050 | 0.017297934 | 0.023971904 |
| 0.083 | 0.007788584 | 0.011391612 | 0.016204440 | 0.022485833 |
| 0.082 | 0.007269925 | 0.010646633 | 0.015164248 | 0.021069828 |
| 0.081 | 0.006778600 | 0.009939784 | 0.014175685 | 0.019721858 |
| 0.080 | 0.006313607 | 0.009269749 | 0.013237085 | 0.018439891 |
| 0.079 | 0.005873958 | 0.008635226 | 0.012346794 | 0.017221896 |
| 0.078 | 0.005458682 | 0.008034931 | 0.011503170 | 0.016065849 |
| 0.077 | 0.005066826 | 0.007467595 | 0.010704586 | 0.014969732 |
| 0.076 | 0.004697452 | 0.006931967 | 0.009949430 | 0.013931536 |
| 0.075 | 0.004349639 | 0.006426816 | 0.009236107 | 0.012949264 |
| 0.074 | 0.004022487 | 0.005950928 | 0.008563042 | 0.012020936 |
| 0.073 | 0.003715112 | 0.005503111 | 0.007928681 | 0.011144585 |
| 0.072 | 0.003426649 | 0.005082197 | 0.007331490 | 0.010318266 |
| 0.071 | 0.003156254 | 0.004687036 | 0.006769961 | 0.009540055 |
| 0.070 | 0.002903102 | 0.004316504 | 0.006242607 | 0.008808050 |
| 0.069 | 0.002666387 | 0.003969499 | 0.005747970 | 0.008120376 |
| 0.068 | 0.002445326 | 0.003644944 | 0.005284617 | 0.007475184 |
| 0.067 | 0.002239153 | 0.003341787 | 0.004851146 | 0.006870656 |
| 0.066 | 0.002047128 | 0.003059001 | 0.004446180 | 0.006305005 |
| 0.065 | 0.001868527 | 0.002795586 | 0.004068377 | 0.005776474 |
| 0.064 | 0.001702649 | 0.002550568 | 0.003716422 | 0.005283344 |
| 0.063 | 0.001548817 | 0.002322998 | 0.003389035 | 0.004823930 |
| 0.062 | 0.001406373 | 0.002111958 | 0.003084966 | 0.004396584 |
| 0.061 | 0.001274680 | 0.001916555 | 0.002803002 | 0.003999699 |
| 0.060 | 0.001153124 | 0.001735923 | 0.002541960 | 0.003631706 |
| 0.059 | 0.001041112 | 0.001569227 | 0.002300695 | 0.003291076 |
| 0.058 | 0.000938074 | 0.001415657 | 0.002078096 | 0.002976326 |
| 0.057 | 0.000843460 | 0.001274434 | 0.001873087 | 0.002686010 |
| 0.056 | 0.000756741 | 0.001144804 | 0.001684628 | 0.002418733 |
| 0.055 | 0.000677411 | 0.001026044 | 0.001511717 | 0.002173138 |
| 0.054 | 0.000604983 | 0.000917458 | 0.001353385 | 0.001947916 |
| 0.053 | 0.000538993 | 0.000818380 | 0.001208703 | 0.001741804 |
| 0.052 | 0.000478996 | 0.000728168 | 0.001076775 | 0.001553585 |
| 0.051 | 0.000424569 | 0.000646211 | 0.000956744 | 0.001382086 |

TABLE 1a:  CLUSTERING PROBABILITY  P(n;N,p)

n = 8

| P | N =20 | N =21 | N =22 | N =23 |
|-------|------------|------------|------------|------------|
| 0.050 | 0.000375307 | 0.000571925 | 0.000847789 | 0.001226182 |
| 0.049 | 0.000330827 | 0.000504752 | 0.000749123 | 0.001084795 |
| 0.048 | 0.000290764 | 0.000444163 | 0.000659997 | 0.000956893 |
| 0.047 | 0.000254772 | 0.000389652 | 0.000579697 | 0.000841488 |
| 0.046 | 0.000222525 | 0.000340742 | 0.000507544 | 0.000737642 |
| 0.045 | 0.000193713 | 0.000296981 | 0.000442894 | 0.000644459 |
| 0.044 | 0.000168047 | 0.000257941 | 0.000385136 | 0.000561090 |
| 0.043 | 0.000145252 | 0.000223220 | 0.000333694 | 0.000486730 |
| 0.042 | 0.000125073 | 0.000192439 | 0.000288023 | 0.000420619 |
| 0.041 | 0.000107268 | 0.000165241 | 0.000247613 | 0.000362038 |
| 0.040 | 0.000091614 | 0.000141295 | 0.000211982 | 0.000310313 |
| 0.039 | 0.000077901 | 0.000120289 | 0.000180682 | 0.000264811 |
| 0.038 | 0.000065935 | 0.000101933 | 0.000153293 | 0.000224937 |
| 0.037 | 0.000055536 | 0.000085959 | 0.000129424 | 0.000190138 |
| 0.036 | 0.000046538 | 0.000072117 | 0.000108711 | 0.000159899 |
| 0.035 | 0.000038786 | 0.000060176 | 0.000090820 | 0.000133741 |
| 0.034 | 0.000032141 | 0.000049926 | 0.000075438 | 0.000111222 |
| 0.033 | 0.000026473 | 0.000041170 | 0.000062282 | 0.000091934 |
| 0.032 | 0.000021665 | 0.000033732 | 0.000051089 | 0.000075502 |
| 0.031 | 0.000017608 | 0.000027448 | 0.000041622 | 0.000061583 |
| 0.030 | 0.000014207 | 0.000022173 | 0.000033662 | 0.000049864 |
| 0.029 | 0.000011374 | 0.000017772 | 0.000027012 | 0.000040061 |
| 0.028 | 0.000009030 | 0.000014126 | 0.000021496 | 0.000031918 |
| 0.027 | 0.000007106 | 0.000011128 | 0.000016954 | 0.000025203 |
| 0.026 | 0.000005537 | 0.000008682 | 0.000013243 | 0.000019710 |
| 0.025 | 0.000004271 | 0.000006704 | 0.000010238 | 0.000015255 |
| 0.024 | 0.000003257 | 0.000005119 | 0.000007826 | 0.000011675 |
| 0.023 | 0.000002454 | 0.000003861 | 0.000005910 | 0.000008827 |
| 0.022 | 0.000001825 | 0.000002874 | 0.000004404 | 0.000006586 |
| 0.021 | 0.000001337 | 0.000002109 | 0.000003235 | 0.000004843 |
| 0.020 | 0.000000964 | 0.000001523 | 0.000002339 | 0.000003505 |
| 0.019 | 0.000000683 | 0.000001080 | 0.000001661 | 0.000002493 |
| 0.018 | 0.000000475 | 0.000000752 | 0.000001157 | 0.000001738 |
| 0.017 | 0.000000323 | 0.000000512 | 0.000000789 | 0.000001187 |
| 0.016 | 0.000000214 | 0.000000340 | 0.000000525 | 0.000000790 |
| 0.015 | 0.000000139 | 0.000000220 | 0.000000340 | 0.000000512 |
| 0.014 | 0.000000087 | 0.000000138 | 0.000000213 | 0.000000322 |
| 0.013 | 0.000000052 | 0.000000083 | 0.000000129 | 0.000000195 |
| 0.012 | 0.000000030 | 0.000000048 | 0.000000075 | 0.000000113 |
| 0.011 | 0.000000017 | 0.000000027 | 0.000000041 | 0.000000063 |
| 0.010 | 0.000000009 | 0.000000014 | 0.000000022 | 0.000000033 |
| 0.009 | 0.000000004 | 0.000000007 | 0.000000011 | 0.000000016 |
| 0.008 | 0.000000002 | 0.000000003 | 0.000000005 | 0.000000007 |
| 0.007 | 0.000000001 | 0.000000001 | 0.000000002 | 0.000000003 |
| 0.006 | 0.000000000 | 0.000000000 | 0.000000001 | 0.000000001 |

TABLE 1a: CLUSTERING PROBABILITY P(n;N,p)

n = 9

| P | N = 18 | N = 19 | N = 20 | N = 21 |
|---|---|---|---|---|
| 0.450 | 0.998585182 | 0.999944494 | 0.999998942 | 0.999999990 |
| 0.449 | 0.998435890 | 0.999935122 | 0.999998689 | 0.999999987 |
| 0.448 | 0.998269981 | 0.999924434 | 0.999998384 | 0.999999983 |
| 0.447 | 0.998090761 | 0.999912282 | 0.999998017 | 0.999999978 |
| 0.446 | 0.997897522 | 0.999898506 | 0.999997577 | 0.999999972 |
| 0.445 | 0.997689541 | 0.999882935 | 0.999997051 | 0.999999964 |
| 0.444 | 0.997466078 | 0.999865381 | 0.999996426 | 0.999999954 |
| 0.443 | 0.997226381 | 0.999845647 | 0.999995686 | 0.999999941 |
| 0.442 | 0.996969684 | 0.999823518 | 0.999994811 | 0.999999925 |
| 0.441 | 0.996695209 | 0.999798766 | 0.999993782 | 0.999999906 |
| 0.440 | 0.996402165 | 0.999771145 | 0.999992575 | 0.999999881 |
| 0.439 | 0.996089751 | 0.999740395 | 0.999991164 | 0.999999851 |
| 0.438 | 0.995757157 | 0.999706240 | 0.999989520 | 0.999999814 |
| 0.437 | 0.995403561 | 0.999668385 | 0.999987609 | 0.999999769 |
| 0.436 | 0.995028135 | 0.999626517 | 0.999985395 | 0.999999714 |
| 0.435 | 0.994630041 | 0.999580307 | 0.999982837 | 0.999999647 |
| 0.434 | 0.994208436 | 0.999529406 | 0.999979889 | 0.999999566 |
| 0.433 | 0.993762471 | 0.999473447 | 0.999976503 | 0.999999469 |
| 0.432 | 0.993291292 | 0.999412042 | 0.999972621 | 0.999999351 |
| 0.431 | 0.992794041 | 0.999344785 | 0.999968183 | 0.999999211 |
| 0.430 | 0.992269855 | 0.999271249 | 0.999963122 | 0.999999043 |
| 0.429 | 0.991717872 | 0.999190987 | 0.999957363 | 0.999998843 |
| 0.428 | 0.991137226 | 0.999103531 | 0.999950826 | 0.999998605 |
| 0.427 | 0.990527052 | 0.999008392 | 0.999943423 | 0.999998324 |
| 0.426 | 0.989886485 | 0.998905061 | 0.999935056 | 0.999997992 |
| 0.425 | 0.989214663 | 0.998793007 | 0.999925621 | 0.999997601 |
| 0.424 | 0.988510724 | 0.998671678 | 0.999915004 | 0.999997142 |
| 0.423 | 0.987773810 | 0.998540500 | 0.999903081 | 0.999996604 |
| 0.422 | 0.987003069 | 0.998398878 | 0.999889718 | 0.999995976 |
| 0.421 | 0.986197652 | 0.998246197 | 0.999874769 | 0.999995243 |
| 0.420 | 0.985356716 | 0.998081818 | 0.999858078 | 0.999994391 |
| 0.419 | 0.984479428 | 0.997905082 | 0.999839478 | 0.999993402 |
| 0.418 | 0.983564958 | 0.997715311 | 0.999818786 | 0.999992258 |
| 0.417 | 0.982612488 | 0.997511803 | 0.999795808 | 0.999990936 |
| 0.416 | 0.981621209 | 0.997293838 | 0.999770336 | 0.999989413 |
| 0.415 | 0.980590321 | 0.997060674 | 0.999742147 | 0.999987662 |
| 0.414 | 0.979519037 | 0.996811550 | 0.999711003 | 0.999985652 |
| 0.413 | 0.978406579 | 0.996545685 | 0.999676649 | 0.999983350 |
| 0.412 | 0.977252186 | 0.996262280 | 0.999638814 | 0.999980719 |
| 0.411 | 0.976055106 | 0.995960517 | 0.999597212 | 0.999977718 |
| 0.410 | 0.974814604 | 0.995639559 | 0.999551535 | 0.999974301 |
| 0.409 | 0.973529959 | 0.995298553 | 0.999501459 | 0.999970419 |
| 0.408 | 0.972200466 | 0.994936630 | 0.999446643 | 0.999966015 |
| 0.407 | 0.970825436 | 0.994552902 | 0.999386721 | 0.999961029 |
| 0.406 | 0.969404196 | 0.994146468 | 0.999321311 | 0.999955394 |
| 0.405 | 0.967936093 | 0.993716413 | 0.999250009 | 0.999949036 |
| 0.404 | 0.966420490 | 0.993261806 | 0.999172389 | 0.999941876 |
| 0.403 | 0.964856769 | 0.992781705 | 0.999088003 | 0.999933825 |
| 0.402 | 0.963244334 | 0.992275157 | 0.998996381 | 0.999924787 |
| 0.401 | 0.961582605 | 0.991741194 | 0.998897031 | 0.999914658 |

## TABLE 1a:  CLUSTERING PROBABILITY  $P(n;N,p)$

$n = 9$

| P | N ≡ 18 | N ≡ 19 | N ≡ 20 | N ≡ 21 |
|---|---|---|---|---|
| 0.400 | 0.955871024 | 0.991178842 | 0.998789434 | 0.999903325 |
| 0.399 | 0.958109056 | 0.990587116 | 0.998673052 | 0.999890663 |
| 0.398 | 0.956256184 | 0.989965023 | 0.998547319 | 0.999876540 |
| 0.397 | 0.954431915 | 0.989311565 | 0.998411647 | 0.999860809 |
| 0.396 | 0.952515778 | 0.988625736 | 0.998265421 | 0.999843314 |
| 0.395 | 0.950547323 | 0.987906526 | 0.998108002 | 0.999823887 |
| 0.394 | 0.948526125 | 0.987152921 | 0.997938726 | 0.999802344 |
| 0.393 | 0.946451781 | 0.986363907 | 0.997756902 | 0.999778489 |
| 0.392 | 0.944323913 | 0.985538465 | 0.997561817 | 0.999752111 |
| 0.391 | 0.942142165 | 0.984675578 | 0.997352728 | 0.999722982 |
| 0.390 | 0.939906207 | 0.983774232 | 0.997128869 | 0.999690861 |
| 0.389 | 0.937615731 | 0.982833411 | 0.996889448 | 0.999655485 |
| 0.388 | 0.935270458 | 0.981852108 | 0.996633647 | 0.999616578 |
| 0.387 | 0.932870129 | 0.980829316 | 0.996360624 | 0.999573842 |
| 0.386 | 0.930414512 | 0.979764038 | 0.996069511 | 0.999526961 |
| 0.385 | 0.927903401 | 0.978655282 | 0.995759416 | 0.999475596 |
| 0.384 | 0.925336614 | 0.977502066 | 0.995429422 | 0.999419390 |
| 0.383 | 0.922713995 | 0.976303417 | 0.995078590 | 0.999357960 |
| 0.382 | 0.920035412 | 0.975058374 | 0.994705959 | 0.999290905 |
| 0.381 | 0.917300761 | 0.973765988 | 0.994310532 | 0.999217794 |
| 0.380 | 0.914509961 | 0.972425323 | 0.993891312 | 0.999138176 |
| 0.379 | 0.911662957 | 0.971035458 | 0.993447267 | 0.999051573 |
| 0.378 | 0.908759721 | 0.969595489 | 0.992977346 | 0.998957480 |
| 0.377 | 0.905800249 | 0.968104529 | 0.992480480 | 0.998855366 |
| 0.376 | 0.902784562 | 0.966561707 | 0.991955582 | 0.998744671 |
| 0.375 | 0.899712707 | 0.964966175 | 0.991401544 | 0.998624807 |
| 0.374 | 0.896584756 | 0.963317103 | 0.990817247 | 0.998495157 |
| 0.373 | 0.893400806 | 0.961613683 | 0.990201551 | 0.998355075 |
| 0.372 | 0.890160980 | 0.959855130 | 0.989553305 | 0.998203883 |
| 0.371 | 0.886865423 | 0.958040682 | 0.988871344 | 0.998040873 |
| 0.370 | 0.883514308 | 0.956169602 | 0.988154493 | 0.997865306 |
| 0.369 | 0.880107831 | 0.954241179 | 0.987401563 | 0.997676410 |
| 0.368 | 0.876644210 | 0.952254729 | 0.986611360 | 0.997473382 |
| 0.367 | 0.873129690 | 0.950209592 | 0.985782680 | 0.997255387 |
| 0.366 | 0.869558539 | 0.948105140 | 0.984914316 | 0.997021556 |
| 0.365 | 0.865933049 | 0.945940773 | 0.984005053 | 0.996770989 |
| 0.364 | 0.862253533 | 0.943715918 | 0.983053676 | 0.996502752 |
| 0.363 | 0.858520329 | 0.941430036 | 0.982058968 | 0.996215879 |
| 0.362 | 0.854733797 | 0.939082617 | 0.981019713 | 0.995909371 |
| 0.361 | 0.850894321 | 0.936673182 | 0.979934697 | 0.995582199 |
| 0.360 | 0.847002305 | 0.934201288 | 0.978802710 | 0.995233299 |
| 0.359 | 0.843058175 | 0.931666519 | 0.977622548 | 0.994861578 |
| 0.358 | 0.839062379 | 0.929068498 | 0.976393014 | 0.994465911 |
| 0.357 | 0.835015388 | 0.926406877 | 0.975112921 | 0.994045144 |
| 0.356 | 0.830917690 | 0.923681346 | 0.973781093 | 0.993598092 |
| 0.355 | 0.826769797 | 0.920891626 | 0.972396365 | 0.993123543 |
| 0.354 | 0.822572239 | 0.918037475 | 0.970957590 | 0.992620258 |
| 0.353 | 0.818325566 | 0.915118685 | 0.969463634 | 0.992086969 |
| 0.352 | 0.814030348 | 0.912135084 | 0.967913382 | 0.991522387 |
| 0.351 | 0.809687174 | 0.909086534 | 0.966305740 | 0.990925195 |

NEFF and NAUS

TABLE 1a:  CLUSTERING PROBABILITY  P(n;N,p)

n = 9

| P | N =18 | N =19 | N =20 | N =21 |
|------|-------------|-------------|-------------|-------------|
| 0.350 | 0.805296651 | 0.905972935 | 0.964639634 | 0.990294056 |
| 0.349 | 0.800859404 | 0.902794220 | 0.962914014 | 0.989627613 |
| 0.348 | 0.796376077 | 0.899550359 | 0.961127853 | 0.988924488 |
| 0.347 | 0.791847330 | 0.896241358 | 0.959280153 | 0.988183287 |
| 0.346 | 0.787273841 | 0.892867257 | 0.957369941 | 0.987402599 |
| 0.345 | 0.782656304 | 0.889428134 | 0.955396275 | 0.986581002 |
| 0.344 | 0.777995430 | 0.885924101 | 0.953358243 | 0.985717059 |
| 0.343 | 0.773291945 | 0.882355305 | 0.951254967 | 0.984809327 |
| 0.342 | 0.768546590 | 0.878721928 | 0.949085600 | 0.983856353 |
| 0.341 | 0.763760122 | 0.875024189 | 0.946849330 | 0.982856680 |
| 0.340 | 0.758933312 | 0.871262340 | 0.944545384 | 0.981808849 |
| 0.339 | 0.754066945 | 0.867436668 | 0.942173022 | 0.980711398 |
| 0.338 | 0.749161820 | 0.863547493 | 0.939731545 | 0.979562870 |
| 0.337 | 0.744218748 | 0.859595170 | 0.937220292 | 0.978361810 |
| 0.336 | 0.739238554 | 0.855580087 | 0.934638644 | 0.977106770 |
| 0.335 | 0.734222076 | 0.851502666 | 0.931986021 | 0.975796311 |
| 0.334 | 0.729170162 | 0.847363361 | 0.929261887 | 0.974429006 |
| 0.333 | 0.724083672 | 0.843162656 | 0.926465748 | 0.973003441 |
| 0.332 | 0.718963478 | 0.838901070 | 0.923597151 | 0.971518217 |
| 0.331 | 0.713810463 | 0.834579152 | 0.920655692 | 0.969971957 |
| 0.330 | 0.708625518 | 0.830197481 | 0.917641008 | 0.968363301 |
| 0.329 | 0.703409546 | 0.825756667 | 0.914552782 | 0.966690913 |
| 0.328 | 0.698163457 | 0.821257350 | 0.911390744 | 0.964953484 |
| 0.327 | 0.692888173 | 0.816700198 | 0.908154667 | 0.963149730 |
| 0.326 | 0.687584621 | 0.812085909 | 0.904844374 | 0.961278398 |
| 0.325 | 0.682253738 | 0.807415207 | 0.901459731 | 0.959338266 |
| 0.324 | 0.676896470 | 0.802688846 | 0.898000654 | 0.957328146 |
| 0.323 | 0.671513766 | 0.797907605 | 0.894467103 | 0.955246886 |
| 0.322 | 0.666106585 | 0.793072288 | 0.890859087 | 0.953093370 |
| 0.321 | 0.660675892 | 0.788183728 | 0.887176660 | 0.950866523 |
| 0.320 | 0.655222656 | 0.783242778 | 0.883419926 | 0.948565310 |
| 0.319 | 0.649747854 | 0.778250321 | 0.879589032 | 0.946188740 |
| 0.318 | 0.644252466 | 0.773207258 | 0.875684173 | 0.943735866 |
| 0.317 | 0.638737478 | 0.768114517 | 0.871705593 | 0.941205785 |
| 0.316 | 0.633203878 | 0.762973045 | 0.867653578 | 0.938597645 |
| 0.315 | 0.627652661 | 0.757783813 | 0.863528463 | 0.935910640 |
| 0.314 | 0.622084822 | 0.752547812 | 0.859330627 | 0.933144016 |
| 0.313 | 0.616501361 | 0.747266051 | 0.855060495 | 0.930297069 |
| 0.312 | 0.610903281 | 0.741939562 | 0.850718535 | 0.927369149 |
| 0.311 | 0.605291584 | 0.736569393 | 0.846305261 | 0.924359658 |
| 0.310 | 0.599667278 | 0.731156611 | 0.841821230 | 0.921268054 |
| 0.309 | 0.594031368 | 0.725702301 | 0.837267042 | 0.918093849 |
| 0.308 | 0.588384863 | 0.720207563 | 0.832643340 | 0.914836613 |
| 0.307 | 0.582728772 | 0.714673513 | 0.827950807 | 0.911495972 |
| 0.306 | 0.577064102 | 0.709101284 | 0.823190171 | 0.908071609 |
| 0.305 | 0.571391862 | 0.703492022 | 0.818362196 | 0.904563266 |
| 0.304 | 0.565713059 | 0.697846886 | 0.813467689 | 0.900970744 |
| 0.303 | 0.560028700 | 0.692167049 | 0.808507495 | 0.897293903 |
| 0.302 | 0.554339789 | 0.686453698 | 0.803482496 | 0.893532662 |
| 0.301 | 0.548647331 | 0.680708027 | 0.798393613 | 0.889686998 |

TABLE 1a:   CLUSTERING PROBABILITY   P(n;N,p)

n = 9

| P | N =18 | N =19 | N =20 | N =21 |
|---|---|---|---|---|
| 0.300 | 0.542952324 | 0.674931246 | 0.793241804 | 0.885756951 |
| 0.299 | 0.537255768 | 0.669124572 | 0.788028060 | 0.881742619 |
| 0.298 | 0.531558658 | 0.663289233 | 0.782753408 | 0.877644159 |
| 0.297 | 0.525861986 | 0.657426464 | 0.777418912 | 0.873461788 |
| 0.296 | 0.520166739 | 0.651537510 | 0.772025663 | 0.869195785 |
| 0.295 | 0.514473903 | 0.645623623 | 0.766574790 | 0.864846485 |
| 0.294 | 0.508784456 | 0.639686060 | 0.761067448 | 0.860414282 |
| 0.293 | 0.503099374 | 0.633726087 | 0.755504826 | 0.855899632 |
| 0.292 | 0.497419626 | 0.627744973 | 0.749888139 | 0.851303047 |
| 0.291 | 0.491746176 | 0.621743993 | 0.744218632 | 0.846625094 |
| 0.290 | 0.486079984 | 0.615724426 | 0.738497577 | 0.841866402 |
| 0.289 | 0.480422002 | 0.609687551 | 0.732726271 | 0.837027653 |
| 0.288 | 0.474773175 | 0.603634655 | 0.726906037 | 0.832109586 |
| 0.287 | 0.469134443 | 0.597567025 | 0.721038221 | 0.827112994 |
| 0.286 | 0.463506737 | 0.591485949 | 0.715124193 | 0.822038725 |
| 0.285 | 0.457890982 | 0.585392714 | 0.709165344 | 0.816887681 |
| 0.284 | 0.452288096 | 0.579288611 | 0.703163086 | 0.811660814 |
| 0.283 | 0.446698986 | 0.573174928 | 0.697118851 | 0.806359129 |
| 0.282 | 0.441124554 | 0.567052952 | 0.691034091 | 0.800983662 |
| 0.281 | 0.435565690 | 0.560923970 | 0.684910273 | 0.795535576 |
| 0.280 | 0.430023279 | 0.554789263 | 0.678748882 | 0.790015965 |
| 0.279 | 0.424498194 | 0.548650115 | 0.672551419 | 0.784426048 |
| 0.278 | 0.418991298 | 0.542507800 | 0.666319400 | 0.778767071 |
| 0.277 | 0.413503446 | 0.536363593 | 0.660054352 | 0.773040323 |
| 0.276 | 0.408035483 | 0.530218762 | 0.653757817 | 0.767247140 |
| 0.275 | 0.402588243 | 0.524074571 | 0.647431346 | 0.761388896 |
| 0.274 | 0.397162548 | 0.517932277 | 0.641076503 | 0.755467008 |
| 0.273 | 0.391759213 | 0.511793130 | 0.634694860 | 0.749482933 |
| 0.272 | 0.386379038 | 0.505658377 | 0.628287996 | 0.743438166 |
| 0.271 | 0.381022813 | 0.499529254 | 0.621857500 | 0.737334236 |
| 0.270 | 0.375691318 | 0.493406990 | 0.615404965 | 0.731172713 |
| 0.269 | 0.370385319 | 0.487292806 | 0.608931990 | 0.724955195 |
| 0.268 | 0.365105572 | 0.481187915 | 0.602440178 | 0.718683317 |
| 0.267 | 0.359852820 | 0.475093519 | 0.595931136 | 0.712358742 |
| 0.266 | 0.354627792 | 0.469010811 | 0.589406473 | 0.705983165 |
| 0.265 | 0.349431206 | 0.462940973 | 0.582867798 | 0.699558309 |
| 0.264 | 0.344263768 | 0.456885178 | 0.576316723 | 0.693085921 |
| 0.263 | 0.339126170 | 0.450844586 | 0.569754857 | 0.686567777 |
| 0.262 | 0.334019090 | 0.444820346 | 0.563183808 | 0.680005674 |
| 0.261 | 0.328943193 | 0.438813594 | 0.556605184 | 0.673401432 |
| 0.260 | 0.323899133 | 0.432825455 | 0.550020586 | 0.666756894 |
| 0.259 | 0.318887546 | 0.426857041 | 0.543431614 | 0.660073918 |
| 0.258 | 0.313909057 | 0.420909448 | 0.536839860 | 0.653354383 |
| 0.257 | 0.308964277 | 0.414983761 | 0.530246913 | 0.646600183 |
| 0.256 | 0.304053801 | 0.409081050 | 0.523654353 | 0.639813227 |
| 0.255 | 0.299178212 | 0.403202371 | 0.517063753 | 0.632995438 |
| 0.254 | 0.294338077 | 0.397348763 | 0.510476678 | 0.626148750 |
| 0.253 | 0.289533949 | 0.391521253 | 0.503894684 | 0.619275108 |
| 0.252 | 0.284766366 | 0.385720850 | 0.497319316 | 0.612376465 |
| 0.251 | 0.280035852 | 0.379948548 | 0.490752108 | 0.605454783 |

TABLE 1a:   CLUSTERING PROBABILITY   P(n;N,p)

n = 9

| P | N =18 | N =19 | N =20 | N =21 |
|-------|-------------|-------------|-------------|-------------|
| 0.250 | 0.275342915 | 0.374205324 | 0.484194583 | 0.598512028 |
| 0.249 | 0.270688049 | 0.368492139 | 0.477648253 | 0.591550172 |
| 0.248 | 0.266071733 | 0.362809937 | 0.471114615 | 0.584571191 |
| 0.247 | 0.261494430 | 0.357159645 | 0.464595153 | 0.577577061 |
| 0.246 | 0.256956587 | 0.351542172 | 0.458091336 | 0.570569758 |
| 0.245 | 0.252458639 | 0.345958411 | 0.451604618 | 0.563551261 |
| 0.244 | 0.248001003 | 0.340409234 | 0.445136439 | 0.556523542 |
| 0.243 | 0.243584080 | 0.334895497 | 0.438688219 | 0.549488572 |
| 0.242 | 0.239208258 | 0.329418037 | 0.432261365 | 0.542448317 |
| 0.241 | 0.234873908 | 0.323977673 | 0.425857263 | 0.535404737 |
| 0.240 | 0.230581385 | 0.318575202 | 0.419477282 | 0.528359783 |
| 0.239 | 0.226331031 | 0.313211407 | 0.413122772 | 0.521315399 |
| 0.238 | 0.222123168 | 0.307887046 | 0.406795065 | 0.514273519 |
| 0.237 | 0.217958108 | 0.302602861 | 0.400495471 | 0.507236064 |
| 0.236 | 0.213836142 | 0.297359575 | 0.394225279 | 0.500204946 |
| 0.235 | 0.209757549 | 0.292157887 | 0.387985760 | 0.493182061 |
| 0.234 | 0.205722592 | 0.286998481 | 0.381778161 | 0.486169292 |
| 0.233 | 0.201731517 | 0.281882017 | 0.375603707 | 0.479168505 |
| 0.232 | 0.197784555 | 0.276809136 | 0.369463602 | 0.472181551 |
| 0.231 | 0.193881923 | 0.271780458 | 0.363359027 | 0.465210261 |
| 0.230 | 0.190023820 | 0.266796585 | 0.357291138 | 0.458256449 |
| 0.229 | 0.186210432 | 0.261858095 | 0.351261069 | 0.451321910 |
| 0.228 | 0.182441928 | 0.256965546 | 0.345269929 | 0.444408416 |
| 0.227 | 0.178718463 | 0.252119477 | 0.339318804 | 0.437517719 |
| 0.226 | 0.175040175 | 0.247320405 | 0.333408752 | 0.430651547 |
| 0.225 | 0.171407188 | 0.242568825 | 0.327540809 | 0.423811607 |
| 0.224 | 0.167819612 | 0.237865212 | 0.321715984 | 0.416999578 |
| 0.223 | 0.164277539 | 0.233210021 | 0.315935262 | 0.410217118 |
| 0.222 | 0.160781050 | 0.228603683 | 0.310199599 | 0.403465856 |
| 0.221 | 0.157330207 | 0.224046612 | 0.304509927 | 0.396747394 |
| 0.220 | 0.153925061 | 0.219539197 | 0.298867150 | 0.390063309 |
| 0.219 | 0.150565646 | 0.215081808 | 0.293272147 | 0.383415149 |
| 0.218 | 0.147251983 | 0.210674794 | 0.287725769 | 0.376804431 |
| 0.217 | 0.143984077 | 0.206318482 | 0.282228839 | 0.370232645 |
| 0.216 | 0.140761920 | 0.202013179 | 0.276782153 | 0.363701249 |
| 0.215 | 0.137585490 | 0.197759170 | 0.271386481 | 0.357211672 |
| 0.214 | 0.134454750 | 0.193556720 | 0.266042563 | 0.350765309 |
| 0.213 | 0.131369650 | 0.189406072 | 0.260751113 | 0.344363525 |
| 0.212 | 0.128330127 | 0.185307451 | 0.255512817 | 0.338007652 |
| 0.211 | 0.125336103 | 0.181261058 | 0.250328331 | 0.331698990 |
| 0.210 | 0.122387487 | 0.177267076 | 0.245198285 | 0.325438803 |
| 0.209 | 0.119484175 | 0.173325665 | 0.240123281 | 0.319228324 |
| 0.208 | 0.116626051 | 0.169436968 | 0.235103890 | 0.313068749 |
| 0.207 | 0.113812985 | 0.165601104 | 0.230140659 | 0.306961242 |
| 0.206 | 0.111044834 | 0.161818175 | 0.225234103 | 0.300906929 |
| 0.205 | 0.108321444 | 0.158088262 | 0.220384711 | 0.294906904 |
| 0.204 | 0.105642648 | 0.154411425 | 0.215592943 | 0.288962222 |
| 0.203 | 0.103008266 | 0.150787708 | 0.210859230 | 0.283073904 |
| 0.202 | 0.100418108 | 0.147217131 | 0.206183978 | 0.277242934 |
| 0.201 | 0.097871971 | 0.143699698 | 0.201567561 | 0.271470261 |

TABLE 1a:  CLUSTERING PROBABILITY  P(n;N,p)

n = 9

| P | N = 18 | N = 19 | N = 20 | N = 21 |
|---|---|---|---|---|
| 0.200 | 0.095369640 | 0.140235393 | 0.197010328 | 0.265756795 |
| 0.199 | 0.092910891 | 0.136824181 | 0.192512598 | 0.260103411 |
| 0.198 | 0.090455486 | 0.133466008 | 0.188074665 | 0.254510946 |
| 0.197 | 0.088123180 | 0.130160802 | 0.183696792 | 0.248980201 |
| 0.196 | 0.085793713 | 0.126908473 | 0.179379217 | 0.243511939 |
| 0.195 | 0.083506819 | 0.123708914 | 0.175122150 | 0.238106886 |
| 0.194 | 0.081262218 | 0.120561998 | 0.170925774 | 0.232765732 |
| 0.193 | 0.079059623 | 0.117467582 | 0.166790244 | 0.227489127 |
| 0.192 | 0.076898736 | 0.114425505 | 0.162715691 | 0.222277686 |
| 0.191 | 0.074779249 | 0.111435591 | 0.158702217 | 0.217131986 |
| 0.190 | 0.072700847 | 0.108497645 | 0.154749898 | 0.212052567 |
| 0.189 | 0.070663204 | 0.105611456 | 0.150858785 | 0.207039931 |
| 0.188 | 0.068665987 | 0.102776799 | 0.147028902 | 0.202094546 |
| 0.187 | 0.066708852 | 0.099993430 | 0.143260250 | 0.197216840 |
| 0.186 | 0.064791451 | 0.097261093 | 0.139552801 | 0.192407205 |
| 0.185 | 0.062913424 | 0.094579513 | 0.135906505 | 0.187665998 |
| 0.184 | 0.061074405 | 0.091948402 | 0.132321287 | 0.182993538 |
| 0.183 | 0.059274021 | 0.089367459 | 0.128797046 | 0.178390109 |
| 0.182 | 0.057511893 | 0.086836366 | 0.125333660 | 0.173855958 |
| 0.181 | 0.055787631 | 0.084354793 | 0.121930981 | 0.169391298 |
| 0.180 | 0.054100843 | 0.081922395 | 0.118588838 | 0.164996306 |
| 0.179 | 0.052451128 | 0.079538815 | 0.115307039 | 0.160671124 |
| 0.178 | 0.050838080 | 0.077203682 | 0.112085367 | 0.156415859 |
| 0.177 | 0.049261287 | 0.074916613 | 0.108923584 | 0.152230585 |
| 0.176 | 0.047720331 | 0.072677212 | 0.105821431 | 0.148115342 |
| 0.175 | 0.046214789 | 0.070485073 | 0.102778628 | 0.144070133 |
| 0.174 | 0.044744234 | 0.068339777 | 0.099794871 | 0.140094934 |
| 0.173 | 0.043308232 | 0.066240893 | 0.096869840 | 0.136189682 |
| 0.172 | 0.041906347 | 0.064187982 | 0.094003193 | 0.132354288 |
| 0.171 | 0.040538137 | 0.062180592 | 0.091194566 | 0.128588625 |
| 0.170 | 0.039203158 | 0.060218261 | 0.088443580 | 0.124892540 |
| 0.169 | 0.037900959 | 0.058300520 | 0.085749835 | 0.121265846 |
| 0.168 | 0.036631090 | 0.056426888 | 0.083112914 | 0.117708327 |
| 0.167 | 0.035393094 | 0.054596876 | 0.080532381 | 0.114219737 |
| 0.166 | 0.034186514 | 0.052809987 | 0.078007784 | 0.110799801 |
| 0.165 | 0.033010888 | 0.051065714 | 0.075538654 | 0.107448214 |
| 0.164 | 0.031865753 | 0.049363544 | 0.073124505 | 0.104164646 |
| 0.163 | 0.030750644 | 0.047702957 | 0.070764836 | 0.100948737 |
| 0.162 | 0.029665093 | 0.046083424 | 0.068459131 | 0.097800100 |
| 0.161 | 0.028608632 | 0.044504411 | 0.066206859 | 0.094718324 |
| 0.160 | 0.027580791 | 0.042965376 | 0.064007474 | 0.091702969 |
| 0.159 | 0.026581099 | 0.041465774 | 0.061860418 | 0.088753574 |
| 0.158 | 0.025609083 | 0.040005052 | 0.059765119 | 0.085869650 |
| 0.157 | 0.024664271 | 0.038582652 | 0.057720992 | 0.083050686 |
| 0.156 | 0.023746191 | 0.037198012 | 0.055727440 | 0.080296149 |
| 0.155 | 0.022854368 | 0.035850567 | 0.053783855 | 0.077605482 |
| 0.154 | 0.021988331 | 0.034539745 | 0.051889619 | 0.074978106 |
| 0.153 | 0.021147607 | 0.033264973 | 0.050044100 | 0.072413424 |
| 0.152 | 0.020331723 | 0.032025673 | 0.048246660 | 0.069910815 |
| 0.151 | 0.019540210 | 0.030821266 | 0.046496650 | 0.067469640 |

TABLE 1a:   CLUSTERING PROBABILITY   $P(n;N,p)$

$\underline{n = 9}$

| __P_____ | __N_=18_____ | __N_=19_____ | __N_=20_____ | __N_=21____ |
|-----------|------------------|------------------|------------------|-------------|
| 0.150 | 0.018772596 | 0.029651168 | 0.044793410 | 0.065089243 |
| 0.149 | 0.018028414 | 0.028514795 | 0.043136277 | 0.062768947 |
| 0.148 | 0.017307196 | 0.027411560 | 0.041524574 | 0.060508060 |
| 0.147 | 0.016608477 | 0.026340877 | 0.039957621 | 0.058305871 |
| 0.146 | 0.015931793 | 0.025302154 | 0.038434730 | 0.056161655 |
| 0.145 | 0.015276683 | 0.024294804 | 0.036955207 | 0.054074672 |
| 0.144 | 0.014642688 | 0.023318236 | 0.035518352 | 0.052044165 |
| 0.143 | 0.014029351 | 0.022371860 | 0.034123460 | 0.050069368 |
| 0.142 | 0.013436218 | 0.021455087 | 0.032769822 | 0.048149497 |
| 0.141 | 0.012862839 | 0.020567327 | 0.031456724 | 0.046283759 |
| 0.140 | 0.012308766 | 0.019707993 | 0.030183449 | 0.044471348 |
| 0.139 | 0.011773554 | 0.018876498 | 0.028949276 | 0.042711449 |
| 0.138 | 0.011256763 | 0.018072257 | 0.027753484 | 0.041003237 |
| 0.137 | 0.010757955 | 0.017294687 | 0.026595348 | 0.039345875 |
| 0.136 | 0.010276696 | 0.016543207 | 0.025474141 | 0.037738521 |
| 0.135 | 0.009812557 | 0.015817240 | 0.024389137 | 0.036180323 |
| 0.134 | 0.009365112 | 0.015116210 | 0.023339608 | 0.034670424 |
| 0.133 | 0.008933940 | 0.014439545 | 0.022324827 | 0.033207959 |
| 0.132 | 0.008518624 | 0.013786675 | 0.021344066 | 0.031792059 |
| 0.131 | 0.008118752 | 0.013157038 | 0.020396599 | 0.030421849 |
| 0.130 | 0.007733916 | 0.012550070 | 0.019481701 | 0.029096451 |
| 0.129 | 0.007363714 | 0.011965217 | 0.018598650 | 0.027814983 |
| 0.128 | 0.007007747 | 0.011401924 | 0.017746724 | 0.026576560 |
| 0.127 | 0.006665622 | 0.010859646 | 0.016925207 | 0.025380297 |
| 0.126 | 0.006336953 | 0.010337840 | 0.016133382 | 0.024225306 |
| 0.125 | 0.006021356 | 0.009835968 | 0.015370540 | 0.023110698 |
| 0.124 | 0.005718455 | 0.009353499 | 0.014635973 | 0.022035587 |
| 0.123 | 0.005427879 | 0.008889907 | 0.013928978 | 0.020999084 |
| 0.122 | 0.005149260 | 0.008444672 | 0.013248658 | 0.020000304 |
| 0.121 | 0.004882239 | 0.008017279 | 0.012594920 | 0.019038364 |
| 0.120 | 0.004626461 | 0.007607221 | 0.011966476 | 0.018112382 |
| 0.119 | 0.004381578 | 0.007213997 | 0.011362846 | 0.017221482 |
| 0.118 | 0.004147247 | 0.006837112 | 0.010783354 | 0.016364788 |
| 0.117 | 0.003923130 | 0.006476079 | 0.010227333 | 0.015541433 |
| 0.116 | 0.003708897 | 0.006130416 | 0.009694120 | 0.014750553 |
| 0.115 | 0.003504222 | 0.005799650 | 0.009183063 | 0.013991288 |
| 0.114 | 0.003308788 | 0.005483314 | 0.008693514 | 0.013262787 |
| 0.113 | 0.003122281 | 0.005180950 | 0.008224835 | 0.012564204 |
| 0.112 | 0.002944395 | 0.004892106 | 0.007776396 | 0.011894701 |
| 0.111 | 0.002774829 | 0.004616338 | 0.007347574 | 0.011253447 |
| 0.110 | 0.002613291 | 0.004353211 | 0.006937756 | 0.010639622 |
| 0.109 | 0.002459490 | 0.004102297 | 0.006546338 | 0.010052410 |
| 0.108 | 0.002313147 | 0.003863174 | 0.006172725 | 0.009491008 |
| 0.107 | 0.002173986 | 0.003635432 | 0.005816330 | 0.008954621 |
| 0.106 | 0.002041738 | 0.003418666 | 0.005476577 | 0.008442465 |
| 0.105 | 0.001916140 | 0.003212481 | 0.005152900 | 0.007953764 |
| 0.104 | 0.001796936 | 0.003016489 | 0.004844742 | 0.007487757 |
| 0.103 | 0.001683876 | 0.002830311 | 0.004551557 | 0.007043690 |
| 0.102 | 0.001576715 | 0.002653576 | 0.004272809 | 0.006620823 |
| 0.101 | 0.001475217 | 0.002485921 | 0.004007971 | 0.006218426 |

TABLE 1a:   CLUSTERING PROBABILITY   P(n;N,p)

n = 9

| P | N =18 | N =19 | N =20 | N =21 |
|---|---|---|---|---|
| 0.100 | 0.001379148 | 0.002326994 | 0.003756530 | 0.005835784 |
| 0.099 | 0.001288286 | 0.002176447 | 0.003517979 | 0.005472192 |
| 0.098 | 0.001202409 | 0.002033944 | 0.003291827 | 0.005126958 |
| 0.097 | 0.001121305 | 0.001899156 | 0.003077589 | 0.004799404 |
| 0.096 | 0.001044767 | 0.001771763 | 0.002874794 | 0.004488864 |
| 0.095 | 0.000972594 | 0.001651454 | 0.002682981 | 0.004194686 |
| 0.094 | 0.000904590 | 0.001537923 | 0.002501701 | 0.003916233 |
| 0.093 | 0.000840567 | 0.001430877 | 0.002330514 | 0.003652878 |
| 0.092 | 0.000780342 | 0.001330028 | 0.002168993 | 0.003404013 |
| 0.091 | 0.000723736 | 0.001235099 | 0.002016721 | 0.003169040 |
| 0.090 | 0.000670578 | 0.001145818 | 0.001873294 | 0.002947378 |
| 0.089 | 0.000620702 | 0.001061923 | 0.001738317 | 0.002738457 |
| 0.088 | 0.000573947 | 0.000983162 | 0.001611408 | 0.002541724 |
| 0.087 | 0.000530158 | 0.000909287 | 0.001492193 | 0.002356641 |
| 0.086 | 0.000489185 | 0.000840060 | 0.001380313 | 0.002182682 |
| 0.085 | 0.000450884 | 0.000775252 | 0.001275417 | 0.002019337 |
| 0.084 | 0.000415116 | 0.000714640 | 0.001177167 | 0.001866112 |
| 0.083 | 0.000381747 | 0.000658009 | 0.001085234 | 0.001722523 |
| 0.082 | 0.000350647 | 0.000605153 | 0.000999301 | 0.001588106 |
| 0.081 | 0.000321693 | 0.000555872 | 0.000919062 | 0.001462407 |
| 0.080 | 0.000294766 | 0.000509973 | 0.000844220 | 0.001344989 |
| 0.079 | 0.000269751 | 0.000467272 | 0.000774490 | 0.001235429 |
| 0.078 | 0.000246540 | 0.000427591 | 0.000709596 | 0.001133317 |
| 0.077 | 0.000225026 | 0.000390759 | 0.000649274 | 0.001038258 |
| 0.076 | 0.000205110 | 0.000356613 | 0.000593268 | 0.000949870 |
| 0.075 | 0.000186695 | 0.000324994 | 0.000541332 | 0.000867787 |
| 0.074 | 0.000169689 | 0.000295753 | 0.000493233 | 0.000791655 |
| 0.073 | 0.000154005 | 0.000268746 | 0.000448742 | 0.000721133 |
| 0.072 | 0.000139559 | 0.000243835 | 0.000407646 | 0.000655895 |
| 0.071 | 0.000126271 | 0.000220888 | 0.000369735 | 0.000595627 |
| 0.070 | 0.000114065 | 0.000199780 | 0.000334812 | 0.000540029 |
| 0.069 | 0.000102870 | 0.000180392 | 0.000302688 | 0.000488812 |
| 0.068 | 0.000092617 | 0.000162609 | 0.000273182 | 0.000441701 |
| 0.067 | 0.000083240 | 0.000146323 | 0.000246121 | 0.000398432 |
| 0.066 | 0.000074678 | 0.000131432 | 0.000221342 | 0.000358756 |
| 0.065 | 0.000066873 | 0.000117838 | 0.000198689 | 0.000322431 |
| 0.064 | 0.000059769 | 0.000105448 | 0.000178014 | 0.000289231 |
| 0.063 | 0.000053315 | 0.000094175 | 0.000159177 | 0.000258937 |
| 0.062 | 0.000047462 | 0.000083937 | 0.000142043 | 0.000231346 |
| 0.061 | 0.000042163 | 0.000074655 | 0.000126488 | 0.000206261 |
| 0.060 | 0.000037374 | 0.000066256 | 0.000112393 | 0.000183497 |
| 0.059 | 0.000033055 | 0.000058670 | 0.000099645 | 0.000162880 |
| 0.058 | 0.000029168 | 0.000051833 | 0.000088138 | 0.000144245 |
| 0.057 | 0.000025677 | 0.000045683 | 0.000077774 | 0.000127436 |
| 0.056 | 0.000022547 | 0.000040163 | 0.000068458 | 0.000112306 |
| 0.055 | 0.000019748 | 0.000035220 | 0.000060104 | 0.000098719 |
| 0.054 | 0.000017251 | 0.000030803 | 0.000052629 | 0.000086545 |
| 0.053 | 0.000015028 | 0.000026865 | 0.000045956 | 0.000075663 |
| 0.052 | 0.000013054 | 0.000023364 | 0.000040015 | 0.000065960 |
| 0.051 | 0.000011306 | 0.000020259 | 0.000034738 | 0.000057330 |

TABLE 1a:   CLUSTERING PROBABILITY   P(n;N,p)

n = 9

| P | N =18 | N =19 | N =20 | N =21 |
|---|---|---|---|---|
| 0.050 | 0.000009761 | 0.000017512 | 0.000030064 | 0.000049675 |
| 0.049 | 0.000008401 | 0.000015089 | 0.000025935 | 0.000042904 |
| 0.048 | 0.000007206 | 0.000012958 | 0.000022299 | 0.000036932 |
| 0.047 | 0.000006159 | 0.000011089 | 0.000019105 | 0.000031680 |
| 0.046 | 0.000005246 | 0.000009456 | 0.000016310 | 0.000027077 |
| 0.045 | 0.000004451 | 0.000008032 | 0.000013871 | 0.000023054 |
| 0.044 | 0.000003761 | 0.000006796 | 0.000011749 | 0.000019552 |
| 0.043 | 0.000003165 | 0.000005726 | 0.000009911 | 0.000016513 |
| 0.042 | 0.000002652 | 0.000004803 | 0.000008325 | 0.000013885 |
| 0.041 | 0.000002212 | 0.000004011 | 0.000006960 | 0.000011623 |
| 0.040 | 0.000001837 | 0.000003334 | 0.000005792 | 0.000009683 |
| 0.039 | 0.000001517 | 0.000002757 | 0.000004795 | 0.000008026 |
| 0.038 | 0.000001247 | 0.000002268 | 0.000003949 | 0.000006618 |
| 0.037 | 0.000001019 | 0.000001855 | 0.000003234 | 0.000005427 |
| 0.036 | 0.000000827 | 0.000001509 | 0.000002634 | 0.000004424 |
| 0.035 | 0.000000668 | 0.000001220 | 0.000002131 | 0.000003584 |
| 0.034 | 0.000000536 | 0.000000979 | 0.000001713 | 0.000002885 |
| 0.033 | 0.000000427 | 0.000000781 | 0.000001368 | 0.000002306 |
| 0.032 | 0.000000337 | 0.000000618 | 0.000001084 | 0.000001829 |
| 0.031 | 0.000000265 | 0.000000486 | 0.000000852 | 0.000001440 |
| 0.030 | 0.000000206 | 0.000000378 | 0.000000665 | 0.000001124 |
| 0.029 | 0.000000159 | 0.000000292 | 0.000000514 | 0.000000870 |
| 0.028 | 0.000000121 | 0.000000223 | 0.000000393 | 0.000000667 |
| 0.027 | 0.000000092 | 0.000000169 | 0.000000298 | 0.000000506 |
| 0.026 | 0.000000069 | 0.000000126 | 0.000000223 | 0.000000380 |
| 0.025 | 0.000000051 | 0.000000094 | 0.000000165 | 0.000000281 |
| 0.024 | 0.000000037 | 0.000000068 | 0.000000121 | 0.000000206 |
| 0.023 | 0.000000027 | 0.000000049 | 0.000000087 | 0.000000149 |
| 0.022 | 0.000000019 | 0.000000035 | 0.000000062 | 0.000000106 |
| 0.021 | 0.000000013 | 0.000000024 | 0.000000043 | 0.000000074 |
| 0.020 | 0.000000009 | 0.000000017 | 0.000000030 | 0.000000051 |
| 0.019 | 0.000000006 | 0.000000011 | 0.000000020 | 0.000000034 |
| 0.018 | 0.000000004 | 0.000000007 | 0.000000013 | 0.000000023 |
| 0.017 | 0.000000003 | 0.000000005 | 0.000000008 | 0.000000014 |
| 0.016 | 0.000000002 | 0.000000003 | 0.000000005 | 0.000000009 |
| 0.015 | 0.000000001 | 0.000000002 | 0.000000003 | 0.000000005 |
| 0.014 | 0.000000001 | 0.000000001 | 0.000000002 | 0.000000003 |
| 0.013 | 0.000000000 | 0.000000001 | 0.000000001 | 0.000000002 |
| 0.012 | 0.000000000 | 0.000000000 | 0.000000001 | 0.000000001 |
| 0.011 | 0.000000000 | 0.000000000 | 0.000000000 | 0.000000000 |

TABLE 1a: CLUSTERING PROBABILITY $P(n;N,p)$

$n = 9$

| P | N =22 | N =23 | N =24 | N =25 |
|---|---|---|---|---|
| 0.350 | 0.998238383 | 0.999822811 | 0.999993399 | |
| 0.349 | 0.998071586 | 0.999800014 | 0.999992258 | |
| 0.348 | 0.997891148 | 0.999774542 | 0.999990929 | |
| 0.347 | 0.997696161 | 0.999746113 | 0.999989383 | |
| 0.346 | 0.997485675 | 0.999714422 | 0.999987586 | |
| 0.345 | 0.997258697 | 0.999679136 | 0.999985500 | |
| 0.344 | 0.997014192 | 0.999639892 | 0.999983080 | |
| 0.343 | 0.996751082 | 0.999596298 | 0.999980276 | |
| 0.342 | 0.996468246 | 0.999547929 | 0.999977029 | |
| 0.341 | 0.996164521 | 0.999494324 | 0.999973274 | |
| 0.340 | 0.995838701 | 0.999434987 | 0.999968934 | |
| 0.339 | 0.995489539 | 0.999369383 | 0.999963923 | |
| 0.338 | 0.995115746 | 0.999296936 | 0.999958143 | |
| 0.337 | 0.994715995 | 0.999217028 | 0.999951481 | |
| 0.336 | 0.994288917 | 0.999128997 | 0.999943810 | |
| 0.335 | 0.993833108 | 0.999032134 | 0.999934983 | |
| 0.334 | 0.993347127 | 0.998925683 | 0.999924837 | |
| 0.333 | 0.992829498 | 0.998808839 | 0.999913183 | 1.000000000 |
| 0.332 | 0.992278712 | 0.998680749 | 0.999899808 | 1.000000000 |
| 0.331 | 0.991693229 | 0.998540505 | 0.999884472 | 0.999999995 |
| 0.330 | 0.991071482 | 0.998387150 | 0.999866903 | 0.999999980 |
| 0.329 | 0.990411874 | 0.998219676 | 0.999846798 | 0.999999941 |
| 0.328 | 0.989712785 | 0.998037021 | 0.999823820 | 0.999999857 |
| 0.327 | 0.988972574 | 0.997838069 | 0.999797593 | 0.999999703 |
| 0.326 | 0.988189577 | 0.997621654 | 0.999767705 | 0.999999441 |
| 0.325 | 0.987362115 | 0.997386556 | 0.999733701 | 0.999999024 |
| 0.324 | 0.986488495 | 0.997131502 | 0.999695084 | 0.999998392 |
| 0.323 | 0.985567008 | 0.996855169 | 0.999651312 | 0.999997474 |
| 0.322 | 0.984595938 | 0.996556182 | 0.999601796 | 0.999996180 |
| 0.321 | 0.983573563 | 0.996233114 | 0.999545898 | 0.999994405 |
| 0.320 | 0.982498156 | 0.995884490 | 0.999482928 | 0.999992022 |
| 0.319 | 0.981367989 | 0.995508787 | 0.999412145 | 0.999988885 |
| 0.318 | 0.980181334 | 0.995104435 | 0.999332754 | 0.999984823 |
| 0.317 | 0.978936471 | 0.994669817 | 0.999243903 | 0.999979639 |
| 0.316 | 0.977631685 | 0.994203273 | 0.999144684 | 0.999973108 |
| 0.315 | 0.976265272 | 0.993703102 | 0.999034129 | 0.999964973 |
| 0.314 | 0.974835543 | 0.993167560 | 0.998911213 | 0.999954945 |
| 0.313 | 0.973340823 | 0.992594869 | 0.998774849 | 0.999942699 |
| 0.312 | 0.971779457 | 0.991983211 | 0.998623889 | 0.999927870 |
| 0.311 | 0.970149812 | 0.991330739 | 0.998457124 | 0.999910053 |
| 0.310 | 0.968450281 | 0.990635573 | 0.998273281 | 0.999888799 |
| 0.309 | 0.966679284 | 0.989895805 | 0.998071026 | 0.999863612 |
| 0.308 | 0.964835271 | 0.989109501 | 0.997848963 | 0.999833948 |
| 0.307 | 0.962916727 | 0.988274707 | 0.997605833 | 0.999799210 |
| 0.306 | 0.960922170 | 0.987389450 | 0.997339514 | 0.999758747 |
| 0.305 | 0.958850161 | 0.986451737 | 0.997049025 | 0.999711851 |
| 0.304 | 0.956699298 | 0.985459567 | 0.996732524 | 0.999657757 |
| 0.303 | 0.954468225 | 0.984410927 | 0.996388308 | 0.999595635 |
| 0.302 | 0.952155632 | 0.983303797 | 0.996014618 | 0.999524593 |
| 0.301 | 0.949760258 | 0.982136158 | 0.995609637 | 0.999443675 |

TABLE 1a:   CLUSTERING PROBABILITY   P(n;N,p)

n = 9

| P | N =22 | N =23 | N =24 | N =25 |
|-------|-----------|-----------|-----------|-----------|
| 0.300 | 0.947280890 | 0.980905989 | 0.995171494 | 0.999351855 |
| 0.299 | 0.944716372 | 0.979611273 | 0.994698264 | 0.999248037 |
| 0.298 | 0.942065601 | 0.978250004 | 0.994187973 | 0.999131057 |
| 0.297 | 0.939327530 | 0.976820185 | 0.993638596 | 0.998999676 |
| 0.296 | 0.936501174 | 0.975319835 | 0.993048065 | 0.998852583 |
| 0.295 | 0.933585605 | 0.973746994 | 0.992414266 | 0.998688392 |
| 0.294 | 0.930579962 | 0.972099722 | 0.991735047 | 0.998505642 |
| 0.293 | 0.927483444 | 0.970376106 | 0.991008218 | 0.998302797 |
| 0.292 | 0.924295318 | 0.968574264 | 0.990231557 | 0.998078245 |
| 0.291 | 0.921014918 | 0.966692346 | 0.989402808 | 0.997830298 |
| 0.290 | 0.917641645 | 0.964728540 | 0.988519693 | 0.997557194 |
| 0.289 | 0.914174970 | 0.962681075 | 0.987579908 | 0.997257097 |
| 0.288 | 0.910614436 | 0.960548222 | 0.986581132 | 0.996928094 |
| 0.287 | 0.906959655 | 0.958328302 | 0.985521028 | 0.996568205 |
| 0.286 | 0.903210313 | 0.956019683 | 0.984397250 | 0.996175376 |
| 0.285 | 0.899366169 | 0.953620791 | 0.983207444 | 0.995747484 |
| 0.284 | 0.895427055 | 0.951130105 | 0.981949256 | 0.995282342 |
| 0.283 | 0.891392878 | 0.948546167 | 0.980620332 | 0.994777697 |
| 0.282 | 0.887263618 | 0.945867580 | 0.979218329 | 0.994231236 |
| 0.281 | 0.883039331 | 0.943093015 | 0.977740912 | 0.993640585 |
| 0.280 | 0.878720149 | 0.940221207 | 0.976185764 | 0.993003319 |
| 0.279 | 0.874306277 | 0.937250966 | 0.974550588 | 0.992316959 |
| 0.278 | 0.869797996 | 0.934181175 | 0.972833114 | 0.991578978 |
| 0.277 | 0.865195662 | 0.931010791 | 0.971031100 | 0.990786809 |
| 0.276 | 0.860499705 | 0.927738850 | 0.969142341 | 0.989937842 |
| 0.275 | 0.855710631 | 0.924364468 | 0.967164669 | 0.989029436 |
| 0.274 | 0.850829017 | 0.920886842 | 0.965095962 | 0.988058916 |
| 0.273 | 0.845855515 | 0.917305254 | 0.962934146 | 0.987023588 |
| 0.272 | 0.840790849 | 0.913619071 | 0.960677197 | 0.985920733 |
| 0.271 | 0.835635817 | 0.909827747 | 0.958323153 | 0.984747621 |
| 0.270 | 0.830391285 | 0.905930822 | 0.955870109 | 0.983501512 |
| 0.269 | 0.825058191 | 0.901927930 | 0.953316227 | 0.982179664 |
| 0.268 | 0.819637541 | 0.897818790 | 0.950659738 | 0.980779335 |
| 0.267 | 0.814130411 | 0.893603215 | 0.947898949 | 0.979297793 |
| 0.266 | 0.808537942 | 0.889281111 | 0.945032240 | 0.977732321 |
| 0.265 | 0.802861342 | 0.884852475 | 0.942058074 | 0.976080219 |
| 0.264 | 0.797101883 | 0.880317396 | 0.938974999 | 0.974338815 |
| 0.263 | 0.791260899 | 0.875676059 | 0.935781649 | 0.972505468 |
| 0.262 | 0.785339789 | 0.870928741 | 0.932476750 | 0.970577575 |
| 0.261 | 0.779340007 | 0.866075812 | 0.929059120 | 0.968552577 |
| 0.260 | 0.773263071 | 0.861117737 | 0.925527676 | 0.966427963 |
| 0.259 | 0.767110552 | 0.856055071 | 0.921881432 | 0.964201278 |
| 0.258 | 0.760884080 | 0.850888464 | 0.918119505 | 0.961870129 |
| 0.257 | 0.754585334 | 0.845618658 | 0.914241115 | 0.959432187 |
| 0.256 | 0.748216050 | 0.840246485 | 0.910245589 | 0.956885197 |
| 0.255 | 0.741778010 | 0.834772868 | 0.906132361 | 0.954226982 |
| 0.254 | 0.735273048 | 0.829198820 | 0.901900974 | 0.951455444 |
| 0.253 | 0.728703040 | 0.823525442 | 0.897551084 | 0.948568577 |
| 0.252 | 0.722069912 | 0.817753923 | 0.893082457 | 0.945564465 |
| 0.251 | 0.715375628 | 0.811885535 | 0.888494973 | 0.942441291 |

TABLE 1a:  CLUSTERING PROBABILITY  P(n;N,p)

n = 9

| P | N =22 | N =23 | N =24 | N =25 |
|------|------------|------------|------------|------------|
| 0.250 | 0.708622194 | 0.805921639 | 0.883788628 | 0.939197338 |
| 0.249 | 0.701811657 | 0.799863674 | 0.878963530 | 0.935830996 |
| 0.248 | 0.694946099 | 0.793713165 | 0.874019904 | 0.932340766 |
| 0.247 | 0.688027634 | 0.787471713 | 0.868958090 | 0.928725264 |
| 0.246 | 0.681058415 | 0.781140999 | 0.863778543 | 0.924983220 |
| 0.245 | 0.674040619 | 0.774722778 | 0.858481837 | 0.921113490 |
| 0.244 | 0.666976456 | 0.768218880 | 0.853068655 | 0.917115052 |
| 0.243 | 0.659868162 | 0.761631205 | 0.847539801 | 0.912987012 |
| 0.242 | 0.652717995 | 0.754961726 | 0.841896188 | 0.908728607 |
| 0.241 | 0.645528238 | 0.748212479 | 0.836138846 | 0.904339204 |
| 0.240 | 0.638301194 | 0.741385567 | 0.830268913 | 0.899818308 |
| 0.239 | 0.631039184 | 0.734483156 | 0.824287641 | 0.895165558 |
| 0.238 | 0.623744544 | 0.727507472 | 0.818196391 | 0.890380733 |
| 0.237 | 0.616419626 | 0.720460797 | 0.811996629 | 0.885463751 |
| 0.236 | 0.609066794 | 0.713345470 | 0.805689930 | 0.880414670 |
| 0.235 | 0.601688422 | 0.706163881 | 0.799277972 | 0.875233691 |
| 0.234 | 0.594286891 | 0.698918472 | 0.792762935 | 0.869921155 |
| 0.233 | 0.586864591 | 0.691611729 | 0.786145499 | 0.864477546 |
| 0.232 | 0.579423913 | 0.684246186 | 0.779428842 | 0.858903492 |
| 0.231 | 0.571967253 | 0.676824417 | 0.772614635 | 0.853199760 |
| 0.230 | 0.564497006 | 0.669349036 | 0.765705045 | 0.847367260 |
| 0.229 | 0.557015566 | 0.661822691 | 0.758702326 | 0.841407045 |
| 0.228 | 0.549525322 | 0.654248067 | 0.751608821 | 0.835320303 |
| 0.227 | 0.542028662 | 0.646627877 | 0.744426954 | 0.829108363 |
| 0.226 | 0.534527961 | 0.638964864 | 0.737159234 | 0.822772692 |
| 0.225 | 0.527025590 | 0.631261796 | 0.729808245 | 0.816314890 |
| 0.224 | 0.519523908 | 0.623521462 | 0.722376648 | 0.809736691 |
| 0.223 | 0.512025261 | 0.615746672 | 0.714867173 | 0.803039960 |
| 0.222 | 0.504531981 | 0.607940253 | 0.707282621 | 0.796226691 |
| 0.221 | 0.497046386 | 0.600105045 | 0.699625855 | 0.789299002 |
| 0.220 | 0.489570773 | 0.592243902 | 0.691899801 | 0.782259136 |
| 0.219 | 0.482107425 | 0.584359684 | 0.684107442 | 0.775109457 |
| 0.218 | 0.474658600 | 0.576455259 | 0.676251817 | 0.767852442 |
| 0.217 | 0.467226538 | 0.568533496 | 0.668336013 | 0.760490686 |
| 0.216 | 0.459813453 | 0.560597268 | 0.660363165 | 0.753026892 |
| 0.215 | 0.452421535 | 0.552649444 | 0.652336452 | 0.745463868 |
| 0.214 | 0.445052947 | 0.544692890 | 0.644259092 | 0.737804526 |
| 0.213 | 0.437709825 | 0.536730463 | 0.636134338 | 0.730051877 |
| 0.212 | 0.430394277 | 0.528765013 | 0.627965478 | 0.722209026 |
| 0.211 | 0.423108379 | 0.520799376 | 0.619755826 | 0.714279167 |
| 0.210 | 0.415854178 | 0.512836376 | 0.611508722 | 0.706265582 |
| 0.209 | 0.408633684 | 0.504878818 | 0.603227526 | 0.698171634 |
| 0.208 | 0.401448879 | 0.496929490 | 0.594915618 | 0.690000761 |
| 0.207 | 0.394301705 | 0.488991157 | 0.586576389 | 0.681756476 |
| 0.206 | 0.387194072 | 0.481066563 | 0.578213241 | 0.673442359 |
| 0.205 | 0.380127850 | 0.473158424 | 0.569829585 | 0.665062052 |
| 0.204 | 0.373104373 | 0.465269430 | 0.561428831 | 0.656619257 |
| 0.203 | 0.366126936 | 0.457402240 | 0.553014392 | 0.648117729 |
| 0.202 | 0.359195793 | 0.449559483 | 0.544589675 | 0.639561271 |
| 0.201 | 0.352313159 | 0.441743752 | 0.536158082 | 0.630953731 |

TABLE 1a:  CLUSTERING PROBABILITY  P(n;N,p)

n = 9

| P | N = 22 | N = 23 | N = 24 | N = 25 |
|---|---|---|---|---|
| 0.200 | 0.345480707 | 0.433957607 | 0.527723000 | 0.622298995 |
| 0.199 | 0.338700067 | 0.426203568 | 0.519287807 | 0.613600985 |
| 0.198 | 0.331972828 | 0.418484119 | 0.510855860 | 0.604863649 |
| 0.197 | 0.325300534 | 0.410801701 | 0.502430497 | 0.596090963 |
| 0.196 | 0.318684685 | 0.403158712 | 0.494015032 | 0.587286920 |
| 0.195 | 0.312126736 | 0.395557509 | 0.485612753 | 0.578455530 |
| 0.194 | 0.305628100 | 0.388000401 | 0.477226918 | 0.569600811 |
| 0.193 | 0.299190139 | 0.380489650 | 0.468860751 | 0.560726787 |
| 0.192 | 0.292814172 | 0.373027473 | 0.460517442 | 0.551837485 |
| 0.191 | 0.286501471 | 0.365616033 | 0.452200142 | 0.542936924 |
| 0.190 | 0.280253262 | 0.358257445 | 0.443911963 | 0.534029117 |
| 0.189 | 0.274070721 | 0.350953771 | 0.435655971 | 0.525118064 |
| 0.188 | 0.267954978 | 0.343707020 | 0.427435186 | 0.516207747 |
| 0.187 | 0.261907116 | 0.336519148 | 0.419252583 | 0.507302125 |
| 0.186 | 0.255928168 | 0.329392053 | 0.411111081 | 0.498405134 |
| 0.185 | 0.250019121 | 0.322327581 | 0.403013552 | 0.489520677 |
| 0.184 | 0.244180911 | 0.315327518 | 0.394962807 | 0.480652625 |
| 0.183 | 0.238414428 | 0.308393593 | 0.386961603 | 0.471804809 |
| 0.182 | 0.232720513 | 0.301527478 | 0.379012639 | 0.462981019 |
| 0.181 | 0.227099957 | 0.294730783 | 0.371118549 | 0.454184998 |
| 0.180 | 0.221553504 | 0.288005061 | 0.363281908 | 0.445420443 |
| 0.179 | 0.216081849 | 0.281351804 | 0.355505223 | 0.436690993 |
| 0.178 | 0.210685640 | 0.274772443 | 0.347790939 | 0.428000235 |
| 0.177 | 0.205365474 | 0.268268349 | 0.340141429 | 0.419351694 |
| 0.176 | 0.200121903 | 0.261840831 | 0.332559001 | 0.410748834 |
| 0.175 | 0.194955429 | 0.255491135 | 0.325045889 | 0.402195050 |
| 0.174 | 0.189866509 | 0.249220446 | 0.317604258 | 0.393693673 |
| 0.173 | 0.184855550 | 0.243029888 | 0.310236199 | 0.385247957 |
| 0.172 | 0.179922914 | 0.236920522 | 0.302943731 | 0.376861088 |
| 0.171 | 0.175068915 | 0.230893345 | 0.295728796 | 0.368536170 |
| 0.170 | 0.170293822 | 0.224949295 | 0.288593263 | 0.360276230 |
| 0.169 | 0.165597857 | 0.219089246 | 0.281538922 | 0.352084216 |
| 0.168 | 0.160981198 | 0.213314010 | 0.274567488 | 0.343962988 |
| 0.167 | 0.156443977 | 0.207624336 | 0.267680599 | 0.335915325 |
| 0.166 | 0.151986283 | 0.202020914 | 0.260879814 | 0.327943917 |
| 0.165 | 0.147608158 | 0.196504370 | 0.254166613 | 0.320051364 |
| 0.164 | 0.143309604 | 0.191075270 | 0.247542400 | 0.312240177 |
| 0.163 | 0.139090579 | 0.185734120 | 0.241008498 | 0.304512775 |
| 0.162 | 0.134950999 | 0.180481364 | 0.234566152 | 0.296871484 |
| 0.161 | 0.130890738 | 0.175317389 | 0.228216526 | 0.289318534 |
| 0.160 | 0.126909629 | 0.170242519 | 0.221960708 | 0.281856062 |
| 0.159 | 0.123007467 | 0.165257022 | 0.215799707 | 0.274486107 |
| 0.158 | 0.119184005 | 0.160361107 | 0.209734450 | 0.267210611 |
| 0.157 | 0.115438959 | 0.155554925 | 0.203765789 | 0.260031420 |
| 0.156 | 0.111772005 | 0.150838572 | 0.197894496 | 0.252950279 |
| 0.155 | 0.108182784 | 0.146212086 | 0.192121268 | 0.245968836 |
| 0.154 | 0.104670900 | 0.141675451 | 0.186446721 | 0.239088640 |
| 0.153 | 0.101235920 | 0.137228596 | 0.180871399 | 0.232311141 |
| 0.152 | 0.097877379 | 0.132871397 | 0.175395766 | 0.225637689 |
| 0.151 | 0.094594776 | 0.128603677 | 0.170020213 | 0.219069537 |

TABLE 1a:  CLUSTERING PROBABILITY  P(n;N,p)

n = 9

| P | N =22 | N =23 | N =24 | N =25 |
|-------|-------------|-------------|-------------|-------------|
| 0.150 | 0.091387577 | 0.124425208 | 0.164745056 | 0.212607835 |
| 0.149 | 0.088255218 | 0.120335709 | 0.159570538 | 0.206253640 |
| 0.148 | 0.085197101 | 0.116334852 | 0.154496830 | 0.200007907 |
| 0.147 | 0.082212599 | 0.112422259 | 0.149524028 | 0.193871494 |
| 0.146 | 0.079301056 | 0.108597504 | 0.144652162 | 0.187845164 |
| 0.145 | 0.076461787 | 0.104860116 | 0.139881188 | 0.181929581 |
| 0.144 | 0.073694080 | 0.101209576 | 0.135210998 | 0.176125315 |
| 0.143 | 0.070997197 | 0.097645323 | 0.130641412 | 0.170432842 |
| 0.142 | 0.068370372 | 0.094166752 | 0.126172189 | 0.164852545 |
| 0.141 | 0.065812818 | 0.090773216 | 0.121803018 | 0.159384713 |
| 0.140 | 0.063323722 | 0.087464027 | 0.117533529 | 0.154029545 |
| 0.139 | 0.060902249 | 0.084238459 | 0.113363288 | 0.148787149 |
| 0.138 | 0.058547542 | 0.081095746 | 0.109291800 | 0.143657548 |
| 0.137 | 0.056258726 | 0.078035086 | 0.105318513 | 0.138640674 |
| 0.136 | 0.054034903 | 0.075055641 | 0.101442815 | 0.133736376 |
| 0.135 | 0.051875159 | 0.072156539 | 0.097664039 | 0.128944419 |
| 0.134 | 0.049778562 | 0.069336874 | 0.093981464 | 0.124264484 |
| 0.133 | 0.047744163 | 0.066595711 | 0.090394316 | 0.119696175 |
| 0.132 | 0.045770998 | 0.063932081 | 0.086901769 | 0.115239015 |
| 0.131 | 0.043858089 | 0.061344989 | 0.083502949 | 0.110892450 |
| 0.130 | 0.042004443 | 0.058833410 | 0.080196932 | 0.106655854 |
| 0.129 | 0.040209058 | 0.056396296 | 0.076982750 | 0.102528526 |
| 0.128 | 0.038470918 | 0.054032572 | 0.073859389 | 0.098509695 |
| 0.127 | 0.036788997 | 0.051741139 | 0.070825794 | 0.094598521 |
| 0.126 | 0.035162261 | 0.049520877 | 0.067880869 | 0.090794099 |
| 0.125 | 0.033589666 | 0.047370646 | 0.065023477 | 0.087095458 |
| 0.124 | 0.032070163 | 0.045289285 | 0.062252446 | 0.083501567 |
| 0.123 | 0.030602695 | 0.043275617 | 0.059566569 | 0.080011333 |
| 0.122 | 0.029186200 | 0.041328447 | 0.056964604 | 0.076623607 |
| 0.121 | 0.027819613 | 0.039446566 | 0.054445278 | 0.073337186 |
| 0.120 | 0.026501864 | 0.037628751 | 0.052007289 | 0.070150811 |
| 0.119 | 0.025231881 | 0.035873766 | 0.049649305 | 0.067063176 |
| 0.118 | 0.024008592 | 0.034180365 | 0.047369970 | 0.064072926 |
| 0.117 | 0.022830922 | 0.032547291 | 0.045167904 | 0.061178659 |
| 0.116 | 0.021697799 | 0.030973281 | 0.043041703 | 0.058378931 |
| 0.115 | 0.020608150 | 0.029457063 | 0.040989945 | 0.055672259 |
| 0.114 | 0.019560907 | 0.027997360 | 0.039011186 | 0.053057119 |
| 0.113 | 0.018555002 | 0.026592891 | 0.037103968 | 0.050531953 |
| 0.112 | 0.017589372 | 0.025242372 | 0.035266817 | 0.048095169 |
| 0.111 | 0.016662961 | 0.023944516 | 0.033498246 | 0.045745145 |
| 0.110 | 0.015774714 | 0.022698038 | 0.031796757 | 0.043480231 |
| 0.109 | 0.014923587 | 0.021501651 | 0.030160843 | 0.041298748 |
| 0.108 | 0.014108541 | 0.020354071 | 0.028588988 | 0.039198999 |
| 0.107 | 0.013328543 | 0.019254018 | 0.027079669 | 0.037179260 |
| 0.106 | 0.012582573 | 0.018200215 | 0.025631362 | 0.035237793 |
| 0.105 | 0.011869616 | 0.017191390 | 0.024242539 | 0.033372842 |
| 0.104 | 0.011188669 | 0.016226280 | 0.022911668 | 0.031582637 |
| 0.103 | 0.010538741 | 0.015303626 | 0.021637222 | 0.029865397 |
| 0.102 | 0.009918849 | 0.014422181 | 0.020417675 | 0.028219331 |
| 0.101 | 0.009328025 | 0.013580706 | 0.019251503 | 0.026642642 |

TABLE 1a:   CLUSTERING PROBABILITY  P(n;N,p)

$n = 9$

| P | N = 22 | N = 23 | N = 24 | N = 25 |
|---|---|---|---|---|
| 0.100 | 0.008765312 | 0.012777972 | 0.018137190 | 0.025133528 |
| 0.099 | 0.008229765 | 0.012012763 | 0.017073226 | 0.023690186 |
| 0.098 | 0.007720456 | 0.011283875 | 0.016058108 | 0.022310809 |
| 0.097 | 0.007236467 | 0.010590117 | 0.015090345 | 0.020993596 |
| 0.096 | 0.006776898 | 0.009930313 | 0.014168455 | 0.019736748 |
| 0.095 | 0.006340861 | 0.009303302 | 0.013290971 | 0.018538472 |
| 0.094 | 0.005927487 | 0.008707938 | 0.012456438 | 0.017396984 |
| 0.093 | 0.005535921 | 0.008143094 | 0.011663416 | 0.016310508 |
| 0.092 | 0.005165323 | 0.007607657 | 0.010910484 | 0.015277282 |
| 0.091 | 0.004814871 | 0.007100534 | 0.010196235 | 0.014295557 |
| 0.090 | 0.004483761 | 0.006620652 | 0.009519282 | 0.013363598 |
| 0.089 | 0.004171204 | 0.006166955 | 0.008878259 | 0.012479690 |
| 0.088 | 0.003876430 | 0.005738408 | 0.008271818 | 0.011642135 |
| 0.087 | 0.003598687 | 0.005333996 | 0.007698635 | 0.010849254 |
| 0.086 | 0.003337240 | 0.004952725 | 0.007157409 | 0.010099394 |
| 0.085 | 0.003091372 | 0.004593623 | 0.006646859 | 0.009390921 |
| 0.084 | 0.002860385 | 0.004255738 | 0.006165731 | 0.008722229 |
| 0.083 | 0.002643599 | 0.003938144 | 0.005712797 | 0.008091736 |
| 0.082 | 0.002440354 | 0.003639932 | 0.005286851 | 0.007497889 |
| 0.081 | 0.002250006 | 0.003360220 | 0.004886717 | 0.006939162 |
| 0.080 | 0.002071931 | 0.003098148 | 0.004511243 | 0.006414060 |
| 0.079 | 0.001905524 | 0.002852879 | 0.004159308 | 0.005921118 |
| 0.078 | 0.001750199 | 0.002623598 | 0.003829815 | 0.005458902 |
| 0.077 | 0.001605388 | 0.002409516 | 0.003521698 | 0.005026011 |
| 0.076 | 0.001470540 | 0.002209866 | 0.003233919 | 0.004621078 |
| 0.075 | 0.001345125 | 0.002023905 | 0.002965469 | 0.004242769 |
| 0.074 | 0.001228631 | 0.001850915 | 0.002715369 | 0.003889785 |
| 0.073 | 0.001120563 | 0.001690200 | 0.002482667 | 0.003560863 |
| 0.072 | 0.001020445 | 0.001541087 | 0.002266444 | 0.003254773 |
| 0.071 | 0.000927820 | 0.001402930 | 0.002065810 | 0.002970325 |
| 0.070 | 0.000842246 | 0.001275104 | 0.001879902 | 0.002706362 |
| 0.069 | 0.000763301 | 0.001157006 | 0.001707890 | 0.002461766 |
| 0.068 | 0.000690579 | 0.001048059 | 0.001548974 | 0.002235455 |
| 0.067 | 0.000623693 | 0.000947708 | 0.001402380 | 0.002026384 |
| 0.066 | 0.000562269 | 0.000855420 | 0.001267367 | 0.001833546 |
| 0.065 | 0.000505954 | 0.000770684 | 0.001143222 | 0.001655970 |
| 0.064 | 0.000454409 | 0.000693012 | 0.001029260 | 0.001492722 |
| 0.063 | 0.000407310 | 0.000621938 | 0.000924828 | 0.001342905 |
| 0.062 | 0.000364349 | 0.000557016 | 0.000829298 | 0.001205659 |
| 0.061 | 0.000325236 | 0.000497823 | 0.000742071 | 0.001080161 |
| 0.060 | 0.000289691 | 0.000443954 | 0.000662576 | 0.000965621 |
| 0.059 | 0.000257453 | 0.000395027 | 0.000590269 | 0.000861288 |
| 0.058 | 0.000228273 | 0.000350677 | 0.000524633 | 0.000766443 |
| 0.057 | 0.000201915 | 0.000310559 | 0.000465176 | 0.000680405 |
| 0.056 | 0.000178157 | 0.000274348 | 0.000411433 | 0.000602524 |
| 0.055 | 0.000156791 | 0.000241737 | 0.000362963 | 0.000532183 |
| 0.054 | 0.000137621 | 0.000212434 | 0.000319350 | 0.000468801 |
| 0.053 | 0.000120460 | 0.000186168 | 0.000280200 | 0.000411824 |
| 0.052 | 0.000105138 | 0.000162682 | 0.000245145 | 0.000360735 |
| 0.051 | 0.000091491 | 0.000141735 | 0.000213836 | 0.000315041 |

TABLE 1a:   CLUSTERING PROBABILITY   P(n;N,p)

n = 9

| P | N =22 | N =23 | N =24 | N =25 |
|------|------------|------------|------------|------------|
| 0.050 | 0.000079369 | 0.000123104 | 0.000185949 | 0.000274283 |
| 0.049 | 0.000068633 | 0.000106578 | 0.000161178 | 0.000238029 |
| 0.048 | 0.000059149 | 0.000091961 | 0.000139239 | 0.000205874 |
| 0.047 | 0.000050799 | 0.000079072 | 0.000119866 | 0.000177441 |
| 0.046 | 0.000043469 | 0.000067742 | 0.000102813 | 0.000152379 |
| 0.045 | 0.000037055 | 0.000057815 | 0.000087851 | 0.000130358 |
| 0.044 | 0.000031462 | 0.000049148 | 0.000074769 | 0.000111078 |
| 0.043 | 0.000026603 | 0.000041606 | 0.000063371 | 0.000094256 |
| 0.042 | 0.000022397 | 0.000035069 | 0.000053477 | 0.000079634 |
| 0.041 | 0.000018770 | 0.000029425 | 0.000044923 | 0.000066975 |
| 0.040 | 0.000015655 | 0.000024571 | 0.000037557 | 0.000056059 |
| 0.039 | 0.000012992 | 0.000020415 | 0.000031241 | 0.000046687 |
| 0.038 | 0.000010725 | 0.000016873 | 0.000025850 | 0.000038676 |
| 0.037 | 0.000008805 | 0.000013867 | 0.000021271 | 0.000031863 |
| 0.036 | 0.000007186 | 0.000011331 | 0.000017401 | 0.000026096 |
| 0.035 | 0.000005828 | 0.000009201 | 0.000014147 | 0.000021240 |
| 0.034 | 0.000004696 | 0.000007423 | 0.000011426 | 0.000017175 |
| 0.033 | 0.000003758 | 0.000005947 | 0.000009165 | 0.000013792 |
| 0.032 | 0.000002985 | 0.000004729 | 0.000007297 | 0.000010994 |
| 0.031 | 0.000002353 | 0.000003732 | 0.000005764 | 0.000008695 |
| 0.030 | 0.000001839 | 0.000002920 | 0.000004516 | 0.000006820 |
| 0.029 | 0.000001425 | 0.000002265 | 0.000003506 | 0.000005301 |
| 0.028 | 0.000001093 | 0.000001740 | 0.000002697 | 0.000004082 |
| 0.027 | 0.000000830 | 0.000001323 | 0.000002053 | 0.000003111 |
| 0.026 | 0.000000624 | 0.000000995 | 0.000001546 | 0.000002345 |
| 0.025 | 0.000000463 | 0.000000739 | 0.000001150 | 0.000001747 |
| 0.024 | 0.000000339 | 0.000000542 | 0.000000845 | 0.000001285 |
| 0.023 | 0.000000245 | 0.000000393 | 0.000000612 | 0.000000932 |
| 0.022 | 0.000000175 | 0.000000280 | 0.000000437 | 0.000000666 |
| 0.021 | 0.000000122 | 0.000000196 | 0.000000306 | 0.000000468 |
| 0.020 | 0.000000084 | 0.000000135 | 0.000000211 | 0.000000323 |
| 0.019 | 0.000000057 | 0.000000091 | 0.000000143 | 0.000000218 |
| 0.018 | 0.000000037 | 0.000000060 | 0.000000094 | 0.000000144 |
| 0.017 | 0.000000024 | 0.000000039 | 0.000000061 | 0.000000093 |
| 0.016 | 0.000000015 | 0.000000024 | 0.000000038 | 0.000000059 |
| 0.015 | 0.000000009 | 0.000000015 | 0.000000023 | 0.000000036 |
| 0.014 | 0.000000005 | 0.000000009 | 0.000000014 | 0.000000021 |
| 0.013 | 0.000000003 | 0.000000005 | 0.000000008 | 0.000000012 |
| 0.012 | 0.000000002 | 0.000000003 | 0.000000004 | 0.000000006 |
| 0.011 | 0.000000001 | 0.000000001 | 0.000000002 | 0.000000003 |
| 0.010 | 0.000000000 | 0.000000001 | 0.000000001 | 0.000000002 |
| 0.009 | 0.000000000 | 0.000000000 | 0.000000000 | 0.000000001 |
| 0.008 | 0.000000000 | 0.000000000 | 0.000000000 | 0.000000000 |

TABLE 2:   POISSON CLUSTER PROBABILITY   $P'(n;\lambda,p)$

|  |  |  |  | n = 3 |
| p | $\lambda = 1$ | $\lambda = 2$ | $\lambda = 3$ | $\lambda = 4$ |
|---|---|---|---|---|
| 0.50 | 0.047728736 | 0.221822083 | 0.443785071 | 0.639792383 |
| 0.49 | 0.046651158 | 0.218185365 | 0.438691497 | 0.634846151 |
| 0.48 | 0.045564413 | 0.214490354 | 0.433484018 | 0.629761934 |
| 0.47 | 0.044468582 | 0.210733235 | 0.428150058 | 0.624519825 |
| 0.46 | 0.043363798 | 0.206910431 | 0.422677815 | 0.619100511 |
| 0.45 | 0.042250250 | 0.203018844 | 0.417056024 | 0.613485634 |
| 0.44 | 0.041128188 | 0.199055672 | 0.411274254 | 0.607657552 |
| 0.43 | 0.039997920 | 0.195018411 | 0.405322671 | 0.601599455 |
| 0.42 | 0.038859807 | 0.190905094 | 0.399192214 | 0.595295370 |
| 0.41 | 0.037714284 | 0.186714053 | 0.392874539 | 0.588729858 |
| 0.40 | 0.036561839 | 0.182443976 | 0.386361778 | 0.581888258 |
| 0.39 | 0.035403021 | 0.178093970 | 0.379646957 | 0.574756324 |
| 0.38 | 0.034238447 | 0.173663437 | 0.372723460 | 0.567320108 |
| 0.37 | 0.033068795 | 0.169152260 | 0.365585148 | 0.559565902 |
| 0.36 | 0.031894810 | 0.164560616 | 0.358226478 | 0.551480055 |
| 0.35 | 0.030717302 | 0.159888983 | 0.350642085 | 0.543048620 |
| 0.34 | 0.029537141 | 0.155138314 | 0.342826962 | 0.534257233 |
| 0.33 | 0.028355263 | 0.150309563 | 0.334776103 | 0.525090933 |
| 0.32 | 0.027172673 | 0.145404339 | 0.326484561 | 0.515533447 |
| 0.31 | 0.025990438 | 0.140424192 | 0.317947030 | 0.505566537 |
| 0.30 | 0.024809692 | 0.135371029 | 0.309157789 | 0.495169759 |
| 0.29 | 0.023631651 | 0.130247295 | 0.300111353 | 0.484320998 |
| 0.28 | 0.022457603 | 0.125055671 | 0.290802240 | 0.472996652 |
| 0.27 | 0.021288920 | 0.119799495 | 0.281225622 | 0.461172163 |
| 0.26 | 0.020127077 | 0.114482760 | 0.271377563 | 0.448822558 |
| 0.25 | 0.018973622 | 0.109110057 | 0.261255026 | 0.435922801 |
| 0.24 | 0.017830215 | 0.103686929 | 0.250856578 | 0.422447979 |
| 0.23 | 0.016698625 | 0.098219693 | 0.240182281 | 0.408373654 |
| 0.22 | 0.015580717 | 0.092715740 | 0.229234278 | 0.393676043 |
| 0.21 | 0.014478482 | 0.087183535 | 0.218016863 | 0.378332734 |
| 0.20 | 0.013394024 | 0.081632733 | 0.206537247 | 0.362322986 |
| 0.19 | 0.012329582 | 0.076074481 | 0.194805980 | 0.345629215 |
| 0.18 | 0.011287514 | 0.070521355 | 0.182837725 | 0.328237951 |
| 0.17 | 0.010270327 | 0.064987481 | 0.170651913 | 0.310141683 |
| 0.16 | 0.009280670 | 0.059489153 | 0.158273757 | 0.291340828 |
| 0.15 | 0.008321334 | 0.054044355 | 0.145735264 | 0.271846116 |
| 0.14 | 0.007395279 | 0.048673533 | 0.133076429 | 0.251682043 |
| 0.13 | 0.006505616 | 0.043399543 | 0.120346725 | 0.230890572 |
| 0.12 | 0.005655631 | 0.038247932 | 0.107606590 | 0.209536195 |
| 0.11 | 0.004848778 | 0.033247169 | 0.094929159 | 0.187711895 |
| 0.10 | 0.004088696 | 0.028428920 | 0.082402587 | 0.165546238 |
| 0.09 | 0.003379216 | 0.023828268 | 0.070131779 | 0.143212140 |
| 0.08 | 0.002724345 | 0.019484002 | 0.058241192 | 0.120937169 |
| 0.07 | 0.002128298 | 0.015438825 | 0.046877019 | 0.099015057 |
| 0.06 | 0.001595488 | 0.011739619 | 0.036209960 | 0.077818811 |
| 0.05 | 0.001130534 | 0.008437622 | 0.026437558 | 0.057814848 |
| 0.04 | 0.000738263 | 0.005588606 | 0.017786384 | 0.039576501 |
| 0.03 | 0.000423715 | 0.003252988 | 0.010513622 | 0.02379615 |
| 0.02 | 0.000192141 | 0.001495835 | 0.004907735 | 0.01129248 |
| 0.01 | 0.000049009 | 0.000386818 | 0.001287659 | 0.00300926 |

TABLE 2:  POISSON CLUSTER PROBABILITY  $P'(n;\lambda,p)$

$n = 3$

| p | $\lambda = 5$ | $\lambda = 6$ | $\lambda = 7$ | $\lambda = 8$ |
|---|---|---|---|---|
| 0.50 | 0.783227563 | 0.876682043 | 0.932895899 | 0.964776397 |
| 0.49 | 0.779310763 | 0.873961270 | 0.931171536 | 0.963755310 |
| 0.48 | 0.775265455 | 0.871139109 | 0.929375708 | 0.962687969 |
| 0.47 | 0.771068990 | 0.868194759 | 0.927492142 | 0.961562693 |
| 0.46 | 0.766699612 | 0.865108311 | 0.925504744 | 0.960368037 |
| 0.45 | 0.762136102 | 0.861860037 | 0.923397958 | 0.959092736 |
| 0.44 | 0.757358074 | 0.858431041 | 0.921156526 | 0.957725763 |
| 0.43 | 0.752345741 | 0.854802668 | 0.918765306 | 0.956256092 |
| 0.42 | 0.747079909 | 0.850956678 | 0.916209400 | 0.954672873 |
| 0.41 | 0.741541922 | 0.846875012 | 0.913473845 | 0.952964902 |
| 0.40 | 0.735713363 | 0.842539608 | 0.910543501 | 0.951120913 |
| 0.39 | 0.729575992 | 0.837932229 | 0.907402694 | 0.949128926 |
| 0.38 | 0.723111391 | 0.833033979 | 0.904035270 | 0.946976423 |
| 0.37 | 0.716300964 | 0.827825427 | 0.900423884 | 0.944649935 |
| 0.36 | 0.709125340 | 0.822285891 | 0.896550000 | 0.942134738 |
| 0.35 | 0.701564193 | 0.816393137 | 0.892393529 | 0.939414680 |
| 0.34 | 0.693596005 | 0.810123265 | 0.887932241 | 0.936471701 |
| 0.33 | 0.685197532 | 0.803449810 | 0.883141398 | 0.933285594 |
| 0.32 | 0.676343024 | 0.796343267 | 0.877993047 | 0.929833114 |
| 0.31 | 0.667002738 | 0.788768411 | 0.872453153 | 0.926085651 |
| 0.30 | 0.657142282 | 0.780683815 | 0.866480708 | 0.922007859 |
| 0.29 | 0.646722972 | 0.772041619 | 0.860027611 | 0.917557716 |
| 0.28 | 0.635702431 | 0.762788057 | 0.853038609 | 0.912686229 |
| 0.27 | 0.624034564 | 0.752863467 | 0.845451415 | 0.907336950 |
| 0.26 | 0.611670434 | 0.742202401 | 0.837196112 | 0.901445925 |
| 0.25 | 0.598558068 | 0.730733633 | 0.828195035 | 0.894940555 |
| 0.24 | 0.584642768 | 0.718379378 | 0.818361223 | 0.887737870 |
| 0.23 | 0.569866598 | 0.705054224 | 0.807596743 | 0.879742384 |
| 0.22 | 0.554168284 | 0.690663874 | 0.795789599 | 0.870842338 |
| 0.21 | 0.537482858 | 0.675103605 | 0.782811880 | 0.860906184 |
| 0.20 | 0.519742429 | 0.658257842 | 0.768516898 | 0.849779010 |
| 0.19 | 0.500876606 | 0.639999628 | 0.752737343 | 0.837278724 |
| 0.18 | 0.480814040 | 0.620190680 | 0.735282958 | 0.823191285 |
| 0.17 | 0.459483922 | 0.598681450 | 0.715937972 | 0.807265043 |
| 0.16 | 0.436818361 | 0.575312257 | 0.694458604 | 0.789203882 |
| 0.15 | 0.412755907 | 0.549914956 | 0.670570910 | 0.768659770 |
| 0.14 | 0.387246132 | 0.522316933 | 0.643970370 | 0.745225310 |
| 0.13 | 0.360256195 | 0.492347419 | 0.614323497 | 0.718426287 |
| 0.12 | 0.331779540 | 0.459847391 | 0.581273258 | 0.687717021 |
| 0.11 | 0.301847994 | 0.424685240 | 0.544451237 | 0.652480006 |
| 0.10 | 0.270547390 | 0.386779904 | 0.503500581 | 0.612035811 |
| 0.09 | 0.238038242 | 0.346134543 | 0.458115160 | 0.565672338 |
| 0.08 | 0.204582334 | 0.302885354 | 0.408104897 | 0.512707353 |
| 0.07 | 0.170576572 | 0.257369518 | 0.35349923 | 0.4526090 |
| 0.06 | 0.136594176 | 0.21021813 | 0.2947062 | 0.385211 |
| 0.05 | 0.10343361 | 0.1624777 | 0.232746 | 0.31107 |
| 0.04 | 0.0721728 | 0.115762 | 0.16958 | 0.2321 |
| 0.03 | 0.0442229 | 0.072421 | 0.10852 | 0.1522 |
| 0.02 | 0.021371 | 0.03571 | 0.0547 | 0.079 |
| 0.01 | 0.005792 | 0.00986 | 0.0154 | 0.023 |

TABLE 2:   POISSON CLUSTER PROBABILITY   $P'(n;\lambda,p)$

<u>n = 3</u>

| p | $\lambda = 9$ | $\lambda = 10$ | $\lambda = 11$ | $\lambda = 12$ |
|------|-----------|------------|------------|------------|
| 0.50 | 0.982053459 | 0.991082668 | 0.995662987 | 0.997929394 |
| 0.49 | 0.981479764 | 0.990773499 | 0.995501995 | 0.997847855 |
| 0.48 | 0.980877936 | 0.990448117 | 0.995332003 | 0.997761548 |
| 0.47 | 0.980240345 | 0.990101755 | 0.995150268 | 0.997668862 |
| 0.46 | 0.979559362 | 0.989729702 | 0.994953871 | 0.997568130 |
| 0.45 | 0.978827477 | 0.989327192 | 0.994740069 | 0.997457862 |
| 0.44 | 0.978037417 | 0.988889754 | 0.994506240 | 0.997336447 |
| 0.43 | 0.977181792 | 0.988412678 | 0.994249463 | 0.997202337 |
| 0.42 | 0.976253092 | 0.987891257 | 0.993966997 | 0.997053802 |
| 0.41 | 0.975243866 | 0.987320662 | 0.993655860 | 0.996889234 |
| 0.40 | 0.974146247 | 0.986695766 | 0.993312955 | 0.996706724 |
| 0.39 | 0.972951889 | 0.986011267 | 0.992934942 | 0.996504307 |
| 0.38 | 0.971651971 | 0.985261202 | 0.992518127 | 0.996279895 |
| 0.37 | 0.970236778 | 0.984439194 | 0.992058456 | 0.996030927 |
| 0.36 | 0.968695760 | 0.983538151 | 0.991551459 | 0.995754719 |
| 0.35 | 0.967017055 | 0.982549906 | 0.990991890 | 0.995448053 |
| 0.34 | 0.965187371 | 0.981465399 | 0.990373969 | 0.995107472 |
| 0.33 | 0.963191688 | 0.980274379 | 0.989690959 | 0.994728684 |
| 0.32 | 0.961012602 | 0.978964627 | 0.988934934 | 0.994306862 |
| 0.31 | 0.958628178 | 0.977520823 | 0.988095701 | 0.993835265 |
| 0.30 | 0.956011236 | 0.975923359 | 0.987160146 | 0.993306458 |
| 0.29 | 0.953128457 | 0.974148095 | 0.986111760 | 0.992708862 |
| 0.28 | 0.949940681 | 0.972166240 | 0.984930754 | 0.992029786 |
| 0.27 | 0.946402371 | 0.969943762 | 0.983593345 | 0.991253555 |
| 0.26 | 0.942460895 | 0.967441022 | 0.982071698 | 0.990361691 |
| 0.25 | 0.938055634 | 0.964611650 | 0.980332613 | 0.989331782 |
| 0.24 | 0.933116198 | 0.961400986 | 0.978336573 | 0.988136768 |
| 0.23 | 0.927559853 | 0.957743585 | 0.976035416 | 0.986743271 |
| 0.22 | 0.921287775 | 0.953559697 | 0.973369539 | 0.985109329 |
| 0.21 | 0.914180756 | 0.948751390 | 0.970264077 | 0.983181298 |
| 0.20 | 0.906095147 | 0.943197668 | 0.966624975 | 0.980890393 |
| 0.19 | 0.896857440 | 0.936749339 | 0.962333620 | 0.978148341 |
| 0.18 | 0.886258304 | 0.929221749 | 0.957240105 | 0.974841237 |
| 0.17 | 0.874043763 | 0.920385003 | 0.951153159 | 0.970820546 |
| 0.16 | 0.859905005 | 0.909951150 | 0.943826795 | 0.965890527 |
| 0.15 | 0.843465447 | 0.897557676 | 0.934942305 | 0.959790766 |
| 0.14 | 0.824265897 | 0.882746696 | 0.924084246 | 0.952171624 |
| 0.13 | 0.801746607 | 0.864938021 | 0.910707414 | 0.942559779 |
| 0.12 | 0.775228381 | 0.843395650 | 0.894093156 | 0.930308580 |
| 0.11 | 0.743893087 | 0.817187726 | 0.873290718 | 0.914528251 |
| 0.10 | 0.706769288 | 0.785140872 | 0.847041965 | 0.893987656 |
| 0.09 | 0.662732124 | 0.745796800 | 0.813689709 | 0.866979897 |
| 0.08 | 0.61053538 | 0.69738734 | 0.7710788 | 0.8311483 |
| 0.07 | 0.5489115 | 0.637867 | 0.716483 | 0.78329 |
| 0.06 | 0.47680 | 0.56509 | 0.64665 | 0.7192 |
| 0.05 | 0.39380 | 0.4773 | 0.5582 | 0.634 |
| 0.04 | 0.3011 | 0.374 | 0.449 | 0.522 |
| 0.03 | 0.203 | 0.259 | 0.32 | 0.38 |
| 0.02 | 0.108 | 0.14 | 0.18 | 0.23 |
| 0.01 | 0.032 | 0.05 | 0.06 | |

TABLE 2:   POISSON CLUSTER PROBABILITY   P'(n;λ,p)

| p | λ = 1 | λ = 2 | λ = 3 | n = 4<br>λ = 4 |
|---|---|---|---|---|
| 0.50 | 0.007452004 | 0.068630099 | 0.201802433 | 0.375233173 |
| 0.49 | 0.007161140 | 0.066486239 | 0.196894884 | 0.368334591 |
| 0.48 | 0.006873086 | 0.064340770 | 0.191937208 | 0.361306369 |
| 0.47 | 0.006588075 | 0.062194582 | 0.186927319 | 0.354137480 |
| 0.46 | 0.006306335 | 0.060048562 | 0.181863904 | 0.346818626 |
| 0.45 | 0.006028108 | 0.057903945 | 0.176746309 | 0.339341879 |
| 0.44 | 0.005753633 | 0.055762086 | 0.171574771 | 0.331701100 |
| 0.43 | 0.005483147 | 0.053624522 | 0.166350365 | 0.323891521 |
| 0.42 | 0.005216900 | 0.051492956 | 0.161074698 | 0.315909922 |
| 0.41 | 0.004955132 | 0.049369242 | 0.155750334 | 0.307754457 |
| 0.40 | 0.004698087 | 0.047255393 | 0.150380254 | 0.299424589 |
| 0.39 | 0.004446011 | 0.045153555 | 0.144968331 | 0.290920973 |
| 0.38 | 0.004199143 | 0.043066017 | 0.139519036 | 0.282245457 |
| 0.37 | 0.003957722 | 0.040995181 | 0.134037316 | 0.273401022 |
| 0.36 | 0.003721993 | 0.038943570 | 0.128528774 | 0.264391661 |
| 0.35 | 0.003492178 | 0.036913812 | 0.122999668 | 0.255222499 |
| 0.34 | 0.003268507 | 0.034908634 | 0.117456675 | 0.245899618 |
| 0.33 | 0.003051201 | 0.032930847 | 0.111907125 | 0.236430228 |
| 0.32 | 0.002840474 | 0.030983347 | 0.106358886 | 0.226822615 |
| 0.31 | 0.002636532 | 0.029069107 | 0.100820363 | 0.217086315 |
| 0.30 | 0.002439573 | 0.027191162 | 0.095300674 | 0.207232118 |
| 0.29 | 0.002249782 | 0.025352627 | 0.089809477 | 0.197272718 |
| 0.28 | 0.002067336 | 0.023556657 | 0.084357321 | 0.187222660 |
| 0.27 | 0.001892399 | 0.021806452 | 0.078955352 | 0.177098751 |
| 0.26 | 0.001725119 | 0.020105250 | 0.073615491 | 0.166920304 |
| 0.25 | 0.001565632 | 0.018456295 | 0.068350554 | 0.156709254 |
| 0.24 | 0.001414056 | 0.016862821 | 0.063173831 | 0.146490335 |
| 0.23 | 0.001270492 | 0.015328038 | 0.058099594 | 0.136291385 |
| 0.22 | 0.001135020 | 0.013855107 | 0.053142551 | 0.126143336 |
| 0.21 | 0.001007702 | 0.012447093 | 0.048318125 | 0.116080284 |
| 0.20 | 0.000888575 | 0.011106960 | 0.043642223 | 0.106139839 |
| 0.19 | 0.000777652 | 0.009837512 | 0.039131135 | 0.096362710 |
| 0.18 | 0.000674921 | 0.008641366 | 0.034801405 | 0.086793065 |
| 0.17 | 0.000580341 | 0.007520918 | 0.030669656 | 0.077478170 |
| 0.16 | 0.000493840 | 0.006478265 | 0.026752360 | 0.068468034 |
| 0.15 | 0.000415317 | 0.005515184 | 0.023065597 | 0.059815083 |
| 0.14 | 0.000344635 | 0.004633054 | 0.019624736 | 0.051573277 |
| 0.13 | 0.000281619 | 0.003832816 | 0.016444076 | 0.043797549 |
| 0.12 | 0.000226059 | 0.003114873 | 0.013536453 | 0.036542419 |
| 0.11 | 0.000177701 | 0.002479048 | 0.010912735 | 0.029860597 |
| 0.10 | 0.000136249 | 0.001924492 | 0.008581311 | 0.023801155 |
| 0.09 | 0.000101361 | 0.001449604 | 0.006547485 | 0.018407367 |
| 0.08 | 0.000072646 | 0.001051943 | 0.004812792 | 0.013714083 |
| 0.07 | 0.000049661 | 0.000728130 | 0.003374280 | 0.009744473 |
| 0.06 | 0.000031912 | 0.000473753 | 0.002223687 | 0.006507820 |
| 0.05 | 0.000018844 | 0.000283258 | 0.001346577 | 0.003993110 |
| 0.04 | 0.000009844 | 0.000149835 | 0.000721386 | 0.002167220 |
| 0.03 | 0.000004237 | 0.000065305 | 0.000318407 | 0.000966897 |
| 0.02 | 0.000001281 | 0.000019990 | 0.000098698 | 0.000304200 |
| 0.01 | 0.000000163 | 0.000002558 | 0.000012780 | 0.000040300 |

TABLE 2:   POISSON CLUSTER PROBABILITY   P'(n;λ,p)

n = 4

| p | λ = 5 | λ = 6 | λ = 7 | λ = 8 |
|---|---|---|---|---|
| 0.50 | 0.548083007 | 0.694028974 | 0.803964257 | 0.880090714 |
| 0.49 | 0.540699899 | 0.687402666 | 0.798709989 | 0.876291990 |
| 0.48 | 0.533120275 | 0.680552304 | 0.793243170 | 0.872316003 |
| 0.47 | 0.525321662 | 0.673446774 | 0.787529349 | 0.868130207 |
| 0.46 | 0.517283976 | 0.666057169 | 0.781536043 | 0.863703728 |
| 0.45 | 0.508989036 | 0.658357024 | 0.775232911 | 0.859007120 |
| 0.44 | 0.500420928 | 0.650321841 | 0.768591404 | 0.854012191 |
| 0.43 | 0.491565704 | 0.641929150 | 0.761584222 | 0.848691761 |
| 0.42 | 0.482411146 | 0.633157969 | 0.754185379 | 0.843018889 |
| 0.41 | 0.472946823 | 0.623988807 | 0.746369362 | 0.836966634 |
| 0.40 | 0.463163614 | 0.614403129 | 0.738111019 | 0.830507636 |
| 0.39 | 0.453053892 | 0.604383230 | 0.729385078 | 0.823613524 |
| 0.38 | 0.442611158 | 0.593911886 | 0.720165670 | 0.816254377 |
| 0.37 | 0.431829810 | 0.582972050 | 0.710426152 | 0.808398604 |
| 0.36 | 0.420705318 | 0.571546853 | 0.700138628 | 0.800012112 |
| 0.35 | 0.409234107 | 0.559619367 | 0.689273715 | 0.791058123 |
| 0.34 | 0.397413313 | 0.547172248 | 0.677800238 | 0.781496704 |
| 0.33 | 0.385241091 | 0.534188151 | 0.665684998 | 0.771283925 |
| 0.32 | 0.372716367 | 0.520648956 | 0.652892113 | 0.760371327 |
| 0.31 | 0.359839022 | 0.506535769 | 0.639381766 | 0.748703778 |
| 0.30 | 0.346610188 | 0.491829216 | 0.625110924 | 0.736219168 |
| 0.29 | 0.333033204 | 0.476510704 | 0.610034049 | 0.722849429 |
| 0.28 | 0.319113910 | 0.460563183 | 0.594104409 | 0.708521426 |
| 0.27 | 0.304861724 | 0.443972647 | 0.577275276 | 0.693157852 |
| 0.26 | 0.290290058 | 0.426728785 | 0.559501290 | 0.676678121 |
| 0.25 | 0.275417149 | 0.408826351 | 0.540739179 | 0.658999026 |
| 0.24 | 0.260266364 | 0.390266299 | 0.520949483 | 0.640035272 |
| 0.23 | 0.244867325 | 0.371056616 | 0.500097334 | 0.619700313 |
| 0.22 | 0.229256094 | 0.351214230 | 0.478154719 | 0.597907722 |
| 0.21 | 0.213476419 | 0.330766380 | 0.455102921 | 0.574573517 |
| 0.20 | 0.197580159 | 0.309753180 | 0.430935800 | 0.549619973 |
| 0.19 | 0.181628406 | 0.288229644 | 0.405664027 | 0.522980452 |
| 0.18 | 0.165692091 | 0.266268551 | 0.379319727 | 0.494605601 |
| 0.17 | 0.149852574 | 0.243962765 | 0.351962209 | 0.464471281 |
| 0.16 | 0.134202182 | 0.221428216 | 0.323684335 | 0.432588696 |
| 0.15 | 0.118844211 | 0.198806584 | 0.294619381 | 0.399016440 |
| 0.14 | 0.103892803 | 0.176267147 | 0.264949143 | 0.363875508 |
| 0.13 | 0.089472055 | 0.154008806 | 0.234911203 | 0.327366114 |
| 0.12 | 0.075714409 | 0.13226014 | 0.2048059 | 0.2897866 |
| 0.11 | 0.062758327 | 0.11127824 | 0.1750017 | 0.251552 |
| 0.10 | 0.05074453 | 0.0913449 | 0.145937 | 0.21321 |
| 0.09 | 0.03981110 | 0.072760 | 0.11812 | 0.17547 |
| 0.08 | 0.0300870 | 0.055828 | 0.09210 | 0.1392 |
| 0.07 | 0.0216838 | 0.040847 | 0.06847 | 0.1052 |
| 0.06 | 0.0146853 | 0.02808 | 0.0478 | 0.0748 |
| 0.05 | 0.009136 | 0.01772 | 0.0307 | 0.049 |
| 0.04 | 0.005026 | 0.00989 | 0.0174 | 0.028 |
| 0.03 | 0.002277 | 0.00455 | 0.0081 | 0.013 |
| 0.02 | 0.000724 | 0.00147 | 0.0027 | 0.005 |
| 0.01 | 0.000097 | 0.00020 | 0.0004 | 0.001 |

TABLE 2:  POISSON CLUSTER PROBABILITY  $P'(n;\lambda,p)$

n = 4

| p | $\lambda = 9$ | $\lambda = 10$ | $\lambda = 11$ | $\lambda = 12$ |
|---|---|---|---|---|
| 0.50 | 0.929485619 | 0.959909618 | 0.977864087 | 0.988086343 |
| 0.49 | 0.926929295 | 0.958285511 | 0.976879716 | 0.987512827 |
| 0.48 | 0.924238861 | 0.956567228 | 0.975833178 | 0.986900210 |
| 0.47 | 0.921386898 | 0.954733670 | 0.974709332 | 0.986238360 |
| 0.46 | 0.918347120 | 0.952764690 | 0.973493695 | 0.985517323 |
| 0.45 | 0.915094495 | 0.950640559 | 0.972171903 | 0.984727442 |
| 0.44 | 0.911604524 | 0.948342085 | 0.970729947 | 0.983859003 |
| 0.43 | 0.907853127 | 0.945849895 | 0.969153523 | 0.982902050 |
| 0.42 | 0.903816104 | 0.943144381 | 0.967427909 | 0.981846154 |
| 0.41 | 0.899468720 | 0.940205038 | 0.965537429 | 0.980680227 |
| 0.40 | 0.894785345 | 0.937010288 | 0.963465273 | 0.979392052 |
| 0.39 | 0.889738560 | 0.933536649 | 0.961193144 | 0.977968216 |
| 0.38 | 0.884299278 | 0.929758728 | 0.958700657 | 0.976393580 |
| 0.37 | 0.878435731 | 0.925648391 | 0.955965161 | 0.974651098 |
| 0.36 | 0.872113168 | 0.921174347 | 0.952961087 | 0.972721338 |
| 0.35 | 0.865293503 | 0.916301847 | 0.949659526 | 0.970582187 |
| 0.34 | 0.857934296 | 0.910991609 | 0.946027637 | 0.968208015 |
| 0.33 | 0.849988461 | 0.905199409 | 0.942027688 | 0.965569377 |
| 0.32 | 0.841402769 | 0.898874283 | 0.937616050 | 0.962631524 |
| 0.31 | 0.832115233 | 0.891956091 | 0.932739794 | 0.959351778 |
| 0.30 | 0.822054505 | 0.884373844 | 0.927335739 | 0.955678165 |
| 0.29 | 0.811140060 | 0.876045763 | 0.921329260 | 0.951548517 |
| 0.28 | 0.799282789 | 0.866879106 | 0.914634466 | 0.946889639 |
| 0.27 | 0.786385298 | 0.856769979 | 0.907152832 | 0.941616356 |
| 0.26 | 0.772342026 | 0.845602691 | 0.898772120 | 0.935629487 |
| 0.25 | 0.757039189 | 0.833248317 | 0.889364183 | 0.928813457 |
| 0.24 | 0.740354061 | 0.819563091 | 0.878782094 | 0.921032488 |
| 0.23 | 0.722154438 | 0.804386199 | 0.866856456 | 0.912126303 |
| 0.22 | 0.702298880 | 0.787537813 | 0.853391707 | 0.901904225 |
| 0.21 | 0.680637479 | 0.768818021 | 0.838162243 | 0.890140057 |
| 0.20 | 0.657014787 | 0.748006761 | 0.820909619 | 0.876566052 |
| 0.19 | 0.631273270 | 0.724865496 | 0.801340401 | 0.860866725 |
| 0.18 | 0.603259742 | 0.699140072 | 0.779124618 | 0.842672825 |
| 0.17 | 0.572832942 | 0.670566261 | 0.753896058 | 0.821554780 |
| 0.16 | 0.539875329 | 0.638878942 | 0.725255847 | 0.797018468 |
| 0.15 | 0.504308522 | 0.603827417 | 0.692781866 | 0.768504620 |
| 0.14 | 0.466114283 | 0.565197408 | 0.656046569 | 0.735396087 |
| 0.13 | 0.425361216 | 0.522844017 | 0.614647388 | 0.697037518 |
| 0.12 | 0.382238 | 0.476736 | 0.56826 | 0.65278 |
| 0.11 | 0.33709 | 0.42702 | 0.5167 | 0.6020 |
| 0.10 | 0.29047 | 0.3741 | 0.4600 | 0.544 |
| 0.09 | 0.2432 | 0.3187 | 0.399 | 0.480 |
| 0.08 | 0.1963 | 0.262 | 0.334 | 0.41 |
| 0.07 | 0.151 | 0.206 | 0.27 | 0.33 |
| 0.06 | 0.109 | 0.152 | 0.20 | 0.26 |
| 0.05 | 0.073 | 0.10 | 0.14 | 0.19 |
| 0.04 | 0.043 | 0.06 | 0.09 | |
| 0.03 | 0.021 | 0.03 | 0.05 | |
| 0.02 | 0.008 | 0.01 | 0.02 | |
| 0.01 | 0.00 | 0.00 | 0.01 | |

NEFF and NAUS

TABLE 2: POISSON CLUSTER PROBABILITY P'(n;λ,p)

n = 5

| p | λ = 1 | λ = 2 | λ = 3 | λ = 4 |
|---|---|---|---|---|
| 0.50 | 0.000857059 | 0.016516615 | 0.072447777 | 0.177424252 |
| 0.49 | 0.000846452 | 0.015726436 | 0.069566309 | 0.171683013 |
| 0.48 | 0.000797320 | 0.014950275 | 0.066704333 | 0.165920854 |
| 0.47 | 0.000749690 | 0.014188778 | 0.063863873 | 0.160138786 |
| 0.46 | 0.000703584 | 0.013442598 | 0.061047271 | 0.154338777 |
| 0.45 | 0.000659026 | 0.012712389 | 0.058256943 | 0.148524106 |
| 0.44 | 0.000616032 | 0.011998814 | 0.055495709 | 0.142698944 |
| 0.43 | 0.000574619 | 0.011302523 | 0.052766550 | 0.136868238 |
| 0.42 | 0.000534798 | 0.010624163 | 0.050072625 | 0.131038129 |
| 0.41 | 0.000496578 | 0.009964358 | 0.047417257 | 0.125215352 |
| 0.40 | 0.000459966 | 0.009323720 | 0.044803899 | 0.119407356 |
| 0.39 | 0.000424963 | 0.008702837 | 0.042236101 | 0.113622427 |
| 0.38 | 0.000391568 | 0.008102261 | 0.039717492 | 0.107869267 |
| 0.37 | 0.000359778 | 0.007522523 | 0.037251744 | 0.102157235 |
| 0.36 | 0.000329584 | 0.006964102 | 0.034842566 | 0.096496165 |
| 0.35 | 0.000300975 | 0.006427452 | 0.032493647 | 0.090896428 |
| 0.34 | 0.000273937 | 0.005912963 | 0.030208658 | 0.085368693 |
| 0.33 | 0.000248452 | 0.005420990 | 0.027991220 | 0.079924107 |
| 0.32 | 0.000224499 | 0.004951824 | 0.025844865 | 0.074574053 |
| 0.31 | 0.000202051 | 0.004505698 | 0.023773029 | 0.069330215 |
| 0.30 | 0.000181082 | 0.004082792 | 0.021779012 | 0.064204693 |
| 0.29 | 0.000161560 | 0.003683207 | 0.019865949 | 0.059209622 |
| 0.28 | 0.000143450 | 0.003306986 | 0.018036786 | 0.054357268 |
| 0.27 | 0.000126714 | 0.002954093 | 0.016294230 | 0.049659986 |
| 0.26 | 0.000111312 | 0.002624417 | 0.014640726 | 0.045130018 |
| 0.25 | 0.000097199 | 0.002317769 | 0.013078406 | 0.040779378 |
| 0.24 | 0.000084329 | 0.002033879 | 0.011609059 | 0.036619712 |
| 0.23 | 0.000072652 | 0.001772390 | 0.010234073 | 0.032662123 |
| 0.22 | 0.000062117 | 0.001532863 | 0.008954421 | 0.028917000 |
| 0.21 | 0.000052669 | 0.001314772 | 0.007770598 | 0.025393810 |
| 0.20 | 0.000044252 | 0.001117502 | 0.006682593 | 0.022100888 |
| 0.19 | 0.000036807 | 0.000940355 | 0.005689856 | 0.019045237 |
| 0.18 | 0.000030276 | 0.000782544 | 0.004791256 | 0.016232278 |
| 0.17 | 0.000024596 | 0.000643203 | 0.003985059 | 0.013665650 |
| 0.16 | 0.000019706 | 0.000521383 | 0.003268912 | 0.011346962 |
| 0.15 | 0.000015542 | 0.000416060 | 0.002639817 | 0.009275611 |
| 0.14 | 0.000012040 | 0.000326142 | 0.002094137 | 0.007448576 |
| 0.13 | 0.000009139 | 0.000250473 | 0.001627596 | 0.005860269 |
| 0.12 | 0.000006773 | 0.000187842 | 0.001235300 | 0.00450242 |
| 0.11 | 0.000004882 | 0.000136996 | 0.000911768 | 0.00336400 |
| 0.10 | 0.000003403 | 0.000096648 | 0.000650503 | 0.00243128 |
| 0.09 | 0.000002279 | 0.000065494 | 0.000446462 | 0.00168785 |
| 0.08 | 0.000001452 | 0.000042230 | 0.000291345 | 0.00111489 |
| 0.07 | 0.000000869 | 0.000025566 | 0.000178508 | 0.00069143 |
| 0.06 | 0.000000479 | 0.000014252 | 0.000100711 | 0.00039484 |
| 0.05 | 0.000000235 | 0.000007098 | 0.000050762 | 0.00020144 |
| 0.04 | 0.000000098 | 0.000003002 | 0.000021731 | 0.00008729 |
| 0.03 | 0.000000032 | 0.000000981 | 0.000007186 | 0.00002922 |
| 0.02 | 0.000000006 | 0.000000200 | 0.000001484 | 0.0000061 |
| 0.01 | 0.000000000 | 0.000000013 | 0.000000097 | 0.0000004 |

TABLE 2:   POISSON CLUSTER PROBABILITY   $P'(n;\lambda,p)$

n = 5

| p | $\lambda = 5$ | $\lambda = 6$ | $\lambda = 7$ | $\lambda = 8$ |
|---|---|---|---|---|
| 0.50 | 0.317410827 | 0.468361855 | 0.609012783 | 0.726702273 |
| 0.49 | 0.309247613 | 0.459026373 | 0.599857152 | 0.718695879 |
| 0.48 | 0.300974369 | 0.449479640 | 0.590413451 | 0.710371673 |
| 0.47 | 0.292585194 | 0.439701974 | 0.580650091 | 0.701687872 |
| 0.46 | 0.284076631 | 0.429679394 | 0.570539773 | 0.692607224 |
| 0.45 | 0.275447428 | 0.419400930 | 0.560059428 | 0.683096528 |
| 0.44 | 0.266698658 | 0.408858836 | 0.549189806 | 0.673125982 |
| 0.43 | 0.257833421 | 0.398048401 | 0.537914872 | 0.662669003 |
| 0.42 | 0.248856664 | 0.386967778 | 0.526221633 | 0.651701093 |
| 0.41 | 0.239775181 | 0.375617623 | 0.514099658 | 0.640199840 |
| 0.40 | 0.230597496 | 0.364000797 | 0.501540720 | 0.628144085 |
| 0.39 | 0.221333504 | 0.352122545 | 0.488538861 | 0.615513921 |
| 0.38 | 0.211994827 | 0.339990139 | 0.475089967 | 0.602290392 |
| 0.37 | 0.202594280 | 0.327613115 | 0.461191952 | 0.588455439 |
| 0.36 | 0.193146229 | 0.315002859 | 0.446844757 | 0.573991895 |
| 0.35 | 0.183666408 | 0.302173197 | 0.432050526 | 0.558883667 |
| 0.34 | 0.174171805 | 0.289140165 | 0.416813731 | 0.543115795 |
| 0.33 | 0.164680898 | 0.275922179 | 0.401141524 | 0.526674926 |
| 0.32 | 0.155213654 | 0.262540400 | 0.385044038 | 0.509549379 |
| 0.31 | 0.145791352 | 0.249018788 | 0.368534744 | 0.491729558 |
| 0.30 | 0.136436880 | 0.235384643 | 0.351631582 | 0.473209381 |
| 0.29 | 0.127174675 | 0.221668959 | 0.334357679 | 0.453987837 |
| 0.28 | 0.118030787 | 0.207906961 | 0.316742837 | 0.434071124 |
| 0.27 | 0.109032810 | 0.194138110 | 0.298824191 | 0.413473904 |
| 0.26 | 0.100209773 | 0.180406511 | 0.280646980 | 0.392221510 |
| 0.25 | 0.091591835 | 0.166760802 | 0.262265682 | 0.370351255 |
| 0.24 | 0.083210170 | 0.153254211 | 0.243744075 | 0.347914338 |
| 0.23 | 0.075096786 | 0.139944434 | 0.225156248 | 0.324977458 |
| 0.22 | 0.067283750 | 0.126893103 | 0.206586778 | 0.301624715 |
| 0.21 | 0.059803169 | 0.114165425 | 0.188130796 | 0.277959645 |
| 0.20 | 0.052686457 | 0.101829708 | 0.169894338 | 0.254106700 |
| 0.19 | 0.045963753 | 0.089956105 | 0.151993394 | 0.230212450 |
| 0.18 | 0.039663270 | 0.078615665 | 0.134553075 | 0.206446588 |
| 0.17 | 0.033810526 | 0.067878723 | 0.117705822 | 0.183001220 |
| 0.16 | 0.02842753 | 0.0578132 | 0.101589 | 0.160089 |
| 0.15 | 0.02353191 | 0.0484824 | 0.086342 | 0.13794 |
| 0.14 | 0.01913606 | 0.039943 | 0.07210 | 0.11680 |
| 0.13 | 0.0152463 | 0.032242 | 0.05899 | 0.0969 |
| 0.12 | 0.0118619 | 0.025414 | 0.04714 | 0.0786 |
| 0.11 | 0.0089746 | 0.019480 | 0.03663 | 0.0619 |
| 0.10 | 0.0065679 | 0.01444 | 0.0275 | 0.0472 |
| 0.09 | 0.004617 | 0.01028 | 0.0199 | 0.0346 |
| 0.08 | 0.003088 | 0.00697 | 0.0136 | 0.024 |
| 0.07 | 0.001939 | 0.00443 | 0.0088 | 0.016 |
| 0.06 | 0.001121 | 0.00260 | 0.0052 | 0.010 |
| 0.05 | 0.000579 | 0.00136 | 0.0028 | 0.005 |
| 0.04 | 0.000254 | 0.00061 | 0.0013 | 0.003 |
| 0.03 | 0.000086 | 0.00021 | 0.0005 | 0.001 |
| 0.02 | 0.000019 | 0.00005 | 0.0001 | 0.000 |
| 0.01 | 0.000002 | 0.00001 | 0.0001 | 0.000 |

## TABLE 2:  POISSON CLUSTER PROBABILITY  $P'(n;\lambda,p)$

n = 5

| p | $\lambda = 9$ | $\lambda = 10$ | $\lambda = 11$ | $\lambda = 12$ |
|---|---|---|---|---|
| 0.50 | 0.817326069 | 0.882617414 | 0.927153587 | 0.956168413 |
| 0.49 | 0.810923278 | 0.877852142 | 0.923810720 | 0.953936934 |
| 0.48 | 0.804216564 | 0.872825384 | 0.920260906 | 0.951552331 |
| 0.47 | 0.797159731 | 0.867492557 | 0.916464984 | 0.948982954 |
| 0.46 | 0.789710760 | 0.861812115 | 0.912386239 | 0.946198642 |
| 0.45 | 0.781830907 | 0.855745256 | 0.907989621 | 0.943170488 |
| 0.44 | 0.773484409 | 0.849255025 | 0.903240979 | 0.939869761 |
| 0.43 | 0.764637470 | 0.842305362 | 0.898106515 | 0.936267376 |
| 0.42 | 0.755257607 | 0.834860623 | 0.892551601 | 0.932332873 |
| 0.41 | 0.745312989 | 0.826884389 | 0.886540174 | 0.928034127 |
| 0.40 | 0.734772027 | 0.818339288 | 0.880034208 | 0.923336208 |
| 0.39 | 0.723602712 | 0.809186161 | 0.872992754 | 0.918201208 |
| 0.38 | 0.711772561 | 0.799383759 | 0.865371764 | 0.912587106 |
| 0.37 | 0.699248254 | 0.788888216 | 0.857123375 | 0.906447768 |
| 0.36 | 0.685995579 | 0.777653098 | 0.848195374 | 0.899731755 |
| 0.35 | 0.671979487 | 0.765628636 | 0.838530838 | 0.892382026 |
| 0.34 | 0.657163799 | 0.752761900 | 0.828067303 | 0.884334683 |
| 0.33 | 0.641511679 | 0.738996208 | 0.816736281 | 0.875518203 |
| 0.32 | 0.624985278 | 0.724270523 | 0.804461837 | 0.865851760 |
| 0.31 | 0.607545733 | 0.708518863 | 0.791159093 | 0.855242312 |
| 0.30 | 0.589154601 | 0.691670895 | 0.776734054 | 0.843584239 |
| 0.29 | 0.569775820 | 0.673654139 | 0.761085391 | 0.830759704 |
| 0.28 | 0.549378157 | 0.654396355 | 0.744105875 | 0.816639662 |
| 0.27 | 0.527937472 | 0.633827627 | 0.725684524 | 0.801084816 |
| 0.26 | 0.505439103 | 0.611883104 | 0.705708206 | 0.783945739 |
| 0.25 | 0.481880307 | 0.588505447 | 0.684063613 | 0.765063941 |
| 0.24 | 0.457273006 | 0.563647807 | 0.660639763 | 0.744272768 |
| 0.23 | 0.431646645 | 0.537278056 | 0.635331631 | 0.721399307 |
| 0.22 | 0.405052185 | 0.509383321 | 0.608045101 | 0.696267962 |
| 0.21 | 0.377566040 | 0.479976416 | 0.578704119 | 0.668707073 |
| 0.20 | 0.349294662 | 0.449103534 | 0.547260225 | 0.638557851 |
| 0.19 | 0.320379078 | 0.416852117 | 0.513703763 | 0.605686843 |
| 0.18 | 0.290998816 | 0.383360386 | 0.478077650 | 0.570002079 |
| 0.17 | 0.261375844 | 0.348826468 | 0.440493405 | 0.531473696 |
| 0.16 | 0.23178 | 0.3135 | 0.4012 | 0.4902 |
| 0.15 | 0.2025 | 0.2778 | 0.360 | 0.446 |
| 0.14 | 0.1739 | 0.2421 | 0.319 | 0.400 |
| 0.13 | 0.1464 | 0.207 | 0.276 | 0.35 |
| 0.12 | 0.120 | 0.173 | 0.234 | 0.30 |
| 0.11 | 0.096 | 0.140 | 0.19 | 0.26 |
| 0.10 | 0.075 | 0.111 | 0.16 | 0.21 |
| 0.09 | 0.056 | 0.08 | 0.12 | 0.17 |
| 0.08 | 0.039 | 0.06 | 0.09 | |
| 0.07 | 0.026 | 0.04 | 0.06 | |
| 0.06 | 0.016 | 0.03 | 0.04 | |
| 0.05 | 0.009 | 0.02 | 0.03 | |
| 0.04 | 0.005 | 0.01 | 0.02 | |
| 0.03 | 0.00 | 0.01 | 0.01 | |
| 0.02 | 0.00 | 0.00 | 0.01 | |
| 0.01 | 0.00 | 0.00 | 0.01 | |

TABLE 2:  POISSON CLUSTER PROBABILITY  P'(n;λ,p)

n = 6

| p | λ = 1 | λ = 2 | λ = 3 | λ = 4 |
|---|---|---|---|---|
| 0.50 | 0.000087561 | 0.003232143 | 0.021266777 | 0.069212794 |
| 0.49 | 0.000081073 | 0.003021443 | 0.020064279 | 0.065871358 |
| 0.48 | 0.000074901 | 0.002818576 | 0.018892720 | 0.062578321 |
| 0.47 | 0.000069041 | 0.002623598 | 0.017753046 | 0.059336830 |
| 0.46 | 0.000063489 | 0.002436552 | 0.016646191 | 0.056150150 |
| 0.45 | 0.000058238 | 0.002257463 | 0.015573069 | 0.053021934 |
| 0.44 | 0.000053285 | 0.002086340 | 0.014534559 | 0.049955972 |
| 0.43 | 0.000048621 | 0.001923175 | 0.013531495 | 0.046956196 |
| 0.42 | 0.000044242 | 0.001767943 | 0.012564659 | 0.044026606 |
| 0.41 | 0.000040139 | 0.001620601 | 0.011634763 | 0.041171245 |
| 0.40 | 0.000036304 | 0.001481088 | 0.010742448 | 0.038394138 |
| 0.39 | 0.000032730 | 0.001349325 | 0.009888269 | 0.035699256 |
| 0.38 | 0.000029409 | 0.001225213 | 0.009072684 | 0.033090454 |
| 0.37 | 0.000026330 | 0.001108637 | 0.008296054 | 0.030571450 |
| 0.36 | 0.000023486 | 0.000999463 | 0.007558629 | 0.028145771 |
| 0.35 | 0.000020866 | 0.000897541 | 0.006860550 | 0.025816724 |
| 0.34 | 0.000018461 | 0.000802701 | 0.006201830 | 0.023587342 |
| 0.33 | 0.000016261 | 0.000714757 | 0.005582359 | 0.021460358 |
| 0.32 | 0.000014257 | 0.000633510 | 0.005001899 | 0.019438170 |
| 0.31 | 0.000012437 | 0.000558740 | 0.004460078 | 0.017522793 |
| 0.30 | 0.000010793 | 0.000490218 | 0.003956392 | 0.015715841 |
| 0.29 | 0.000009313 | 0.000427698 | 0.003490202 | 0.014018465 |
| 0.28 | 0.000007988 | 0.000370923 | 0.003060724 | 0.012431350 |
| 0.27 | 0.000006807 | 0.000319626 | 0.002667045 | 0.010954656 |
| 0.26 | 0.000005761 | 0.000273527 | 0.002308119 | 0.009588007 |
| 0.25 | 0.000004839 | 0.000232341 | 0.001982767 | 0.008330461 |
| 0.24 | 0.000004032 | 0.000195774 | 0.001689690 | 0.007180493 |
| 0.23 | 0.000003330 | 0.000163528 | 0.001427467 | 0.006135978 |
| 0.22 | 0.000002724 | 0.000135300 | 0.001194571 | 0.005194195 |
| 0.21 | 0.000002206 | 0.000110789 | 0.000989374 | 0.004351828 |
| 0.20 | 0.000001765 | 0.000089690 | 0.000810163 | 0.00360498 |
| 0.19 | 0.000001395 | 0.000071703 | 0.000655148 | 0.00294917 |
| 0.18 | 0.000001088 | 0.000056530 | 0.000522484 | 0.00237944 |
| 0.17 | 0.000000835 | 0.000043883 | 0.000410283 | 0.00189032 |
| 0.16 | 0.000000630 | 0.000033478 | 0.000316632 | 0.0014759 |
| 0.15 | 0.000000466 | 0.000025044 | 0.000239615 | 0.0011300 |
| 0.14 | 0.000000337 | 0.000018321 | 0.000177331 | 0.0008461 |
| 0.13 | 0.000000237 | 0.000013064 | 0.000127919 | 0.0006175 |
| 0.12 | 0.000000162 | 0.000009043 | 0.000089574 | 0.0004375 |
| 0.11 | 0.000000107 | 0.000006044 | 0.000060574 | 0.0002993 |
| 0.10 | 0.000000068 | 0.000003876 | 0.000039296 | 0.0001965 |
| 0.09 | 0.000000041 | 0.000002363 | 0.000024242 | 0.0001227 |
| 0.08 | 0.000000023 | 0.000001354 | 0.000014054 | 0.0000720 |
| 0.07 | 0.000000012 | 0.000000717 | 0.000007531 | 0.0000390 |
| 0.06 | 0.000000006 | 0.000000343 | 0.000003640 | 0.0000191 |
| 0.05 | 0.000000002 | 0.000000142 | 0.000001528 | 0.0000081 |
| 0.04 | 0.000000001 | 0.000000048 | 0.000000523 | 0.0000029 |
| 0.03 | 0.000000000 | 0.000000012 | 0.000000130 | 0.0000008 |
| 0.02 | 0.000000000 | 0.000000002 | 0.000000018 | 0.0000001 |
| 0.01 | 0.000000000 | 0.000000000 | 0.000000001 | 0.0000001 |

TABLE 2:   POISSON CLUSTER PROBABILITY   P'(n;λ,p)

n = 6

| p | λ = 5 | λ = 6 | λ = 7 | λ = 8 |
|---|---|---|---|---|
| 0.50 | 0.153584898 | 0.268398583 | 0.399344802 | 0.530517757 |
| 0.49 | 0.147362947 | 0.259445488 | 0.388599217 | 0.519255996 |
| 0.48 | 0.141162992 | 0.250429273 | 0.377668619 | 0.507690191 |
| 0.47 | 0.134989262 | 0.241350591 | 0.366543233 | 0.495795727 |
| 0.46 | 0.128847301 | 0.232212901 | 0.355218232 | 0.483554304 |
| 0.45 | 0.122743607 | 0.223022342 | 0.343693137 | 0.470953345 |
| 0.44 | 0.116685748 | 0.213787436 | 0.331971586 | 0.457985640 |
| 0.43 | 0.110682249 | 0.204518914 | 0.320060670 | 0.444648445 |
| 0.42 | 0.104742169 | 0.195229352 | 0.307970941 | 0.430943191 |
| 0.41 | 0.098875284 | 0.185933173 | 0.295715868 | 0.416875005 |
| 0.40 | 0.093091846 | 0.176646292 | 0.283311725 | 0.402452588 |
| 0.39 | 0.087402582 | 0.167386055 | 0.270777464 | 0.387688100 |
| 0.38 | 0.081818342 | 0.158171058 | 0.258134484 | 0.372597098 |
| 0.37 | 0.076350331 | 0.149021149 | 0.245406806 | 0.357198656 |
| 0.36 | 0.071009815 | 0.139957190 | 0.232620895 | 0.341515601 |
| 0.35 | 0.065807939 | 0.131000996 | 0.219805658 | 0.325574458 |
| 0.34 | 0.060755976 | 0.122175157 | 0.206992567 | 0.309405923 |
| 0.33 | 0.055864848 | 0.113503098 | 0.194215477 | 0.293045104 |
| 0.32 | 0.051145289 | 0.105008900 | 0.181510746 | 0.276531756 |
| 0.31 | 0.046607621 | 0.096716940 | 0.168917239 | 0.259910643 |
| 0.30 | 0.042261660 | 0.088652074 | 0.156476259 | 0.243232012 |
| 0.29 | 0.038116593 | 0.080839157 | 0.144231439 | 0.226552129 |
| 0.28 | 0.034180842 | 0.073302805 | 0.132228613 | 0.209933162 |
| 0.27 | 0.030461911 | 0.066067338 | 0.120515227 | 0.193443477 |
| 0.26 | 0.026966237 | 0.059156049 | 0.109140158 | 0.177157104 |
| 0.25 | 0.023699041 | 0.052591003 | 0.098152637 | 0.161153316 |
| 0.24 | 0.020664178 | 0.046392608 | 0.087602019 | 0.145515859 |
| 0.23 | 0.017864000 | 0.040579081 | 0.077536404 | 0.130331755 |
| 0.22 | 0.015299223 | 0.035166018 | 0.068001986 | 0.115690231 |
| 0.21 | 0.012968812 | 0.030165903 | 0.059041727 | 0.101680636 |
| 0.20 | 0.0108699 | 0.025588 | 0.05070 | 0.0884 |
| 0.19 | 0.008998 | 0.02144 | 0.0430 | 0.0759 |
| 0.18 | 0.007346 | 0.01771 | 0.0360 | 0.0643 |
| 0.17 | 0.005905 | 0.01441 | 0.0296 | 0.054 |
| 0.16 | 0.004665 | 0.01153 | 0.0240 | 0.044 |
| 0.15 | 0.003615 | 0.00904 | 0.0191 | 0.035 |
| 0.14 | 0.002739 | 0.00693 | 0.0148 | 0.028 |
| 0.13 | 0.002023 | 0.00519 | 0.0112 | 0.022 |
| 0.12 | 0.001450 | 0.0038 | 0.0083 | 0.016 |
| 0.11 | 0.001004 | 0.0026 | 0.0059 | 0.012 |
| 0.10 | 0.00067 | 0.0018 | 0.004 | 0.008 |
| 0.09 | 0.00042 | 0.0011 | 0.003 | 0.006 |
| 0.08 | 0.00025 | 0.0007 | 0.002 | 0.004 |
| 0.07 | 0.00014 | 0.0004 | 0.001 | 0.002 |
| 0.06 | 0.00007 | 0.0002 | 0.001 | 0.002 |
| 0.05 | 0.00003 | 0.0001 | 0.000 | 0.001 |
| 0.04 | 0.00001 | 0.0000 | 0.000 | 0.001 |
| 0.03 | 0.00000 | 0.0000 | 0.000 | 0.001 |
| 0.02 | 0.00000 | 0.0000 | 0.000 | 0.001 |
| 0.01 | 0.00000 | 0.0000 | 0.000 | 0.001 |

TABLE 2:  POISSON CLUSTER PROBABILITY  $P'(n;\lambda,p)$

n = 6

| p | $\lambda = 9$ | $\lambda = 10$ | $\lambda = 11$ | $\lambda = 12$ |
|---|---|---|---|---|
| 0.50 | 0.649407148 | 0.748879075 | 0.826808929 | 0.884581923 |
| 0.49 | 0.638788223 | 0.739683032 | 0.819384933 | 0.878932059 |
| 0.48 | 0.627783239 | 0.730070114 | 0.811560392 | 0.872930646 |
| 0.47 | 0.616352022 | 0.719988406 | 0.803277850 | 0.866520822 |
| 0.46 | 0.604461551 | 0.709393144 | 0.794486165 | 0.859650910 |
| 0.45 | 0.592085302 | 0.698245585 | 0.785139203 | 0.852272809 |
| 0.44 | 0.579202116 | 0.686511874 | 0.775194466 | 0.844340861 |
| 0.43 | 0.565795541 | 0.674162030 | 0.764611840 | 0.835810125 |
| 0.42 | 0.551853180 | 0.661168993 | 0.753352940 | 0.826635778 |
| 0.41 | 0.537366211 | 0.647508323 | 0.741379917 | 0.816771746 |
| 0.40 | 0.522329032 | 0.633157551 | 0.728655279 | 0.806170523 |
| 0.39 | 0.506739378 | 0.618096292 | 0.715141714 | 0.794782698 |
| 0.38 | 0.490598381 | 0.602306247 | 0.700801969 | 0.782556593 |
| 0.37 | 0.473910570 | 0.585771441 | 0.685598850 | 0.769438386 |
| 0.36 | 0.456684351 | 0.568478584 | 0.669495761 | 0.755371749 |
| 0.35 | 0.438932419 | 0.550417602 | 0.652456999 | 0.740298510 |
| 0.34 | 0.420672178 | 0.531582355 | 0.634448171 | 0.724158406 |
| 0.33 | 0.401926517 | 0.511971235 | 0.615437090 | 0.706889620 |
| 0.32 | 0.382724226 | 0.491588235 | 0.595394433 | 0.688429058 |
| 0.31 | 0.363100946 | 0.470443964 | 0.574294925 | 0.668712974 |
| 0.30 | 0.343100488 | 0.448557854 | 0.552119911 | 0.647679329 |
| 0.29 | 0.322776020 | 0.425960422 | 0.528860509 | 0.625271559 |
| 0.28 | 0.302191257 | 0.402695596 | 0.504521251 | 0.601442575 |
| 0.27 | 0.281421125 | 0.378822923 | 0.479123056 | 0.576158762 |
| 0.26 | 0.260552406 | 0.354419112 | 0.452706814 | 0.549404263 |
| 0.25 | 0.239683926 | 0.329579949 | 0.425336599 | 0.521185935 |
| 0.24 | 0.218926489 | 0.304421306 | 0.397102833 | 0.491538227 |
| 0.23 | 0.198402166 | 0.279080451 | 0.368125975 | 0.460529268 |
| 0.22 | 0.178243339 | 0.253716588 | 0.338559628 | 0.428267717 |
| 0.21 | 0.158591270 | 0.228510320 | 0.308593273 | 0.394909143 |
| 0.20 | 0.140 | 0.204 | 0.279 | 0.36 |
| 0.19 | 0.121 | 0.180 | 0.25 | 0.33 |
| 0.18 | 0.104 | 0.156 | 0.22 | 0.29 |
| 0.17 | 0.088 | 0.13 | 0.19 | 0.26 |
| 0.16 | 0.073 | 0.11 | 0.16 | 0.23 |
| 0.15 | 0.060 | 0.09 | 0.14 | |
| 0.14 | 0.05 | 0.08 | 0.12 | |
| 0.13 | 0.04 | 0.06 | 0.09 | |
| 0.12 | 0.03 | 0.05 | | |
| 0.11 | 0.02 | 0.04 | | |
| 0.10 | 0.02 | 0.03 | | |
| 0.09 | 0.01 | 0.02 | | |
| 0.08 | 0.01 | 0.02 | | |
| 0.07 | 0.01 | 0.01 | | |
| 0.06 | 0.00 | 0.01 | | |
| 0.05 | 0.00 | 0.01 | | |
| 0.04 | 0.00 | 0.01 | | |
| 0.03 | 0.00 | 0.01 | | |
| 0.02 | 0.00 | 0.01 | | |
| 0.01 | 0.00 | 0.01 | | |

TABLE 2:  POISSON CLUSTER PROBABILITY  P'(n;λ,p)

<u>n = 7</u>

| p | λ = 5 | λ = 6 | λ = 7 | λ = 8 |
|---|---|---|---|---|
| 0.50 | 0.0631635795 | 0.131801665 | 0.226796806 | 0.339816153 |
| 0.49 | 0.0599577342 | 0.125375509 | 0.217447937 | 0.328172743 |
| 0.48 | 0.0560071464 | 0.119021177 | 0.208100259 | 0.316405475 |
| 0.47 | 0.0526499718 | 0.112745464 | 0.198760867 | 0.304514885 |
| 0.46 | 0.0499316023 | 0.106556475 | 0.189439356 | 0.292506754 |
| 0.45 | 0.0460074405 | 0.100463033 | 0.180148125 | 0.280391574 |
| 0.44 | 0.0429289227 | 0.094474792 | 0.170901656 | 0.268183947 |
| 0.43 | 0.0399883606 | 0.088601887 | 0.161716223 | 0.255902290 |
| 0.42 | 0.0369423555 | 0.082854927 | 0.152609706 | 0.243567944 |
| 0.41 | 0.034108907 | 0.077244520 | 0.143601179 | 0.231205285 |
| 0.40 | 0.031386759 | 0.071781635 | 0.134710848 | 0.218841136 |
| 0.39 | 0.028779123 | 0.066476822 | 0.125959575 | 0.206504703 |
| 0.38 | 0.026288867 | 0.061340716 | 0.117368937 | 0.194227338 |
| 0.37 | 0.023918476 | 0.056383401 | 0.108960748 | 0.182042181 |
| 0.36 | 0.021669999 | 0.051614601 | 0.100757241 | 0.169984281 |
| 0.35 | 0.019545011 | 0.047043428 | 0.092780411 | 0.158090055 |
| 0.34 | 0.017544597 | 0.042678282 | 0.085052013 | 0.146397293 |
| 0.33 | 0.015669283 | 0.038526740 | 0.077593446 | 0.134944737 |
| 0.32 | 0.013919048 | 0.034595419 | 0.070425332 | 0.123771966 |
| 0.31 | 0.012293275 | 0.030889876 | 0.063567102 | 0.112918794 |
| 0.30 | 0.010790747 | 0.027414486 | 0.057307257 | 0.102425039 |
| 0.29 | 0.009409621 | 0.024172325 | 0.050852314 | 0.092329860 |
| 0.28 | 0.008147445 | 0.021165069 | 0.045027070 | 0.082671225 |
| 0.27 | 0.007001150 | 0.018392913 | 0.039573979 | 0.073485315 |
| 0.26 | 0.005967055 | 0.015854474 | 0.034502890 | 0.064805448 |
| 0.25 | 0.005040906 | 0.013546761 | 0.029820710 | 0.056661587 |
| 0.24 | 0.004217897 | 0.011465125 | 0.025531132 | 0.049079366 |
| 0.23 | 0.003492713 | 0.009603273 | 0.021634385 | 0.042079307 |
| 0.22 | 0.002859583 | 0.007953305 | 0.018127054 | 0.035676066 |
| 0.21 | 0.002312348 | 0.006505769 | 0.015001964 | 0.029877719 |
| 0.20 | 0.0018445 | 0.005250 | 0.01225 | 0.0247 |
| 0.19 | 0.0014494 | 0.004173 | 0.00985 | 0.0201 |
| 0.18 | 0.0011201 | 0.003263 | 0.00779 | 0.0161 |
| 0.17 | 0.0008497 | 0.002504 | 0.00605 | 0.0127 |
| 0.16 | 0.0006313 | 0.001882 | 0.00461 | 0.0098 |
| 0.15 | 0.0004581 | 0.001382 | 0.00342 | 0.0073 |
| 0.14 | 0.0003237 | 0.000988 | 0.00248 | 0.0054 |
| 0.13 | 0.0002218 | 0.00069 | 0.0017 | 0.0039 |
| 0.12 | 0.0001467 | 0.00046 | 0.0012 | 0.0027 |
| 0.11 | 0.0000930 | 0.00029 | 0.0008 | 0.0018 |
| 0.10 | 0.0000561 | 0.00018 | 0.0005 | 0.0011 |
| 0.09 | 0.0000319 | 0.00010 | 0.0003 | 0.0007 |
| 0.08 | 0.0000169 | 0.00006 | 0.0002 | 0.0004 |
| 0.07 | 0.0000081 | 0.00003 | 0.0001 | 0.0003 |
| 0.06 | 0.0000035 | 0.00001 | 0.0000 | 0.0002 |
| 0.05 | 0.0000013 | 0.00001 | 0.0000 | 0.0001 |
| 0.04 | 0.0000004 | 0.00000 | 0.0000 | 0.0001 |
| 0.03 | 0.0000001 | 0.00000 | 0.0000 | 0.0001 |
| 0.02 | 0.0000001 | 0.00000 | 0.0000 | 0.0001 |
| 0.01 | 0.0000001 | 0.00000 | 0.0000 | 0.0001 |

TABLE 2:   POISSON CLUSTER PROBABILITY  P'(n;λ,p)

n = 7

| p | λ = 9 | λ = 10 | λ = 11 | λ = 12 |
|---|-------|--------|--------|--------|
| 0.50 | 0.459512174 | 0.575043619 | 0.678378046 | 0.765081584 |
| 0.49 | 0.446664393 | 0.562182963 | 0.666499257 | 0.754827857 |
| 0.48 | 0.433545291 | 0.548919916 | 0.654131472 | 0.744054079 |
| 0.47 | 0.420142114 | 0.535223961 | 0.641226947 | 0.732698977 |
| 0.46 | 0.406449676 | 0.521074355 | 0.627747953 | 0.720710993 |
| 0.45 | 0.392469823 | 0.506458461 | 0.613665581 | 0.708046794 |
| 0.44 | 0.378210306 | 0.491370916 | 0.598957956 | 0.694669008 |
| 0.43 | 0.363684237 | 0.475812376 | 0.583609343 | 0.680544972 |
| 0.42 | 0.348909378 | 0.459789276 | 0.567609072 | 0.665645957 |
| 0.41 | 0.333908021 | 0.443313003 | 0.550951362 | 0.649946570 |
| 0.40 | 0.318706453 | 0.426400185 | 0.533635259 | 0.633424640 |
| 0.39 | 0.303335011 | 0.409072220 | 0.515664577 | 0.616061568 |
| 0.38 | 0.287827849 | 0.391355932 | 0.497048497 | 0.597842813 |
| 0.37 | 0.272223055 | 0.373283386 | 0.477801979 | 0.578758299 |
| 0.36 | 0.256562531 | 0.354892313 | 0.457946301 | 0.558803439 |
| 0.35 | 0.240891993 | 0.336226463 | 0.437509835 | 0.537980139 |
| 0.34 | 0.225260854 | 0.317336023 | 0.416528821 | 0.516297877 |
| 0.33 | 0.209722221 | 0.298277557 | 0.395048261 | 0.493775010 |
| 0.32 | 0.194332719 | 0.279114842 | 0.373122931 | 0.470440447 |
| 0.31 | 0.179152250 | 0.259918630 | 0.350818157 | 0.446335375 |
| 0.30 | 0.164243758 | 0.240767419 | 0.328211546 | 0.421515882 |
| 0.29 | 0.149672568 | 0.221746862 | 0.305393636 | 0.396055579 |
| 0.28 | 0.135505974 | 0.202950120 | 0.282468855 | 0.370047808 |
| 0.27 | 0.121811867 | 0.184476435 | 0.259555519 | 0.343607306 |
| 0.26 | 0.108657897 | 0.166430235 | 0.236785412 | 0.316871345 |
| 0.25 | 0.096105986 | 0.148919642 | 0.214302540 | 0.290000379 |
| 0.24 | 0.084230781 | 0.132054210 | 0.192261577 | 0.263177276 |
| 0.23 | 0.073078036 | 0.115942597 | 0.170825243 | 0.236606359 |
| 0.22 | 0.062702894 | 0.100689888 | 0.150161028 | 0.210510671 |
| 0.21 | 0.053148113 | 0.086394250 | 0.130436897 | 0.185128093 |
| 0.20 | 0.0445 | 0.073 | 0.112 | 0.16 |
| 0.19 | 0.037 | 0.061 | 0.09 | 0.14 |
| 0.18 | 0.030 | 0.050 | 0.08 | 0.12 |
| 0.17 | 0.024 | 0.041 | 0.07 | 0.10 |
| 0.16 | 0.019 | 0.032 | 0.05 | 0.08 |
| 0.15 | 0.014 | 0.025 | 0.04 | 0.07 |
| 0.14 | 0.011 | 0.02 | 0.03 | 0.05 |
| 0.13 | 0.008 | 0.01 | 0.03 | 0.04 |
| 0.12 | 0.005 | 0.01 | 0.02 | 0.03 |
| 0.11 | 0.004 | 0.01 | 0.01 | 0.03 |
| 0.10 | 0.003 | 0.01 | 0.01 | |
| 0.09 | 0.002 | 0.00 | 0.01 | |
| 0.08 | 0.001 | 0.00 | 0.01 | |
| 0.07 | 0.001 | 0.00 | 0.01 | |
| 0.06 | 0.001 | 0.00 | 0.01 | |
| 0.05 | 0.000 | 0.00 | 0.00 | |

TABLE 2:   POISSON CLUSTER PROBABILITY  P'(n;λ,p)

n = 8

| p | λ = 5 | λ = 6 | λ = 7 | λ = 8 |
|---|---|---|---|---|
| 0.50 | 0.022472180 | 0.056212634 | 0.112488389 | 0.191486418 |
| 0.49 | 0.020813987 | 0.052543126 | 0.106068492 | 0.182055473 |
| 0.48 | 0.019225430 | 0.048935727 | 0.099771559 | 0.172698557 |
| 0.47 | 0.017707136 | 0.045544185 | 0.093606055 | 0.163427532 |
| 0.46 | 0.016259607 | 0.042222254 | 0.087581098 | 0.154256642 |
| 0.45 | 0.014883183 | 0.039023586 | 0.081706464 | 0.145202339 |
| 0.44 | 0.013578013 | 0.035951659 | 0.075992286 | 0.136282504 |
| 0.43 | 0.012344055 | 0.033009686 | 0.070448577 | 0.127516329 |
| 0.42 | 0.011181053 | 0.030200541 | 0.065085351 | 0.118923724 |
| 0.41 | 0.010088518 | 0.027526680 | 0.059912425 | 0.110525131 |
| 0.40 | 0.009065732 | 0.024990100 | 0.054938938 | 0.102341115 |
| 0.39 | 0.008111738 | 0.022592288 | 0.050173558 | 0.094392240 |
| 0.38 | 0.007225338 | 0.020334151 | 0.045624159 | 0.086698592 |
| 0.37 | 0.006405085 | 0.018216021 | 0.041297730 | 0.079279661 |
| 0.36 | 0.005649313 | 0.016237590 | 0.037200246 | 0.072154105 |
| 0.35 | 0.004956119 | 0.014397915 | 0.033336557 | 0.065339208 |
| 0.34 | 0.004323378 | 0.012695387 | 0.029710293 | 0.058851067 |
| 0.33 | 0.003748773 | 0.011127733 | 0.026323758 | 0.052703869 |
| 0.32 | 0.003229786 | 0.009692024 | 0.023177858 | 0.046909954 |
| 0.31 | 0.002763738 | 0.008384682 | 0.020272043 | 0.041479412 |
| 0.30 | 0.002347796 | 0.007201500 | 0.017604243 | 0.036419865 |
| 0.29 | 0.001979006 | 0.006137669 | 0.015170850 | 0.031736240 |
| 0.28 | 0.001654309 | 0.005187839 | 0.012966719 | 0.027430557 |
| 0.27 | 0.001370578 | 0.004346140 | 0.010985170 | 0.023501780 |
| 0.26 | 0.001124637 | 0.003606268 | 0.009218063 | 0.019945703 |
| 0.25 | 0.000913300 | 0.002961539 | 0.007655863 | 0.016754895 |
| 0.24 | 0.000733393 | 0.002404976 | 0.006287765 | 0.013918731 |
| 0.23 | 0.000581790 | 0.001929388 | 0.005101819 | 0.011423480 |
| 0.22 | 0.000455439 | 0.001527464 | 0.004085120 | 0.009252496 |
| 0.21 | 0.000351391 | 0.001191862 | 0.003224001 | 0.007386472 |
| 0.20 | 0.000266830 | 0.00091531 | 0.0025043 | 0.005804 |
| 0.19 | 0.000199090 | 0.00069070 | 0.001911 | 0.00448 |
| 0.18 | 0.000145684 | 0.00051117 | 0.001431 | 0.00339 |
| 0.17 | 0.000104318 | 0.0003702 | 0.001048 | 0.00252 |
| 0.16 | 0.000072903 | 0.0002617 | 0.000749 | 0.00182 |
| 0.15 | 0.000049569 | 0.0001799 | 0.000521 | 0.00128 |
| 0.14 | 0.000032667 | 0.0001199 | 0.000352 | 0.00087 |
| 0.13 | 0.000020770 | 0.0000771 | 0.000229 | 0.00058 |
| 0.12 | 0.000012668 | 0.0000476 | 0.000143 | 0.00037 |
| 0.11 | 0.000007358 | 0.0000280 | 0.000085 | 0.00022 |
| 0.10 | 0.000004033 | 0.0000155 | 0.000048 | 0.00013 |
| 0.09 | 0.000002060 | 0.0000080 | 0.000025 | 0.00007 |
| 0.08 | 0.000000965 | 0.0000038 | 0.000012 | 0.00004 |
| 0.07 | 0.000000405 | 0.0000016 | 0.000005 | 0.00002 |
| 0.06 | 0.000000148 | 0.0000006 | 0.000002 | 0.00001 |
| 0.05 | 0.000000045 | 0.0000002 | 0.000001 | 0.00001 |
| 0.04 | 0.000000011 | 0.0000001 | 0.000001 | 0.00000 |
| 0.03 | 0.000000002 | 0.0000000 | 0.000000 | 0.00000 |
| 0.02 | 0.000000001 | 0.0000000 | 0.000000 | 0.00000 |
| 0.01 | 0.000000001 | 0.0000000 | 0.000000 | 0.00000 |

TABLE 2:  POISSON CLUSTER PROBABILITY  P'(n;λ,p)

n = 8

| p | λ = 9 | λ = 10 | λ = 11 | λ = 12 |
|------|-------------|-------------|-------------|-------------|
| 0.50 | 0.288641036 | 0.396188676 | 0.505411983 | 0.608613014 |
| 0.49 | 0.276546359 | 0.382282078 | 0.490798712 | 0.594378114 |
| 0.48 | 0.264411807 | 0.368177056 | 0.475819826 | 0.579637289 |
| 0.47 | 0.252247393 | 0.353874564 | 0.460460126 | 0.564355373 |
| 0.46 | 0.240068436 | 0.339383781 | 0.444715381 | 0.548509896 |
| 0.45 | 0.227894604 | 0.324721038 | 0.428590953 | 0.532089412 |
| 0.44 | 0.215749204 | 0.309908688 | 0.412100315 | 0.515091598 |
| 0.43 | 0.203658640 | 0.294974387 | 0.395263910 | 0.497521937 |
| 0.42 | 0.191651821 | 0.279950380 | 0.378108561 | 0.479393065 |
| 0.41 | 0.179759562 | 0.264872909 | 0.360667109 | 0.460724592 |
| 0.40 | 0.168014646 | 0.249782026 | 0.342978060 | 0.441542923 |
| 0.39 | 0.156450808 | 0.234721243 | 0.325085580 | 0.421881497 |
| 0.38 | 0.145102918 | 0.219737053 | 0.307039201 | 0.401781261 |
| 0.37 | 0.134006441 | 0.204878807 | 0.288894057 | 0.381290734 |
| 0.36 | 0.123196959 | 0.190198243 | 0.270710707 | 0.360466480 |
| 0.35 | 0.112709880 | 0.175749302 | 0.252554893 | 0.339373648 |
| 0.34 | 0.102580011 | 0.161587358 | 0.234497309 | 0.318086028 |
| 0.33 | 0.092840970 | 0.147768974 | 0.216613472 | 0.296686411 |
| 0.32 | 0.083524942 | 0.134351015 | 0.198983133 | 0.275266707 |
| 0.31 | 0.074661911 | 0.121390283 | 0.181689501 | 0.253927827 |
| 0.30 | 0.066279173 | 0.108942211 | 0.164818823 | 0.232779324 |
| 0.29 | 0.058400828 | 0.097060204 | 0.148458600 | 0.211938798 |
| 0.28 | 0.051046982 | 0.085794508 | 0.132696748 | 0.191530347 |
| 0.27 | 0.044233389 | 0.075190783 | 0.117619276 | 0.171682835 |
| 0.26 | 0.037970781 | 0.065289021 | 0.103308678 | 0.152527273 |
| 0.25 | 0.032264415 | 0.056122236 | 0.089841366 | 0.134193718 |
| 0.24 | 0.027113732 | 0.047715351 | 0.077285349 | 0.116807878 |
| 0.23 | 0.022512093 | 0.040084120 | 0.065698206 | 0.100487173 |
| 0.22 | 0.018446710 | 0.033234235 | 0.055124421 | 0.085336626 |
| 0.21 | 0.014898740 | 0.027160764 | 0.045593474 | 0.071444511 |
| 0.20 | 0.01185 | 0.0219 | 0.037 | 0.059 |
| 0.19 | 0.00926 | 0.0173 | 0.030 | 0.048 |
| 0.18 | 0.00709 | 0.0134 | 0.023 | 0.038 |
| 0.17 | 0.0053 | 0.0102 | 0.018 | 0.030 |
| 0.16 | 0.0039 | 0.0076 | 0.014 | 0.023 |
| 0.15 | 0.0028 | 0.0055 | 0.010 | 0.017 |
| 0.14 | 0.0019 | 0.0039 | 0.007 | 0.01 |
| 0.13 | 0.0013 | 0.0026 | 0.005 | 0.01 |
| 0.12 | 0.0008 | 0.002 | 0.003 | 0.01 |
| 0.11 | 0.0005 | 0.001 | 0.002 | 0.00 |
| 0.10 | 0.0003 | 0.001 | 0.002 | 0.00 |
| 0.09 | 0.0002 | 0.000 | 0.001 | 0.00 |
| 0.08 | 0.0001 | 0.000 | 0.001 | 0.00 |
| 0.07 | 0.0001 | 0.000 | 0.001 | 0.00 |
| 0.06 | 0.0000 | 0.000 | 0.001 | 0.00 |
| 0.05 | 0.0000 | 0.000 | 0.000 | 0.00 |

TABLE 2:   POISSON CLUSTER PROBABILITY   P'(n;λ,p)

n = 8

| p | λ = 13 | λ = 14 | λ = 15 | λ = 16 |
|---|---|---|---|---|
| 0.50 | 0.700300395 | 0.777527928 | 0.839603603 | 0.887474418 |
| 0.49 | 0.687305152 | 0.766313732 | 0.830392838 | 0.880233109 |
| 0.48 | 0.673714519 | 0.754473269 | 0.820577502 | 0.872447312 |
| 0.47 | 0.659473538 | 0.741936326 | 0.810078561 | 0.864036262 |
| 0.46 | 0.644540966 | 0.728645384 | 0.798828304 | 0.854928434 |
| 0.45 | 0.628886223 | 0.714552939 | 0.786766768 | 0.845057607 |
| 0.44 | 0.612487733 | 0.699618697 | 0.773838997 | 0.834360242 |
| 0.43 | 0.595331252 | 0.683808088 | 0.759993434 | 0.822773457 |
| 0.42 | 0.577409208 | 0.667091489 | 0.745180905 | 0.810234070 |
| 0.41 | 0.558720410 | 0.649443984 | 0.729354322 | 0.796677887 |
| 0.40 | 0.539270520 | 0.630845845 | 0.712469041 | 0.782040179 |
| 0.39 | 0.519072235 | 0.611282945 | 0.694483280 | 0.766255617 |
| 0.38 | 0.498146176 | 0.590747774 | 0.675359309 | 0.749259353 |
| 0.37 | 0.476521373 | 0.569240570 | 0.655064464 | 0.730987728 |
| 0.36 | 0.454236746 | 0.546770692 | 0.633572638 | 0.711379707 |
| 0.35 | 0.431341290 | 0.523358047 | 0.610866249 | 0.690378666 |
| 0.34 | 0.407895982 | 0.499034822 | 0.586937964 | 0.667934179 |
| 0.33 | 0.383973956 | 0.473847270 | 0.561793625 | 0.644005001 |
| 0.32 | 0.359662235 | 0.447857857 | 0.535454690 | 0.618561685 |
| 0.31 | 0.335062027 | 0.421147466 | 0.507961571 | 0.591590881 |
| 0.30 | 0.310290277 | 0.393817723 | 0.479378343 | 0.563100874 |
| 0.29 | 0.285479546 | 0.365993679 | 0.449796736 | 0.533128321 |
| 0.28 | 0.260777891 | 0.337824941 | 0.419340611 | 0.501744568 |
| 0.27 | 0.236347616 | 0.309486330 | 0.388168395 | 0.469061553 |
| 0.26 | 0.212362885 | 0.281176984 | 0.356475353 | 0.435236990 |
| 0.25 | 0.189006925 | 0.253118575 | 0.324494302 | 0.400478899 |
| 0.24 | 0.166467726 | 0.225552082 | 0.292494357 | 0.365048528 |
| 0.23 | 0.144933224 | 0.198732674 | 0.260778546 | 0.329261720 |
| 0.22 | 0.124585152 | 0.172923267 | 0.229677975 | 0.29348743 |
| 0.21 | 0.10559267 | 0.14838594 | 0.1995438 | 0.2581421 |
| 0.20 | 0.089 | 0.13 | 0.17 | 0.23 |
| 0.19 | 0.07 | 0.11 | 0.15 | |
| 0.18 | 0.06 | 0.09 | 0.12 | |
| 0.17 | 0.05 | 0.07 | 0.10 | |
| 0.16 | 0.04 | 0.06 | | |
| 0.15 | 0.03 | 0.04 | | |
| 0.14 | 0.02 | 0.03 | | |
| 0.13 | 0.02 | 0.03 | | |
| 0.12 | 0.01 | 0.02 | | |
| 0.11 | 0.01 | 0.02 | | |
| 0.10 | 0.01 | 0.01 | | |
| 0.09 | 0.01 | 0.01 | | |
| 0.08 | 0.00 | 0.01 | | |
| 0.07 | 0.00 | 0.01 | | |
| 0.06 | 0.00 | 0.01 | | |
| 0.05 | 0.00 | 0.01 | | |
| 0.04 | 0.00 | 0.01 | | |
| 0.03 | 0.00 | 0.01 | | |
| 0.02 | 0.00 | 0.01 | | |
| 0.01 | 0.00 | 0.01 | | |

TABLE 2:   POISSON CLUSTER PROBABILITY   $P'(n;\lambda,p)$

$$n = 9$$

| p | $\lambda = 5$ | $\lambda = 6$ | $\lambda = 7$ | $\lambda = 8$ |
|---|---|---|---|---|
| 0.50 | 0.00702644 9 | 0.021108322 | 0.049259774 | 0.095640123 |
| 0.49 | 0.00638631 0 | 0.019369885 | 0.045625396 | 0.089381754 |
| 0.48 | 0.005785633 | 0.017718870 | 0.042132225 | 0.083294690 |
| 0.47 | 0.005223643 | 0.016155228 | 0.038783327 | 0.077387810 |
| 0.46 | 0.004699472 | 0.014678687 | 0.035581559 | 0.071670055 |
| 0.45 | 0.004212160 | 0.013288729 | 0.03252945 8 | 0.066150546 |
| 0.44 | 0.003760662 | 0.011984568 | 0.029629156 | 0.060838260 |
| 0.43 | 0.003343832 | 0.010765150 | 0.026882291 | 0.055741575 |
| 0.42 | 0.002960454 | 0.009629134 | 0.024289943 | 0.050868392 |
| 0.41 | 0.002609229 | 0.008574888 | 0.021852579 | 0.046225779 |
| 0.40 | 0.002288791 | 0.007600497 | 0.019570019 | 0.041819833 |
| 0.39 | 0.001997711 | 0.006703764 | 0.017441388 | 0.037655570 |
| 0.38 | 0.001734510 | 0.005882218 | 0.015465107 | 0.033736799 |
| 0.37 | 0.001497662 | 0.005133148 | 0.013638884 | 0.030066017 |
| 0.36 | 0.001285611 | 0.004453592 | 0.011959713 | 0.026644330 |
| 0.35 | 0.001096779 | 0.003840379 | 0.010423876 | 0.023471393 |
| 0.34 | 0.000929575 | 0.003290146 | 0.009026993 | 0.020545371 |
| 0.33 | 0.000782410 | 0.002799373 | 0.007764019 | 0.017862897 |
| 0.32 | 0.000653704 | 0.002364406 | 0.006629322 | 0.015419096 |
| 0.31 | 0.000541903 | 0.001981495 | 0.005516717 | 0.013207596 |
| 0.30 | 0.000445482 | 0.001646827 | 0.004719526 | 0.011220589 |
| 0.29 | 0.000362962 | 0.001356557 | 0.003930666 | 0.009448901 |
| 0.28 | 0.000292914 | 0.001106852 | 0.003242729 | 0.007882096 |
| 0.27 | 0.000233973 | 0.000893921 | 0.002648048 | 0.006508626 |
| 0.26 | 0.000184843 | 0.000714051 | 0.002138814 | 0.005315967 |
| 0.25 | 0.000144303 | 0.000563642 | 0.00170716 | 0.0042908 |
| 0.24 | 0.000111215 | 0.000439238 | 0.00134526 | 0.0034193 |
| 0.23 | 0.000084526 | 0.000337553 | 0.00104542 | 0.0026872 |
| 0.22 | 0.000063273 | 0.000255501 | 0.00080018 | 0.0020801 |
| 0.21 | 0.000046584 | 0.000190211 | 0.00060240 | 0.001584 |
| 0.20 | 0.000033677 | 0.000139050 | 0.0004453 | 0.001184 |
| 0.19 | 0.000023863 | 0.000099629 | 0.0003227 | 0.000868 |
| 0.18 | 0.000016536 | 0.000069814 | 0.0002287 | 0.000622 |
| 0.17 | 0.000011178 | 0.000047723 | 0.0001581 | 0.000435 |
| 0.16 | 0.000007349 | 0.000031729 | 0.0001063 | 0.000296 |
| 0.15 | 0.000004683 | 0.00002044 | 0.0000693 | 0.000195 |
| 0.14 | 0.000002879 | 0.00001271 | 0.0000436 | 0.000124 |
| 0.13 | 0.000001699 | 0.00000759 | 0.0000263 | 0.000076 |
| 0.12 | 0.000000956 | 0.00000432 | 0.0000151 | 0.000044 |
| 0.11 | 0.000000509 | 0.00000232 | 0.0000083 | 0.000025 |
| 0.10 | 0.000000253 | 0.00000117 | 0.0000042 | 0.000013 |
| 0.09 | 0.000000116 | 0.00000054 | 0.0000020 | 0.000006 |
| 0.08 | 0.000000048 | 0.00000023 | 0.0000009 | 0.000003 |
| 0.07 | 0.000000018 | 0.00000009 | 0.0000003 | 0.000001 |
| 0.06 | 0.000000006 | 0.00000003 | 0.0000001 | 0.000001 |
| 0.05 | 0.000000001 | 0.00000001 | 0.0000001 | 0.000000 |

TABLE 2:   POISSON CLUSTER PROBABILITY   $P'(n;\lambda,p)$

<u>n = 9</u>

| p | $\lambda = 9$ | $\lambda = 10$ | $\lambda = 11$ | $\lambda = 12$ |
|---|---|---|---|---|
| 0.50 | 0.161535442 | 0.244775593 | 0.340303123 | 0.441482306 |
| 0.49 | 0.152262926 | 0.232597828 | 0.325821280 | 0.425639987 |
| 0.48 | 0.143138230 | 0.220475495 | 0.311242044 | 0.409514427 |
| 0.47 | 0.134176135 | 0.208426356 | 0.296579301 | 0.393107653 |
| 0.46 | 0.125393271 | 0.196472704 | 0.281855106 | 0.376433372 |
| 0.45 | 0.116807759 | 0.184640288 | 0.267097950 | 0.359515011 |
| 0.44 | 0.108438194 | 0.172957659 | 0.252341866 | 0.342384100 |
| 0.43 | 0.100303710 | 0.161455214 | 0.237625122 | 0.325079203 |
| 0.42 | 0.092423320 | 0.150164962 | 0.222989619 | 0.307644963 |
| 0.41 | 0.084815562 | 0.139119506 | 0.208480299 | 0.290131569 |
| 0.40 | 0.077498257 | 0.128351867 | 0.194144547 | 0.272594452 |
| 0.39 | 0.070488155 | 0.117895007 | 0.180031478 | 0.255093515 |
| 0.38 | 0.063800573 | 0.107781112 | 0.166191638 | 0.237692893 |
| 0.37 | 0.057449430 | 0.098041475 | 0.152676284 | 0.220460594 |
| 0.36 | 0.051446471 | 0.088705719 | 0.139536679 | 0.203467548 |
| 0.35 | 0.045801543 | 0.079801381 | 0.126823545 | 0.186787307 |
| 0.34 | 0.040522110 | 0.071353674 | 0.114586294 | 0.170494974 |
| 0.33 | 0.035613120 | 0.063384652 | 0.102872312 | 0.154666483 |
| 0.32 | 0.031076834 | 0.055913009 | 0.091725767 | 0.139377117 |
| 0.31 | 0.026912712 | 0.048953414 | 0.081187308 | 0.124700785 |
| 0.30 | 0.023117319 | 0.042516299 | 0.071292520 | 0.110708058 |
| 0.29 | 0.019684285 | 0.036607370 | 0.062071443 | 0.097464919 |
| 0.28 | 0.016604334 | 0.031227324 | 0.053547371 | 0.085030913 |
| 0.27 | 0.013865367 | 0.026371673 | 0.045736246 | 0.073457241 |
| 0.26 | 0.011452600 | 0.022030648 | 0.038645834 | 0.062785327 |
| 0.25 | 0.009349 | 0.01819 | 0.03228 | 0.0531 |
| 0.24 | 0.007535 | 0.01483 | 0.02662 | 0.0443 |
| 0.23 | 0.005989 | 0.01192 | 0.0217 | 0.0364 |
| 0.22 | 0.004689 | 0.00944 | 0.0173 | 0.0295 |
| 0.21 | 0.003611 | 0.00736 | 0.0137 | 0.0236 |
| 0.20 | 0.00273 | 0.00563 | 0.0106 | 0.018 |
| 0.19 | 0.00202 | 0.00422 | 0.0080 | 0.014 |
| 0.18 | 0.00147 | 0.00310 | 0.0060 | 0.011 |
| 0.17 | 0.00104 | 0.0022 | 0.0043 | 0.008 |
| 0.16 | 0.00071 | 0.0015 | 0.0031 | 0.006 |
| 0.15 | 0.00048 | 0.0010 | 0.0021 | 0.004 |
| 0.14 | 0.00031 | 0.0007 | 0.0014 | 0.003 |
| 0.13 | 0.00019 | 0.0004 | 0.0009 | 0.002 |
| 0.12 | 0.00011 | 0.0003 | 0.0006 | 0.001 |
| 0.11 | 0.00006 | 0.0002 | 0.0004 | 0.001 |
| 0.10 | 0.00003 | 0.0001 | 0.0002 | 0.001 |
| 0.09 | 0.00002 | 0.0001 | 0.0002 | 0.000 |
| 0.08 | 0.00001 | 0.0000 | 0.0001 | 0.000 |
| 0.07 | 0.00001 | 0.0000 | 0.0001 | 0.000 |
| 0.06 | 0.00000 | 0.0000 | 0.0001 | 0.000 |
| 0.05 | 0.00000 | 0.0000 | 0.0001 | 0.000 |
| 0.04 | 0.00000 | 0.0000 | 0.0001 | 0.000 |
| 0.03 | 0.00000 | 0.0000 | 0.0001 | 0.000 |
| 0.02 | 0.00000 | 0.0000 | 0.0001 | 0.000 |
| 0.01 | 0.00000 | 0.0000 | 0.0001 | 0.000 |

TABLE 2:   POISSON CLUSTER PROBABILITY   P'(n;λ,p)

n = 9

| p | λ = 13 | λ = 14 | λ = 15 | λ = 16 |
|------|-------------|-------------|-------------|-------------|
| 0.50 | 0.541572392 | 0.634905338 | 0.717547059 | 0.787443340 |
| 0.49 | 0.525436461 | 0.619457960 | 0.703541875 | 0.775343597 |
| 0.48 | 0.508834720 | 0.603397608 | 0.688832164 | 0.762508929 |
| 0.47 | 0.491750181 | 0.586684704 | 0.673356473 | 0.748860598 |
| 0.46 | 0.474180698 | 0.569296420 | 0.657070458 | 0.734335721 |
| 0.45 | 0.456136227 | 0.551223278 | 0.639942825 | 0.718883097 |
| 0.44 | 0.437637091 | 0.532466948 | 0.621952951 | 0.702459991 |
| 0.43 | 0.418712556 | 0.513038814 | 0.603088975 | 0.685030520 |
| 0.42 | 0.399399936 | 0.492959321 | 0.583347738 | 0.666565478 |
| 0.41 | 0.379744589 | 0.472257912 | 0.562734544 | 0.647042155 |
| 0.40 | 0.359799266 | 0.450973213 | 0.541263878 | 0.626445293 |
| 0.39 | 0.339624345 | 0.429153323 | 0.518960118 | 0.604768336 |
| 0.38 | 0.319287539 | 0.406856179 | 0.495858729 | 0.582014680 |
| 0.37 | 0.298863947 | 0.384150326 | 0.472007096 | 0.558199346 |
| 0.36 | 0.278435469 | 0.361114919 | 0.447465956 | 0.533350885 |
| 0.35 | 0.258091033 | 0.337840199 | 0.422310253 | 0.507513404 |
| 0.34 | 0.237925351 | 0.314427614 | 0.396630704 | 0.480748534 |
| 0.33 | 0.218038797 | 0.290989876 | 0.370534003 | 0.453137755 |
| 0.32 | 0.198535979 | 0.267650187 | 0.344144404 | 0.424784660 |
| 0.31 | 0.179524660 | 0.244541883 | 0.317603528 | 0.395816684 |
| 0.30 | 0.161113977 | 0.221806884 | 0.291070759 | 0.366387546 |
| 0.29 | 0.143412530 | 0.199594080 | 0.264722109 | 0.336678326 |
| 0.28 | 0.126525700 | 0.178056300 | 0.238748372 | 0.306897104 |
| 0.27 | 0.110553086 | 0.157347143 | 0.213351905 | 0.277277350 |
| 0.26 | 0.095585406 | 0.137616694 | 0.18874198  | 0.24807429  |
| 0.25 | 0.0817      | 0.119       | 0.165       | 0.220       |
| 0.24 | 0.0690      | 0.102       | 0.143       | 0.19        |
| 0.23 | 0.057       | 0.086       | 0.122       | 0.17        |
| 0.22 | 0.047       | 0.071       | 0.10        | 0.14        |
| 0.21 | 0.038       | 0.058       | 0.09        | 0.12        |
| 0.20 | 0.030       | 0.047       | 0.07        | 0.10        |
| 0.19 | 0.024       | 0.04        | 0.06        | 0.08        |
| 0.18 | 0.018       | 0.03        | 0.04        | 0.07        |
| 0.17 | 0.014       | 0.02        | 0.04        | 0.05        |
| 0.16 | 0.010       | 0.02        | 0.03        | 0.04        |
| 0.15 | 0.007       | 0.01        | 0.02        |             |
| 0.14 | 0.005       | 0.01        | 0.02        |             |
| 0.13 | 0.004       | 0.01        | 0.01        |             |
| 0.12 | 0.003       | 0.01        | 0.01        |             |
| 0.11 | 0.002       | 0.00        | 0.01        |             |
| 0.10 | 0.001       | 0.00        | 0.01        |             |
| 0.09 | 0.001       | 0.00        | 0.01        |             |
| 0.08 | 0.001       | 0.00        | 0.01        |             |
| 0.07 | 0.001       | 0.00        | 0.01        |             |
| 0.06 | 0.001       | 0.00        | 0.01        |             |
| 0.05 | 0.001       | 0.00        | 0.01        |             |
| 0.04 | 0.001       | 0.00        | 0.01        |             |
| 0.03 | 0.001       | 0.00        | 0.01        |             |
| 0.02 | 0.001       | 0.00        | 0.01        |             |
| 0.01 | 0.001       | 0.00        | 0.01        |             |

TABLE 2a:   POISSON CLUSTER PROBABILITY   $P'(n;\lambda,1/L)$

n = 3

| $\lambda$ | L = 2 | L = 3 | L = 4 |
|---|---|---|---|
| 0.100000000 | 0.000078779 | 0.000041495 | 0.000025240 |
| 0.200000000 | 0.000595838 | 0.000318780 | 0.000195695 |
| 0.300000000 | 0.001901399 | 0.001033109 | 0.000640065 |
| 0.400000000 | 0.004261916 | 0.002351361 | 0.001470195 |
| 0.500000000 | 0.007872212 | 0.004409394 | 0.002782254 |
| 0.600000000 | 0.012866230 | 0.007315123 | 0.004657839 |
| 0.700000000 | 0.019326531 | 0.011151376 | 0.007165005 |
| 0.800000000 | 0.027292658 | 0.015978535 | 0.010359243 |
| 0.900000000 | 0.036768503 | 0.021836978 | 0.014284410 |
| 1.000000000 | 0.047728738 | 0.028749354 | 0.018973622 |
| 1.100000000 | 0.060124452 | 0.036722689 | 0.024450106 |
| 1.200000000 | 0.073888037 | 0.045750340 | 0.030728024 |
| 1.300000000 | 0.088937429 | 0.055813801 | 0.037813263 |
| 1.400000000 | 0.105179762 | 0.066884373 | 0.045704199 |
| 1.500000000 | 0.122514497 | 0.078924703 | 0.054392428 |
| 1.600000000 | 0.140836098 | 0.091890197 | 0.063863465 |
| 1.700000000 | 0.160036285 | 0.105730311 | 0.074097423 |
| 1.800000000 | 0.180005938 | 0.120389737 | 0.085069646 |
| 1.900000000 | 0.200636669 | 0.135809467 | 0.096751327 |
| 2.000000000 | 0.221822121 | 0.151927770 | 0.109110085 |
| 2.100000000 | 0.243459020 | 0.168681055 | 0.122110509 |
| 2.200000000 | 0.265448005 | 0.186004656 | 0.135714677 |
| 2.300000000 | 0.287694283 | 0.203833521 | 0.149882634 |
| 2.400000000 | 0.310108109 | 0.222102824 | 0.164572846 |
| 2.500000000 | 0.332605140 | 0.240748498 | 0.179742617 |
| 2.600000000 | 0.355106659 | 0.259707699 | 0.195348472 |
| 2.700000000 | 0.377539706 | 0.278919203 | 0.211346515 |
| 2.800000000 | 0.399837114 | 0.298323741 | 0.227692752 |
| 2.900000000 | 0.421937485 | 0.317864279 | 0.244343382 |
| 3.000000000 | 0.443785096 | 0.337486247 | 0.261255066 |
| 3.100000000 | 0.465329764 | 0.357137721 | 0.278385162 |
| 3.200000000 | 0.486526669 | 0.376769558 | 0.295691937 |
| 3.300000000 | 0.507336146 | 0.396335501 | 0.313134748 |
| 3.400000000 | 0.527723454 | 0.415792239 | 0.330674207 |
| 3.500000000 | 0.547658531 | 0.435099448 | 0.348272315 |
| 3.600000000 | 0.567115730 | 0.454219792 | 0.365892578 |
| 3.700000000 | 0.586073556 | 0.473118912 | 0.383500104 |
| 3.800000000 | 0.604514395 | 0.491765384 | 0.401061679 |
| 3.900000000 | 0.622424242 | 0.510130668 | 0.418545824 |
| 4.000000000 | 0.639792435 | 0.528189031 | 0.435922839 |
| 4.100000000 | 0.656611395 | 0.545917468 | 0.453164834 |
| 4.200000000 | 0.672876365 | 0.563295606 | 0.470245736 |
| 4.300000000 | 0.688585168 | 0.580305600 | 0.487141297 |
| 4.400000000 | 0.703737966 | 0.596932020 | 0.503829082 |
| 4.500000000 | 0.718337034 | 0.613161741 | 0.520288450 |
| 4.600000000 | 0.732386543 | 0.628983820 | 0.536500529 |
| 4.700000000 | 0.745892354 | 0.644389374 | 0.552448175 |
| 4.800000000 | 0.758861328 | 0.659371457 | 0.568115933 |
| 4.900000000 | 0.771303640 | 0.673924942 | 0.583489986 |
| 5.000000000 | 0.783227611 | 0.688046392 | 0.598558107 |

TABLE 2a: POISSON CLUSTER PROBABILITY $P'(n;\lambda,1/L)$

| | | | n = 3 |
|---|---|---|---|
| $\lambda$ | L = 2 | L = 3 | L = 4 |
| 5.100000000 | 0.794644552 | 0.701733943 | 0.613309593 |
| 5.200000000 | 0.805566113 | 0.714987186 | 0.627735212 |
| 5.300000000 | 0.816004646 | 0.727807051 | 0.641827135 |
| 5.400000000 | 0.825973088 | 0.740195695 | 0.655578873 |
| 5.500000000 | 0.835484834 | 0.752156394 | 0.668985208 |
| 5.600000000 | 0.844553643 | 0.763693437 | 0.682042124 |
| 5.700000000 | 0.853193533 | 0.774812030 | 0.694746744 |
| 5.800000000 | 0.861418702 | 0.785518198 | 0.707097258 |
| 5.900000000 | 0.869243444 | 0.795818693 | 0.719092853 |
| 6.000000000 | 0.876682079 | 0.805720913 | 0.730733653 |
| 6.100000000 | 0.883748894 | 0.815232815 | 0.742020647 |
| 6.200000000 | 0.890458082 | 0.824362845 | 0.752955627 |
| 6.300000000 | 0.896823696 | 0.833119859 | 0.763541126 |
| 6.400000000 | 0.902859602 | 0.841513062 | 0.773780354 |
| 6.500000000 | 0.908579445 | 0.849551944 | 0.783677145 |
| 6.600000000 | 0.913996614 | 0.857246221 | 0.793235892 |
| 6.700000000 | 0.919124215 | 0.864605781 | 0.802461498 |
| 6.800000000 | 0.923975045 | 0.871640637 | 0.811359324 |
| 6.900000000 | 0.928561576 | 0.878360880 | 0.819935135 |
| 7.000000000 | 0.932895936 | 0.884776638 | 0.828195055 |
| 7.100000000 | 0.936989899 | 0.890898036 | 0.836145521 |
| 7.200000000 | 0.940854875 | 0.896735166 | 0.843793241 |
| 7.300000000 | 0.944501900 | 0.902298052 | 0.851145153 |
| 7.400000000 | 0.947941635 | 0.907596625 | 0.858208387 |
| 7.500000000 | 0.951184363 | 0.912640696 | 0.864990230 |
| 7.600000000 | 0.954239989 | 0.917439935 | 0.871498092 |
| 7.700000000 | 0.957118039 | 0.922003854 | 0.877739474 |
| 7.800000000 | 0.959827667 | 0.926341783 | 0.883721943 |
| 7.900000000 | 0.962377656 | 0.930462865 | 0.889453100 |
| 8.000000000 | 0.964776424 | 0.934376035 | 0.894940560 |
| 8.100000000 | 0.967032032 | 0.938090015 | 0.900191926 |
| 8.200000000 | 0.969152190 | 0.941613302 | 0.905214770 |
| 8.300000000 | 0.971144266 | 0.944954165 | 0.910016615 |
| 8.400000000 | 0.973015291 | 0.948120634 | 0.914604915 |
| 8.500000000 | 0.974771975 | 0.951120499 | 0.918987041 |
| 8.600000000 | 0.976420710 | 0.953961310 | 0.923170269 |
| 8.700000000 | 0.977967583 | 0.956650370 | 0.927161764 |
| 8.800000000 | 0.979418383 | 0.959194737 | 0.930968571 |
| 8.900000000 | 0.980778618 | 0.961601226 | 0.934597606 |
| 9.000000000 | 0.982053515 | 0.963876408 | 0.938055644 |
| 9.100000000 | 0.983248040 | 0.966026612 | 0.941349318 |
| 9.200000000 | 0.984366901 | 0.968057930 | 0.944485106 |
| 9.300000000 | 0.985414563 | 0.969976219 | 0.947469331 |
| 9.400000000 | 0.986395255 | 0.971787103 | 0.950308154 |
| 9.500000000 | 0.987312980 | 0.973495980 | 0.953007572 |
| 9.600000000 | 0.988171528 | 0.975108028 | 0.955573413 |
| 9.700000000 | 0.988974481 | 0.976628204 | 0.958011339 |
| 9.800000000 | 0.989725225 | 0.978061257 | 0.960326840 |
| 9.900000000 | 0.990426957 | 0.979411726 | 0.962525238 |
| 10.000000000 | 0.991082697 | 0.980683953 | 0.964611682 |

NEFF and NAUS

## TABLE 2a:  POISSON CLUSTER PROBABILITY  P'(n;λ,1/L)

| | | | n = 4 |
|---|---|---|---|
| λ | L = 2 | L = 3 | L = 4 |
| 0.100000000 | 0.000001231 | 0.000000445 | 0.000000205 |
| 0.200000000 | 0.000018627 | 0.000006848 | 0.000003188 |
| 0.300000000 | 0.000089172 | 0.000033331 | 0.000015662 |
| 0.400000000 | 0.000266510 | 0.000101279 | 0.000048034 |
| 0.500000000 | 0.000615324 | 0.000237722 | 0.000113794 |
| 0.600000000 | 0.001206893 | 0.000473907 | 0.000228961 |
| 0.700000000 | 0.002114325 | 0.000844050 | 0.000411581 |
| 0.800000000 | 0.003411523 | 0.001384240 | 0.000681263 |
| 0.900000000 | 0.005168793 | 0.002131489 | 0.001058768 |
| 1.000000000 | 0.007452007 | 0.003122915 | 0.001565632 |
| 1.100000000 | 0.010321022 | 0.004395038 | 0.002223827 |
| 1.200000000 | 0.013828707 | 0.005983182 | 0.003055455 |
| 1.300000000 | 0.018020294 | 0.007920974 | 0.004082474 |
| 1.400000000 | 0.022933015 | 0.010239936 | 0.005326456 |
| 1.500000000 | 0.028595957 | 0.012969140 | 0.006808368 |
| 1.600000000 | 0.035030119 | 0.016134953 | 0.008548381 |
| 1.700000000 | 0.042248608 | 0.019760823 | 0.010565706 |
| 1.800000000 | 0.050256962 | 0.023867144 | 0.012878443 |
| 1.900000000 | 0.059053559 | 0.028471155 | 0.015503462 |
| 2.000000000 | 0.068630099 | 0.033586890 | 0.018456296 |
| 2.100000000 | 0.078972134 | 0.039225171 | 0.021751053 |
| 2.200000000 | 0.090059624 | 0.045393630 | 0.025400352 |
| 2.300000000 | 0.101867521 | 0.052096767 | 0.029415263 |
| 2.400000000 | 0.114386344 | 0.059336029 | 0.033805275 |
| 2.500000000 | 0.127522766 | 0.067109921 | 0.038578265 |
| 2.600000000 | 0.141300174 | 0.075414124 | 0.043740494 |
| 2.700000000 | 0.155659220 | 0.084241644 | 0.049296598 |
| 2.800000000 | 0.170558340 | 0.093582958 | 0.055249610 |
| 2.900000000 | 0.185954257 | 0.103426185 | 0.061600970 |
| 3.000000000 | 0.201802440 | 0.113757255 | 0.068350567 |
| 3.100000000 | 0.218057539 | 0.124560091 | 0.075496767 |
| 3.200000000 | 0.234673786 | 0.135816788 | 0.083036469 |
| 3.300000000 | 0.251605360 | 0.147507799 | 0.090965152 |
| 3.400000000 | 0.268806717 | 0.159612117 | 0.099276935 |
| 3.500000000 | 0.286232890 | 0.172107458 | 0.107964644 |
| 3.600000000 | 0.303839751 | 0.184970443 | 0.117019879 |
| 3.700000000 | 0.321584249 | 0.198176770 | 0.126433084 |
| 3.800000000 | 0.339424607 | 0.211701388 | 0.136193626 |
| 3.900000000 | 0.357320502 | 0.225518655 | 0.146289872 |
| 4.000000000 | 0.375233207 | 0.239602505 | 0.156709265 |
| 4.100000000 | 0.393125712 | 0.253926590 | 0.167438410 |
| 4.200000000 | 0.410962825 | 0.268464429 | 0.178463152 |
| 4.300000000 | 0.428711241 | 0.283189536 | 0.189768656 |
| 4.400000000 | 0.446339601 | 0.298075550 | 0.201339494 |
| 4.500000000 | 0.463818526 | 0.313096352 | 0.213159718 |
| 4.600000000 | 0.481120634 | 0.328226169 | 0.225212945 |
| 4.700000000 | 0.498220546 | 0.343439678 | 0.237482429 |
| 4.800000000 | 0.515094877 | 0.358712093 | 0.249951142 |
| 4.900000000 | 0.531722207 | 0.374019246 | 0.262601844 |
| 5.000000000 | 0.548083058 | 0.389337664 | 0.275417155 |

TABLE 2a: POISSON CLUSTER PROBABILITY P'(n;λ,1/L)

| | | | n = 4 |
|---|---|---|---|
| λ | L = 2 | L = 3 | L = 4 |
| 5.100000000 | 0.564159842 | 0.404644626 | 0.288379619 |
| 5.200000000 | 0.579936816 | 0.419918228 | 0.301471779 |
| 5.300000000 | 0.595430024 | 0.435137424 | 0.314676226 |
| 5.400000000 | 0.610537235 | 0.450282071 | 0.327975670 |
| 5.500000000 | 0.625337871 | 0.465332961 | 0.341352986 |
| 5.600000000 | 0.639792943 | 0.480271845 | 0.354791272 |
| 5.700000000 | 0.653894969 | 0.495081459 | 0.368273895 |
| 5.800000000 | 0.667637904 | 0.509745533 | 0.381784535 |
| 5.900000000 | 0.681017060 | 0.524248801 | 0.395307228 |
| 6.000000000 | 0.694029028 | 0.538577004 | 0.408826406 |
| 6.100000000 | 0.706671601 | 0.552716887 | 0.422326926 |
| 6.200000000 | 0.718943695 | 0.566656195 | 0.435794103 |
| 6.300000000 | 0.730845270 | 0.580383659 | 0.449213740 |
| 6.400000000 | 0.742377261 | 0.593888982 | 0.462572150 |
| 6.500000000 | 0.753541497 | 0.607162825 | 0.475856176 |
| 6.600000000 | 0.764340632 | 0.620196781 | 0.489053211 |
| 6.700000000 | 0.774778076 | 0.632983354 | 0.502151209 |
| 6.800000000 | 0.784857926 | 0.645515933 | 0.515138702 |
| 6.900000000 | 0.794584900 | 0.657788763 | 0.528004804 |
| 7.000000000 | 0.803964276 | 0.669796919 | 0.540739220 |
| 7.100000000 | 0.813001833 | 0.681536266 | 0.553332252 |
| 7.200000000 | 0.821703794 | 0.693003439 | 0.565774792 |
| 7.300000000 | 0.830076768 | 0.704195798 | 0.578058333 |
| 7.400000000 | 0.838127705 | 0.715111402 | 0.590174956 |
| 7.500000000 | 0.845863844 | 0.725748971 | 0.602117331 |
| 7.600000000 | 0.853292669 | 0.736107853 | 0.613878707 |
| 7.700000000 | 0.860421865 | 0.746187987 | 0.625452906 |
| 7.800000000 | 0.867259279 | 0.755989870 | 0.636834314 |
| 7.900000000 | 0.873812886 | 0.765514524 | 0.648017864 |
| 8.000000000 | 0.880090747 | 0.774763458 | 0.658999032 |
| 8.100000000 | 0.886100937 | 0.783738638 | 0.669773816 |
| 8.200000000 | 0.891851756 | 0.792442453 | 0.680338726 |
| 8.300000000 | 0.897351210 | 0.800877684 | 0.690690766 |
| 8.400000000 | 0.902607480 | 0.809047468 | 0.700827419 |
| 8.500000000 | 0.907628655 | 0.816955275 | 0.710746630 |
| 8.600000000 | 0.912422757 | 0.824604872 | 0.720446789 |
| 8.700000000 | 0.916997726 | 0.832000298 | 0.729926711 |
| 8.800000000 | 0.921361401 | 0.839145836 | 0.739185621 |
| 8.900000000 | 0.925521506 | 0.846045986 | 0.748223134 |
| 9.000000000 | 0.929485639 | 0.852705440 | 0.757039236 |
| 9.100000000 | 0.933261256 | 0.859129060 | 0.765634266 |
| 9.200000000 | 0.936855667 | 0.865321850 | 0.774008899 |
| 9.300000000 | 0.940276023 | 0.871238940 | 0.782164124 |
| 9.400000000 | 0.943529309 | 0.877035561 | 0.790101228 |
| 9.500000000 | 0.946622339 | 0.882567028 | 0.797821777 |
| 9.600000000 | 0.949561753 | 0.887888722 | 0.805327598 |
| 9.700000000 | 0.952354009 | 0.893006069 | 0.812620764 |
| 9.800000000 | 0.955005382 | 0.897924528 | 0.819703571 |
| 9.900000000 | 0.957521962 | 0.902649574 | 0.826578528 |
| 10.000000000 | 0.959909655 | 0.907186685 | 0.833248332 |

TABLE 2a:    POISSON CLUSTER PROBABILITY    P'(n;λ,1/L)

n = 5

| λ | L = 2 | L = 3 | L = 4 |
|---|---|---|---|
| 0.100000000 | 0.00000015 | 0.000000004 | 0.000000001 |
| 0.200000000 | 0.00000447 | 0.000000112 | 0.000000039 |
| 0.300000000 | 0.000003214 | 0.000000817 | 0.000000290 |
| 0.400000000 | 0.000012813 | 0.000003313 | 0.000001186 |
| 0.500000000 | 0.000036991 | 0.000009728 | 0.000003516 |
| 0.600000000 | 0.000037076 | 0.000023293 | 0.000008497 |
| 0.700000000 | 0.000178049 | 0.000048444 | 0.000017837 |
| 0.800000000 | 0.000328409 | 0.000090881 | 0.000033772 |
| 0.900000000 | 0.000559888 | 0.000157582 | 0.000059102 |
| 1.000000000 | 0.000897059 | 0.000256776 | 0.000097199 |
| 1.100000000 | 0.001366866 | 0.000397898 | 0.000152017 |
| 1.200000000 | 0.001998114 | 0.000591508 | 0.000228082 |
| 1.300000000 | 0.002820934 | 0.000849196 | 0.000330485 |
| 1.400000000 | 0.003866250 | 0.001183476 | 0.000464854 |
| 1.500000000 | 0.005165253 | 0.001607659 | 0.000637332 |
| 1.600000000 | 0.006748911 | 0.002135731 | 0.000854540 |
| 1.700000000 | 0.008647497 | 0.002782209 | 0.001123544 |
| 1.800000000 | 0.010890170 | 0.003562015 | 0.001451813 |
| 1.900000000 | 0.013504598 | 0.004490330 | 0.001847175 |
| 2.000000000 | 0.016516617 | 0.005582463 | 0.002317769 |
| 2.100000000 | 0.019949953 | 0.006853716 | 0.002872002 |
| 2.200000000 | 0.023825979 | 0.008319258 | 0.003518494 |
| 2.300000000 | 0.028163526 | 0.009993998 | 0.004266035 |
| 2.400000000 | 0.032978736 | 0.011892472 | 0.005123529 |
| 2.500000000 | 0.038284958 | 0.014028732 | 0.006099947 |
| 2.600000000 | 0.044092683 | 0.016416244 | 0.007204279 |
| 2.700000000 | 0.050409518 | 0.019067796 | 0.008445480 |
| 2.800000000 | 0.057240192 | 0.021995409 | 0.009832430 |
| 2.900000000 | 0.064586593 | 0.025210267 | 0.011373879 |
| 3.000000000 | 0.072447833 | 0.028722642 | 0.013078409 |
| 3.100000000 | 0.080820336 | 0.032541842 | 0.014954390 |
| 3.200000000 | 0.089697947 | 0.036676156 | 0.017009934 |
| 3.300000000 | 0.099072057 | 0.041132816 | 0.019252861 |
| 3.400000000 | 0.108931746 | 0.045917963 | 0.021690661 |
| 3.500000000 | 0.119263933 | 0.051036624 | 0.024330459 |
| 3.600000000 | 0.130053543 | 0.056492694 | 0.027178981 |
| 3.700000000 | 0.141283669 | 0.062288926 | 0.030242529 |
| 3.800000000 | 0.152935750 | 0.068426935 | 0.033526948 |
| 3.900000000 | 0.164989744 | 0.074907197 | 0.037037608 |
| 4.000000000 | 0.177424307 | 0.081729069 | 0.040779379 |
| 4.100000000 | 0.190216962 | 0.088890797 | 0.044756609 |
| 4.200000000 | 0.203344278 | 0.096389550 | 0.048973115 |
| 4.300000000 | 0.216782033 | 0.104221443 | 0.053432162 |
| 4.400000000 | 0.230505381 | 0.112381575 | 0.058136457 |
| 4.500000000 | 0.244489008 | 0.120864062 | 0.063088138 |
| 4.600000000 | 0.258707281 | 0.129662085 | 0.068288768 |
| 4.700000000 | 0.273134393 | 0.138767932 | 0.073739334 |
| 4.800000000 | 0.287744500 | 0.148173048 | 0.079440244 |
| 4.900000000 | 0.302511841 | 0.157868086 | 0.085391329 |
| 5.000000000 | 0.317410866 | 0.167842960 | 0.091591847 |

TABLE 2a:   POISSON CLUSTER PROBABILITY   P'(n;λ,1/L)

n = 5

| λ | L = 2 | L = 3 | L = 4 |
|---|---|---|---|
| 5.100000000 | 0.332416337 | 0.178086898 | 0.098040489 |
| 5.200000000 | 0.347503433 | 0.188588501 | 0.104735387 |
| 5.300000000 | 0.362647840 | 0.199335800 | 0.111674123 |
| 5.400000000 | 0.377825834 | 0.210316310 | 0.118853745 |
| 5.500000000 | 0.393014356 | 0.221517093 | 0.126270775 |
| 5.600000000 | 0.408191075 | 0.232924809 | 0.133921228 |
| 5.700000000 | 0.423334451 | 0.244525782 | 0.141800630 |
| 5.800000000 | 0.438423776 | 0.256306046 | 0.149904033 |
| 5.900000000 | 0.453439225 | 0.268251411 | 0.158226037 |
| 6.000000000 | 0.468361881 | 0.280347509 | 0.166760811 |
| 6.100000000 | 0.483173771 | 0.292579851 | 0.175502114 |
| 6.200000000 | 0.497857880 | 0.304933878 | 0.184443319 |
| 6.300000000 | 0.512398172 | 0.317395008 | 0.193577434 |
| 6.400000000 | 0.526779593 | 0.329948686 | 0.202897130 |
| 6.500000000 | 0.540988078 | 0.342580430 | 0.212394764 |
| 6.600000000 | 0.555010550 | 0.355275869 | 0.222062403 |
| 6.700000000 | 0.568834911 | 0.368020790 | 0.231891853 |
| 6.800000000 | 0.582450035 | 0.380801173 | 0.241874632 |
| 6.900000000 | 0.595845753 | 0.393603229 | 0.252002246 |
| 7.000000000 | 0.609012835 | 0.406413431 | 0.262265718 |
| 7.100000000 | 0.621942971 | 0.419218547 | 0.272656108 |
| 7.200000000 | 0.634628747 | 0.432005667 | 0.283164297 |
| 7.300000000 | 0.647063619 | 0.444762234 | 0.293781052 |
| 7.400000000 | 0.659241888 | 0.457476059 | 0.304497060 |
| 7.500000000 | 0.671158668 | 0.470135350 | 0.315302949 |
| 7.600000000 | 0.682809859 | 0.482728726 | 0.326189308 |
| 7.700000000 | 0.694192110 | 0.495245237 | 0.337146719 |
| 7.800000000 | 0.705302792 | 0.507674375 | 0.348165770 |
| 7.900000000 | 0.716139960 | 0.520006086 | 0.359237004 |
| 8.000000000 | 0.726702324 | 0.532230782 | 0.370351337 |
| 8.100000000 | 0.736989211 | 0.544339347 | 0.381499280 |
| 8.200000000 | 0.747000530 | 0.556323144 | 0.392671754 |
| 8.300000000 | 0.756736742 | 0.568174016 | 0.403859715 |
| 8.400000000 | 0.766198825 | 0.579884289 | 0.415054247 |
| 8.500000000 | 0.775388239 | 0.591446776 | 0.426246580 |
| 8.600000000 | 0.784306894 | 0.602854768 | 0.437428107 |
| 8.700000000 | 0.792957116 | 0.614102040 | 0.448590394 |
| 8.800000000 | 0.801341619 | 0.625182837 | 0.459725201 |
| 8.900000000 | 0.809463473 | 0.636091877 | 0.470824487 |
| 9.000000000 | 0.817326071 | 0.646824339 | 0.481880426 |
| 9.100000000 | 0.824933103 | 0.657375857 | 0.492885419 |
| 9.200000000 | 0.832288530 | 0.667742509 | 0.503832098 |
| 9.300000000 | 0.839396551 | 0.677920810 | 0.514713341 |
| 9.400000000 | 0.846261583 | 0.687907698 | 0.525522274 |
| 9.500000000 | 0.852888232 | 0.697700524 | 0.536252283 |
| 9.600000000 | 0.859281273 | 0.707297041 | 0.546897014 |
| 9.700000000 | 0.865445625 | 0.716695388 | 0.557450384 |
| 9.800000000 | 0.871386329 | 0.725894081 | 0.567906580 |
| 9.900000000 | 0.877108530 | 0.734891995 | 0.578260064 |
| 10.000000000 | 0.882617457 | 0.743688351 | 0.588505576 |

TABLE 2a:   POISSON CLUSTER PROBABILITY   P'(n;λ,1/L)

n = 6

| λ | L = 2 | L = 3 | L = 4 |
|---|---|---|---|
| C.100000000 | 0.000000000 | C.000000000 | 0.000000000 |
| C.200000000 | 0.000000009 | 0.000000001 | 0.000000000 |
| 0.300000000 | 0.000000094 | 0.000000016 | 0.000000004 |
| 0.400000000 | 0.000000499 | 0.000000087 | 0.000000024 |
| 0.500000000 | 0.000001802 | 0.000000320 | 0.000000087 |
| C.600000000 | 0.000005092 | 0.000000921 | 0.000000253 |
| 0.700000000 | 0.000012153 | 0.000002237 | 0.000000620 |
| C.800000000 | 0.000025627 | 0.000004799 | 0.000001343 |
| C.900000000 | 0.000049169 | 0.000009368 | 0.000002646 |
| 1.000000000 | 0.000087581 | 0.000016972 | 0.000004839 |
| 1.100000000 | 0.000146808 | 0.000028950 | 0.000008330 |
| 1.200000000 | 0.000234189 | 0.000046982 | 0.000013643 |
| 1.300000000 | 0.000358285 | C.000073121 | 0.000021430 |
| 1.400000000 | 0.000528966 | 0.000109821 | C.000032483 |
| 1.500000000 | C.000757362 | 0.000159954 | 0.000047748 |
| 1.600000000 | 0.001055787 | 0.000226824 | C.000068336 |
| 1.700000000 | C.001437657 | 0.000314180 | C.000095529 |
| 1.800000000 | 0.001917376 | 0.000426217 | 0.000130793 |
| 1.900000000 | 0.002510209 | C.000567571 | C.000175781 |
| 2.000000000 | 0.003232143 | 0.000743318 | C.000232341 |
| 2.100000000 | 0.004039730 | 0.000958958 | C.000302518 |
| 2.200000000 | 0.005129929 | 0.001220399 | C.000388557 |
| 2.300000000 | 0.006339947 | 0.001533936 | C.000492905 |
| 2.400000000 | 0.007747068 | 0.001906228 | C.000618208 |
| 2.500000000 | 0.009368493 | 0.002344265 | C.000767311 |
| 2.600000000 | 0.011221181 | 0.002855341 | C.000943255 |
| 2.700000000 | 0.013321690 | 0.003447017 | 0.001149269 |
| 2.800000000 | 0.015686038 | 0.004127087 | 0.001388769 |
| 2.900000000 | 0.018329559 | 0.004903536 | C.001665348 |
| 3.000000000 | 0.021266778 | 0.005784501 | 0.001982768 |
| 3.100000000 | 0.024511295 | 0.006778231 | 0.002344949 |
| 3.200000000 | 0.028075677 | 0.007893044 | 0.002755963 |
| 3.300000000 | 0.031971372 | C.009137285 | 0.003220021 |
| 3.400000000 | 0.036208624 | 0.010519281 | C.003741458 |
| 3.500000000 | 0.040796407 | 0.012047300 | 0.004324724 |
| 3.600000000 | 0.045742372 | 0.013729512 | 0.004974369 |
| 3.700000000 | 0.051052808 | 0.015573940 | 0.005695031 |
| 3.800000000 | 0.056732607 | 0.017588430 | 0.006491417 |
| 3.900000000 | 0.062785249 | 0.019780603 | C.007368292 |
| 4.000000000 | 0.069212799 | 0.022157821 | 0.008330462 |
| 4.100000000 | 0.076015906 | 0.024727154 | C.009382761 |
| 4.200000000 | 0.083193822 | 0.027495341 | C.010530032 |
| 4.300000000 | 0.090744422 | 0.030468761 | 0.011777110 |
| 4.400000000 | 0.098664240 | 0.033653404 | 0.013128811 |
| 4.500000000 | 0.106948507 | 0.037054841 | 0.014589914 |
| 4.600000000 | 0.115591197 | 0.040678199 | 0.016165141 |
| 4.700000000 | 0.124585086 | 0.044528140 | 0.017859149 |
| 4.800000000 | 0.133921803 | 0.048608840 | 0.019676507 |
| 4.900000000 | 0.143591900 | 0.052923970 | 0.021621686 |
| 5.000000000 | 0.153584917 | 0.057476684 | 0.023699042 |

TABLE 2a:   POISSON CLUSTER PROBABILITY  P'(n;λ,1/L)

n = 6

| λ | L = 2 | L = 3 | L = 4 |
|---|---|---|---|
| 5.100000000 | 0.163889454 | 0.062269600 | 0.025912801 |
| 5.200000000 | 0.174493243 | 0.067304798 | 0.028267046 |
| 5.300000000 | 0.185383225 | 0.072583806 | 0.030765705 |
| 5.400000000 | 0.196545624 | 0.078107598 | 0.033412533 |
| 5.500000000 | 0.207966026 | 0.083876590 | 0.036211105 |
| 5.600000000 | 0.219629457 | 0.089890641 | 0.039164799 |
| 5.700000000 | 0.231520458 | 0.096149054 | 0.042276792 |
| 5.800000000 | 0.243623163 | 0.102650580 | 0.045550042 |
| 5.900000000 | 0.255921370 | 0.109393427 | 0.048987281 |
| 6.000000000 | 0.268398019 | 0.116375264 | 0.052591007 |
| 6.100000000 | 0.281038258 | 0.123593237 | 0.056363477 |
| 6.200000000 | 0.293823517 | 0.131043976 | 0.060306695 |
| 6.300000000 | 0.306737568 | 0.138723614 | 0.064422407 |
| 6.400000000 | 0.319763592 | 0.146627798 | 0.068712099 |
| 6.500000000 | 0.332884839 | 0.154751709 | 0.073176987 |
| 6.600000000 | 0.346084683 | 0.163090082 | 0.077818015 |
| 6.700000000 | 0.359346679 | 0.171637222 | 0.082635850 |
| 6.800000000 | 0.372654607 | 0.180387028 | 0.087630883 |
| 6.900000000 | 0.385992525 | 0.189333014 | 0.092803222 |
| 7.000000000 | 0.399344812 | 0.198468330 | 0.098152697 |
| 7.100000000 | 0.412696200 | 0.207785791 | 0.103678855 |
| 7.200000000 | 0.426031822 | 0.217277893 | 0.109380963 |
| 7.300000000 | 0.439337234 | 0.226936847 | 0.115258009 |
| 7.400000000 | 0.452598450 | 0.236754596 | 0.121308703 |
| 7.500000000 | 0.465801967 | 0.246722845 | 0.127531484 |
| 7.600000000 | 0.478934787 | 0.256833086 | 0.133924518 |
| 7.700000000 | 0.491984437 | 0.267076623 | 0.140485708 |
| 7.800000000 | 0.504938983 | 0.277444596 | 0.147212695 |
| 7.900000000 | 0.517787047 | 0.287928008 | 0.154102864 |
| 8.000000000 | 0.530517816 | 0.298517751 | 0.161153355 |
| 8.100000000 | 0.543121048 | 0.309204630 | 0.168361064 |
| 8.200000000 | 0.555587079 | 0.319979385 | 0.175722652 |
| 8.300000000 | 0.567906826 | 0.330832719 | 0.183234557 |
| 8.400000000 | 0.580071786 | 0.341755316 | 0.190892997 |
| 8.500000000 | 0.592074035 | 0.352737870 | 0.198693983 |
| 8.600000000 | 0.603906226 | 0.363771103 | 0.206633326 |
| 8.700000000 | 0.615561579 | 0.374845784 | 0.214706648 |
| 8.800000000 | 0.627033878 | 0.385952755 | 0.222909391 |
| 8.900000000 | 0.638317460 | 0.397082945 | 0.231236828 |
| 9.000000000 | 0.649407206 | 0.408227395 | 0.239684074 |
| 9.100000000 | 0.660298526 | 0.419377268 | 0.248246097 |
| 9.200000000 | 0.670987350 | 0.430523871 | 0.256917727 |
| 9.300000000 | 0.681470115 | 0.441658669 | 0.265693669 |
| 9.400000000 | 0.691743744 | 0.452773301 | 0.274568516 |
| 9.500000000 | 0.701805636 | 0.463859591 | 0.283536755 |
| 9.600000000 | 0.711653649 | 0.474909565 | 0.292592784 |
| 9.700000000 | 0.721286081 | 0.485915459 | 0.301730922 |
| 9.800000000 | 0.730701654 | 0.496869731 | 0.310945417 |
| 9.900000000 | 0.739899498 | 0.507765071 | 0.320230461 |
| 10.000000000 | 0.748879129 | 0.518594410 | 0.329580202 |

TABLE 2a:   POISSON CLUSTER PROBABILITY   P'(n;λ,l/L)

n = 7

| λ | L = 2 | L = 3 | L = 4 |
|---|---|---|---|
| 0.100000000 | 0.000000000 | 0.000000000 | 0.000000000 |
| 0.200000000 | 0.000000000 | 0.000000000 | 0.000000000 |
| 0.300000000 | 0.000000002 | 0.000000000 | 0.000000000 |
| 0.400000000 | 0.000000016 | 0.000000002 | 0.000000000 |
| 0.500000000 | 0.000000074 | 0.000000009 | 0.000000002 |
| 0.600000000 | 0.000000250 | 0.000000030 | 0.000000006 |
| 0.700000000 | 0.000000696 | 0.000000086 | 0.000000018 |
| 0.800000000 | 0.000001679 | 0.000000212 | 0.000000045 |
| 0.900000000 | 0.000003625 | 0.000000465 | 0.000000099 |
| 1.000000000 | 0.000007176 | 0.000000937 | 0.000000201 |
| 1.100000000 | 0.000013239 | 0.000001760 | 0.000000381 |
| 1.200000000 | 0.000023047 | 0.000003117 | 0.000000681 |
| 1.300000000 | 0.000038210 | 0.000005258 | 0.000001159 |
| 1.400000000 | 0.000060773 | 0.000008510 | 0.000001893 |
| 1.500000000 | 0.000093259 | 0.000013287 | 0.000002983 |
| 1.600000000 | 0.000138716 | 0.000020109 | 0.000004555 |
| 1.700000000 | 0.000200756 | 0.000029610 | 0.000006769 |
| 1.800000000 | 0.000283578 | 0.000042555 | 0.000009818 |
| 1.900000000 | 0.000391994 | 0.000059850 | 0.000013935 |
| 2.000000000 | 0.000531442 | 0.000082554 | 0.000019398 |
| 2.100000000 | 0.000707985 | 0.000111892 | 0.000026533 |
| 2.200000000 | 0.000928310 | 0.000149262 | 0.000035721 |
| 2.300000000 | 0.001199711 | 0.000196250 | 0.000047398 |
| 2.400000000 | 0.001530066 | 0.000254631 | 0.000062064 |
| 2.500000000 | 0.001927809 | 0.000326381 | 0.000080285 |
| 2.600000000 | 0.002401890 | 0.000413680 | 0.000102696 |
| 2.700000000 | 0.002961726 | 0.000518918 | 0.000130008 |
| 2.800000000 | 0.003617156 | 0.000644694 | 0.000163008 |
| 2.900000000 | 0.004378380 | 0.000793822 | 0.000202565 |
| 3.000000000 | 0.005255897 | 0.000969326 | 0.000249631 |
| 3.100000000 | 0.006260445 | 0.001174440 | 0.000305244 |
| 3.200000000 | 0.007402925 | 0.001412604 | 0.000370532 |
| 3.300000000 | 0.008694338 | 0.001687462 | 0.000446714 |
| 3.400000000 | 0.010145709 | 0.002002849 | 0.000535102 |
| 3.500000000 | 0.011768017 | 0.002362790 | 0.000637098 |
| 3.600000000 | 0.013572125 | 0.002771486 | 0.000754204 |
| 3.700000000 | 0.015568705 | 0.003233307 | 0.000888012 |
| 3.800000000 | 0.017768175 | 0.003752779 | 0.001040214 |
| 3.900000000 | 0.020180627 | 0.004334571 | 0.001212591 |
| 4.000000000 | 0.022815765 | 0.004983479 | 0.001407023 |
| 4.100000000 | 0.025682846 | 0.005704418 | 0.001625479 |
| 4.200000000 | 0.028790621 | 0.006502398 | 0.001870021 |
| 4.300000000 | 0.032147284 | 0.007382516 | 0.002142799 |
| 4.400000000 | 0.035760421 | 0.008349931 | 0.002446050 |
| 4.500000000 | 0.039636968 | 0.009409852 | 0.002782096 |
| 4.600000000 | 0.043783174 | 0.010567520 | 0.003153338 |
| 4.700000000 | 0.048204559 | 0.011828186 | 0.003562254 |
| 4.800000000 | 0.052905897 | 0.013197095 | 0.004011397 |
| 4.900000000 | 0.057891179 | 0.014679467 | 0.004503386 |
| 5.000000000 | 0.063163806 | 0.016280479 | 0.005040909 |

TABLE 2a:  POISSON CLUSTER PROBABILITY  P'(n;λ,1/L)

n = 7

| λ | L = 2 | L = 3 | L = 4 |
|---|---|---|---|
| 5.100000000 | 0.068725565 | 0.018005245 | 0.005626708 |
| 5.200000000 | 0.074578626 | 0.019858796 | 0.006263585 |
| 5.300000000 | 0.080723536 | 0.021846067 | 0.006954387 |
| 5.400000000 | 0.087160220 | 0.023971872 | 0.007702006 |
| 5.500000000 | 0.093887783 | 0.026240891 | 0.008509372 |
| 5.600000000 | 0.100904524 | 0.028657653 | 0.009379445 |
| 5.700000000 | 0.108207943 | 0.031226516 | 0.010315212 |
| 5.800000000 | 0.115794762 | 0.033951653 | 0.011319675 |
| 5.900000000 | 0.123660947 | 0.036837037 | 0.012395852 |
| 6.000000000 | 0.131801724 | 0.039886426 | 0.013546763 |
| 6.100000000 | 0.140211614 | 0.043103349 | 0.014775429 |
| 6.200000000 | 0.148884456 | 0.046491092 | 0.016084859 |
| 6.300000000 | 0.157813444 | 0.050052689 | 0.017478051 |
| 6.400000000 | 0.166991155 | 0.053790906 | 0.018957975 |
| 6.500000000 | 0.176409589 | 0.057708235 | 0.020527574 |
| 6.600000000 | 0.186060204 | 0.061806884 | 0.022189755 |
| 6.700000000 | 0.195933955 | 0.066088766 | 0.023947378 |
| 6.800000000 | 0.206021333 | 0.070555495 | 0.025803254 |
| 6.900000000 | 0.216312407 | 0.075208378 | 0.027760137 |
| 7.000000000 | 0.226796862 | 0.080048410 | 0.029820714 |
| 7.100000000 | 0.237464042 | 0.085076272 | 0.031987601 |
| 7.200000000 | 0.248302990 | 0.090292325 | 0.034263336 |
| 7.300000000 | 0.259302490 | 0.095696608 | 0.036650374 |
| 7.400000000 | 0.270451106 | 0.101288842 | 0.039151077 |
| 7.500000000 | 0.281737224 | 0.107068424 | 0.041767709 |
| 7.600000000 | 0.293149090 | 0.113034430 | 0.044502435 |
| 7.700000000 | 0.304674850 | 0.119185619 | 0.047357307 |
| 7.800000000 | 0.316302583 | 0.125520434 | 0.050334265 |
| 7.900000000 | 0.328020345 | 0.132037008 | 0.053435129 |
| 8.000000000 | 0.339816198 | 0.138733165 | 0.056661596 |
| 8.100000000 | 0.351678246 | 0.145606431 | 0.060015232 |
| 8.200000000 | 0.363594666 | 0.152654036 | 0.063497470 |
| 8.300000000 | 0.375553742 | 0.159872922 | 0.067109606 |
| 8.400000000 | 0.387543891 | 0.167259756 | 0.070852795 |
| 8.500000000 | 0.399553691 | 0.174810930 | 0.074728045 |
| 8.600000000 | 0.411571910 | 0.182522580 | 0.078736218 |
| 8.700000000 | 0.423587526 | 0.190390588 | 0.082378025 |
| 8.800000000 | 0.435589752 | 0.198410599 | 0.087154023 |
| 8.900000000 | 0.447568059 | 0.206578026 | 0.091564613 |
| 9.000000000 | 0.459512189 | 0.214888067 | 0.096110041 |
| 9.100000000 | 0.471412179 | 0.223335714 | 0.100790391 |
| 9.200000000 | 0.483258372 | 0.231915767 | 0.105605590 |
| 9.300000000 | 0.495041433 | 0.240622844 | 0.110555402 |
| 9.400000000 | 0.506752362 | 0.249451396 | 0.115639433 |
| 9.500000000 | 0.518382503 | 0.258395723 | 0.120857125 |
| 9.600000000 | 0.529923554 | 0.267449980 | 0.126207760 |
| 9.700000000 | 0.541367577 | 0.276608198 | 0.131690462 |
| 9.800000000 | 0.552706999 | 0.285864293 | 0.137304193 |
| 9.900000000 | 0.563934622 | 0.295212080 | 0.143047759 |
| 10.000000000 | 0.575043622 | 0.304645290 | 0.148919809 |

TABLE 2a:  POISSON CLUSTER PROBABILITY  P'(n;λ,1/L)

n = 8

| λ | L = 2 | L = 3 | L = 4 |
|---|---|---|---|
| 0.100000000 | 0.000000000 | 0.000000000 | 0.000000000 |
| 0.200000000 | 0.000000000 | 0.000000000 | 0.000000000 |
| 0.300000000 | 0.000000000 | 0.000000000 | 0.000000000 |
| 0.400000000 | 0.000000000 | 0.000000000 | 0.000000000 |
| 0.500000000 | 0.000000003 | 0.000000000 | 0.000000000 |
| 0.600000000 | 0.000000011 | 0.000000001 | 0.000000000 |
| 0.700000000 | 0.000000034 | 0.000000003 | 0.000000000 |
| 0.800000000 | 0.000000095 | 0.000000008 | 0.000000001 |
| 0.900000000 | 0.000000230 | 0.000000020 | 0.000000003 |
| 1.000000000 | 0.000000506 | 0.000000044 | 0.000000007 |
| 1.100000000 | 0.000001028 | 0.000000092 | 0.000000015 |
| 1.200000000 | 0.000001953 | 0.000000178 | 0.000000029 |
| 1.300000000 | 0.000003509 | 0.000000325 | 0.000000054 |
| 1.400000000 | 0.000006012 | 0.000000566 | 0.000000095 |
| 1.500000000 | 0.000009887 | 0.000000947 | 0.000000160 |
| 1.600000000 | 0.000015692 | 0.000001530 | 0.000000260 |
| 1.700000000 | 0.000024138 | 0.000002394 | 0.000000411 |
| 1.800000000 | 0.000036112 | 0.000003645 | 0.000000632 |
| 1.900000000 | 0.000052708 | 0.000005413 | 0.000000947 |
| 2.000000000 | 0.000075242 | 0.000007863 | 0.000001388 |
| 2.100000000 | 0.000105281 | 0.000011195 | 0.000001994 |
| 2.200000000 | 0.000144660 | 0.000015652 | 0.000002813 |
| 2.300000000 | 0.000195506 | 0.000021524 | 0.000003904 |
| 2.400000000 | 0.000260254 | 0.000029154 | 0.000005336 |
| 2.500000000 | 0.000341663 | 0.000038943 | 0.000007193 |
| 2.600000000 | 0.000442829 | 0.000051357 | 0.000009573 |
| 2.700000000 | 0.000567192 | 0.000066931 | 0.000012591 |
| 2.800000000 | 0.000718548 | 0.000086273 | 0.000016378 |
| 2.900000000 | 0.000901047 | 0.000110074 | 0.000021087 |
| 3.000000000 | 0.001119196 | 0.000139109 | 0.000026894 |
| 3.100000000 | 0.001377855 | 0.000174245 | 0.000033995 |
| 3.200000000 | 0.001682227 | 0.000216442 | 0.000042615 |
| 3.300000000 | 0.002037851 | 0.000266763 | 0.000053004 |
| 3.400000000 | 0.002450587 | 0.000326372 | 0.000065443 |
| 3.500000000 | 0.002926596 | 0.000396541 | 0.000080243 |
| 3.600000000 | 0.003472322 | 0.000478653 | 0.000097748 |
| 3.700000000 | 0.004094469 | 0.000574206 | 0.000118338 |
| 3.800000000 | 0.004799972 | 0.000684811 | 0.000142429 |
| 3.900000000 | 0.005595974 | 0.000812198 | 0.000170475 |
| 4.000000000 | 0.006489788 | 0.000958215 | 0.000202971 |
| 4.100000000 | 0.007438871 | 0.001124831 | 0.000240455 |
| 4.200000000 | 0.008600786 | 0.001314131 | 0.000283505 |
| 4.300000000 | 0.009833168 | 0.001528322 | 0.000332747 |
| 4.400000000 | 0.011193686 | 0.001769726 | 0.000388852 |
| 4.500000000 | 0.012690008 | 0.002040783 | 0.000452538 |
| 4.600000000 | 0.014329763 | 0.002344046 | 0.000524572 |
| 4.700000000 | 0.016120502 | 0.002682178 | 0.000605773 |
| 4.800000000 | 0.018069660 | 0.003057949 | 0.000697006 |
| 4.900000000 | 0.020184522 | 0.003474233 | 0.000799192 |
| 5.000000000 | 0.022472184 | 0.003934001 | 0.000913300 |

$n = 8$

TABLE 2a:   POISSON CLUSTER PROBABILITY   $P'(n;\lambda,1/L)$

$$\underline{n = 8}$$

| $\lambda$ | L = 2 | L = 3 | L = 4 |
|---|---|---|---|
| 5.100000000 | 0.024939517 | 0.004440318 | 0.001040356 |
| 5.200000000 | 0.027593137 | 0.004996335 | 0.001181434 |
| 5.300000000 | 0.030439367 | 0.005605287 | 0.001337666 |
| 5.400000000 | 0.033484208 | 0.006270482 | 0.001510232 |
| 5.500000000 | 0.036733312 | 0.006995293 | 0.001700370 |
| 5.600000000 | 0.040191949 | 0.007783157 | 0.001909367 |
| 5.700000000 | 0.043864986 | 0.008637560 | 0.002138565 |
| 5.800000000 | 0.047756864 | 0.009562031 | 0.002389356 |
| 5.900000000 | 0.051871573 | 0.010560137 | 0.002663183 |
| 6.000000000 | 0.056212636 | 0.011635468 | 0.002961539 |
| 6.100000000 | 0.060783093 | 0.012791631 | 0.003285968 |
| 6.200000000 | 0.065585487 | 0.014032243 | 0.003638053 |
| 6.300000000 | 0.070621850 | 0.015360916 | 0.004019447 |
| 6.400000000 | 0.075893698 | 0.016781251 | 0.004431815 |
| 6.500000000 | 0.081402022 | 0.018296828 | 0.004876885 |
| 6.600000000 | 0.087147284 | 0.019911196 | 0.005356423 |
| 6.700000000 | 0.093129417 | 0.021627861 | 0.005872233 |
| 6.800000000 | 0.099347824 | 0.023450279 | 0.006426155 |
| 6.900000000 | 0.105801380 | 0.025381844 | 0.007020063 |
| 7.000000000 | 0.112488439 | 0.027425881 | 0.007655866 |
| 7.100000000 | 0.119406842 | 0.029585634 | 0.008335498 |
| 7.200000000 | 0.126553922 | 0.031864255 | 0.009060924 |
| 7.300000000 | 0.133926519 | 0.034264800 | 0.009834131 |
| 7.400000000 | 0.141520991 | 0.036790216 | 0.010657125 |
| 7.500000000 | 0.149333229 | 0.039443332 | 0.011531932 |
| 7.600000000 | 0.157358671 | 0.042226853 | 0.012460593 |
| 7.700000000 | 0.165592322 | 0.045143351 | 0.013445158 |
| 7.800000000 | 0.174028772 | 0.048195255 | 0.014487688 |
| 7.900000000 | 0.182662214 | 0.051384845 | 0.015590247 |
| 8.000000000 | 0.191486467 | 0.054714247 | 0.016754899 |
| 8.100000000 | 0.200494997 | 0.058185424 | 0.017983707 |
| 8.200000000 | 0.209680939 | 0.061800168 | 0.019278728 |
| 8.300000000 | 0.219037120 | 0.065560099 | 0.020642008 |
| 8.400000000 | 0.228556083 | 0.069466655 | 0.022075582 |
| 8.500000000 | 0.238230112 | 0.073521092 | 0.023581465 |
| 8.600000000 | 0.248051255 | 0.077724472 | 0.025161653 |
| 8.700000000 | 0.258011349 | 0.082077670 | 0.026818117 |
| 8.800000000 | 0.268102044 | 0.086581358 | 0.028552800 |
| 8.900000000 | 0.278314830 | 0.091236015 | 0.030367610 |
| 9.000000000 | 0.288641056 | 0.096041914 | 0.032264424 |
| 9.100000000 | 0.299071961 | 0.100999126 | 0.034245077 |
| 9.200000000 | 0.309598694 | 0.106107518 | 0.036311359 |
| 9.300000000 | 0.320212337 | 0.111366751 | 0.038465017 |
| 9.400000000 | 0.330903929 | 0.116776280 | 0.040707745 |
| 9.500000000 | 0.341664490 | 0.122335355 | 0.043041183 |
| 9.600000000 | 0.352485038 | 0.128043021 | 0.045466916 |
| 9.700000000 | 0.363356617 | 0.133898120 | 0.047986466 |
| 9.800000000 | 0.374270311 | 0.139899292 | 0.050601293 |
| 9.900000000 | 0.385217270 | 0.146044980 | 0.053312787 |
| 10.000000000 | 0.396188720 | 0.152233428 | 0.056122271 |

NEFF and NAUS

TABLE 2a:  POISSON CLUSTER PROBABILITY  P'(n;λ,l/L)

| | | | n = 9 |
|---|---|---|---|
| λ | L = 2 | L = 3 | L = 4 |
| 0.100000000 | 0.000000000 | 0.000000000 | 0.000000000 |
| 0.200000000 | 0.000000000 | 0.000000000 | 0.000000000 |
| 0.300000000 | 0.000000000 | 0.000000000 | 0.000000000 |
| 0.400000000 | 0.000000000 | 0.000000000 | 0.000000000 |
| 0.500000000 | 0.000000000 | 0.000000000 | 0.000000000 |
| 0.600000000 | 0.000000000 | 0.000000000 | 0.000000000 |
| 0.700000000 | 0.000000001 | 0.000000000 | 0.000000000 |
| 0.800000000 | 0.000000005 | 0.000000000 | 0.000000000 |
| 0.900000000 | 0.000000013 | 0.000000001 | 0.000000000 |
| 1.000000000 | 0.000000031 | 0.000000002 | 0.000000000 |
| 1.100000000 | 0.000000070 | 0.000000004 | 0.000000001 |
| 1.200000000 | 0.000000145 | 0.000000009 | 0.000000001 |
| 1.300000000 | 0.000000283 | 0.000000018 | 0.000000002 |
| 1.400000000 | 0.000000522 | 0.000000033 | 0.000000004 |
| 1.500000000 | 0.000000920 | 0.000000059 | 0.000000007 |
| 1.600000000 | 0.000001558 | 0.000000102 | 0.000000013 |
| 1.700000000 | 0.000002546 | 0.000000169 | 0.000000022 |
| 1.800000000 | 0.000004035 | 0.000000273 | 0.000000036 |
| 1.900000000 | 0.000006218 | 0.000000428 | 0.000000056 |
| 2.000000000 | 0.000009347 | 0.000000655 | 0.000000087 |
| 2.100000000 | 0.000013736 | 0.000000980 | 0.000000131 |
| 2.200000000 | 0.000019778 | 0.000001436 | 0.000000194 |
| 2.300000000 | 0.000027953 | 0.000002065 | 0.000000281 |
| 2.400000000 | 0.000038840 | 0.000002920 | 0.000000401 |
| 2.500000000 | 0.000053128 | 0.000004065 | 0.000000564 |
| 2.600000000 | 0.000071634 | 0.000005577 | 0.000000780 |
| 2.700000000 | 0.000095306 | 0.000007551 | 0.000001066 |
| 2.800000000 | 0.000125244 | 0.000010097 | 0.000001439 |
| 2.900000000 | 0.000162705 | 0.000013348 | 0.000001919 |
| 3.000000000 | 0.000209121 | 0.000017457 | 0.000002533 |
| 3.100000000 | 0.000266100 | 0.000022604 | 0.000003309 |
| 3.200000000 | 0.000335447 | 0.000028994 | 0.000004283 |
| 3.300000000 | 0.000419163 | 0.000036866 | 0.000005496 |
| 3.400000000 | 0.000519458 | 0.000046488 | 0.000006993 |
| 3.500000000 | 0.000638756 | 0.000058166 | 0.000008830 |
| 3.600000000 | 0.000779700 | 0.000072244 | 0.000011067 |
| 3.700000000 | 0.000945153 | 0.000089108 | 0.000013775 |
| 3.800000000 | 0.001138207 | 0.000109187 | 0.000017033 |
| 3.900000000 | 0.001362173 | 0.000132957 | 0.000020930 |
| 4.000000000 | 0.001620590 | 0.000160945 | 0.000025567 |
| 4.100000000 | 0.001917215 | 0.000193730 | 0.000031056 |
| 4.200000000 | 0.002256020 | 0.000231946 | 0.000037522 |
| 4.300000000 | 0.002641189 | 0.000276284 | 0.000045103 |
| 4.400000000 | 0.003077106 | 0.000327496 | 0.000053953 |
| 4.500000000 | 0.003568346 | 0.000386397 | 0.000064238 |
| 4.600000000 | 0.004119667 | 0.000453864 | 0.000076145 |
| 4.700000000 | 0.004735991 | 0.000530843 | 0.000089874 |
| 4.800000000 | 0.005422398 | 0.000618347 | 0.000105648 |
| 4.900000000 | 0.006184105 | 0.000717460 | 0.000123704 |
| 5.000000000 | 0.007026449 | 0.000829336 | 0.000144303 |

TABLE 2a:   POISSON CLUSTER PROBABILITY   P'(n;λ,1/L)

n = 9

| λ | L = 2 | L = 3 | L = 4 |
|---|---|---|---|
| 5.100000000 | 0.007954874 | 0.000955202 | C.000167726 |
| 5.200000000 | 0.008974908 | 0.001096358 | C.000194277 |
| 5.300000000 | 0.010092142 | 0.001254179 | 0.000224280 |
| 5.400000000 | 0.011312217 | 0.001430113 | C.000258087 |
| 5.500000000 | 0.012640793 | 0.001625684 | 0.000296073 |
| 5.600000000 | 0.014083536 | 0.001842489 | C.000338637 |
| 5.700000000 | 0.015646092 | 0.002082200 | C.000386208 |
| 5.800000000 | 0.017334063 | 0.002346562 | C.000439239 |
| 5.900000000 | 0.019152989 | 0.002637393 | C.000498214 |
| 6.000000000 | 0.021108326 | 0.002956581 | 0.000563642 |
| 6.100000000 | 0.023205417 | 0.003306084 | C.000636066 |
| 6.200000000 | 0.025449479 | 0.003637928 | C.000716056 |
| 6.300000000 | 0.027845578 | 0.004104205 | C.000804213 |
| 6.400000000 | 0.030398606 | 0.004557069 | C.000901169 |
| 6.500000000 | 0.033113266 | C.005048735 | 0.001007590 |
| 6.600000000 | 0.035994046 | 0.005581477 | 0.001124171 |
| 6.700000000 | 0.039045208 | 0.006157620 | C.001251641 |
| 6.800000000 | 0.042270764 | 0.006779543 | C.001390760 |
| 6.900000000 | 0.045674464 | 0.007449670 | 0.001542324 |
| 7.000000000 | 0.049259776 | 0.008170470 | 0.001707158 |
| 7.100000000 | 0.053029878 | 0.008944449 | C.001886123 |
| 7.200000000 | 0.056987638 | 0.009774149 | 0.002080112 |
| 7.300000000 | 0.061135606 | 0.010662142 | 0.002290052 |
| 7.400000000 | 0.065476002 | 0.011611023 | C.002516899 |
| 7.500000000 | 0.070010710 | 0.012623410 | 0.002761647 |
| 7.600000000 | 0.074741262 | 0.013701935 | 0.003025319 |
| 7.700000000 | 0.079668843 | 0.014849238 | C.003308970 |
| 7.800000000 | 0.084794273 | 0.016067965 | 0.003613686 |
| 7.900000000 | 0.090118015 | 0.017360761 | 0.003940586 |
| 8.000000000 | 0.095640165 | 0.018730265 | C.004290818 |
| 8.100000000 | 0.101360450 | 0.020179100 | C.004665558 |
| 8.200000000 | 0.107278235 | 0.021709874 | 0.005066013 |
| 8.300000000 | 0.113392518 | 0.023325170 | 0.005493416 |
| 8.400000000 | 0.119701935 | 0.025027542 | 0.005949028 |
| 8.500000000 | 0.126204765 | 0.026819506 | 0.006434136 |
| 8.600000000 | 0.132898932 | 0.028703538 | 0.006950051 |
| 8.700000000 | 0.139782014 | 0.030682068 | C.007498108 |
| 8.800000000 | 0.146851249 | 0.032757470 | C.008079663 |
| 8.900000000 | 0.154103543 | 0.034932062 | 0.008696095 |
| 9.000000000 | 0.161535479 | 0.037208095 | C.009348802 |
| 9.100000000 | 0.169143330 | 0.039587754 | 0.010039198 |
| 9.200000000 | 0.176923064 | 0.042073144 | 0.010768716 |
| 9.300000000 | 0.184870364 | 0.044666293 | C.011538804 |
| 9.400000000 | 0.192980631 | 0.047369141 | C.012350921 |
| 9.500000000 | 0.201249006 | 0.050183540 | 0.013206540 |
| 9.600000000 | 0.209670376 | 0.053111244 | C.014107142 |
| 9.700000000 | 0.218239394 | 0.056153908 | 0.015054217 |
| 9.800000000 | 0.226950490 | 0.059313082 | 0.016049261 |
| 9.900000000 | 0.235797889 | 0.062590208 | C.017093773 |
| 10.000000000 | 0.244775623 | 0.065986615 | 0.018189256 |

TABLE 3: PIECEWISE POLYNOMIALS FOR P(n;N,p)

n = 3

| N | p range | F | coefficient of $p^{2F}$ | coefficient of $p^{2F+1}$ |
|---|---------|---|----------------------|------------------------|
| 3 | (0,1) | 0 | 0 | 0 |
|   |         | 1 | 3 | -2 |
| 4 | (0,1/2) | 0 | 0 | 0 |
|   |         | 1 | 12 | -24 |
|   |         | 2 | 14 | |
| 4 | (1/2,1) | 0 | -1 | 8 |
|   |         | 1 | -12 | 8 |
|   |         | 2 | -2 | |
| 5 | (0,1/2) | 0 | 0 | 0 |
|   |         | 1 | 30 | -100 |
|   |         | 2 | 120 | -48 |
| 6 | (0,1/3) | 0 | 0 | 0 |
|   |         | 1 | 60 | -280 |
|   |         | 2 | 420 | -12 |
|   |         | 3 | -320 | |
| 6 | (1/3,1/2) | 0 | -4 | 60 |
|   |         | 1 | -300 | 800 |
|   |         | 2 | -1200 | 960 |
|   |         | 3 | -320 | |
| 7 | (0,1/4) | 0 | 0 | 0 |
|   |         | 1 | 105 | -630 |
|   |         | 2 | 910 | 2604 |
|   |         | 3 | -9583 | 8446 |
| 7 | (1/4,1/3) | 0 | 1 | -28 |
|   |         | 1 | 441 | -2870 |
|   |         | 2 | 9870 | -18900 |
|   |         | 3 | 19089 | -7938 |
| 8 | (0,1/4) | 0 | 0 | 0 |
|   |         | 1 | 168 | -1232 |
|   |         | 2 | 1260 | 18480 |
|   |         | 3 | -85064 | 146640 |
|   |         | 4 | -91854 | |
| 8 | (1/4,1/3) | 0 | -13 | 336 |
|   |         | 1 | -3528 | 21168 |
|   |         | 2 | -79380 | 190512 |
|   |         | 3 | -285768 | 244944 |
|   |         | 4 | -91854 | |
| 9 | (0,1/5) | 0 | 0 | 0 |
|   |         | 1 | 252 | -2184 |
|   |         | 2 | 504 | 77616 |
|   |         | 3 | -439908 | 1062072 |
|   |         | 4 | -1112679 | 304904 |
| 9 | (1/5,1/4) | 0 | 7 | -288 |
|   |         | 1 | 5472 | -56784 |
|   |         | 2 | 362754 | -1497384 |
|   |         | 3 | 4022592 | -6812928 |
|   |         | 4 | 6621696 | -2820096 |

TABLE 3: PIECEWISE POLYNOMIALS FOR P(n;N,p)

n = 3

| N | p range | F | coefficient of $p^{2F}$ | coefficient of $p^{2F+1}$ |
|---|---------|---|---|---|
| 10 | (0,1/6) | 0 | 0 | |
| | | 1 | 360 | 0 |
| | | 2 | −3360 | −3600 |
| | | 3 | −1653960 | 247464 |
| | | 4 | −4405320 | 4733040 |
| | | 5 | 11738484 | −5988980 |
| 10 | (1/6,1/5) | 0 | −1 | |
| | | 1 | −1260 | 60 |
| | | 2 | −275520 | 22320 |
| | | 3 | −11451720 | 2207016 |
| | | 4 | −79988040 | 38325360 |
| | | 5 | −48727692 | 94787980 |
| 10 | (1/5,1/4) | 0 | −41 | |
| | | 1 | −30240 | 1680 |
| | | 2 | −2257920 | 322560 |
| | | 3 | −36126720 | 10838016 |
| | | 4 | −123863040 | 82575360 |
| | | 5 | −44040192 | 110100480 |
| 11 | (0,1/6) | 0 | 0 | |
| | | 1 | 495 | 0 |
| | | 2 | −13860 | −5610 |
| | | 3 | −5019630 | 659736 |
| | | 4 | 4097115 | 15057900 |
| | | 5 | 383928336 | −151449540 |
| | | | | −322511040 |
| 11 | (1/6,1/5) | 0 | 34 | |
| | | 1 | 52470 | −1980 |
| | | 2 | 8325900 | −816420 |
| | | 3 | 291362610 | −58616712 |
| | | 4 | 2498326875 | −1024204500 |
| | | 5 | 3875850000 | −4031362500 |
| | | | | −1683000000 |
| 12 | (0,1/7) | 0 | 0 | |
| | | 1 | 660 | 0 |
| | | 2 | −36630 | −8360 |
| | | 3 | −13041336 | 1547568 |
| | | 4 | 139347450 | 36072432 |
| | | 5 | 4725032796 | −1428925960 |
| | | 6 | 4179927862 | −7241178696 |
| 12 | (1/7,1/6) | 0 | −10 | |
| | | 1 | −27984 | 792 |
| | | 2 | −9205020 | 616880 |
| | | 3 | −727406064 | 96627168 |
| | | 4 | −15351496380 | 3949548768 |
| | | 5 | −75175109064 | 41691785520 |
| | | 6 | −39321260484 | 80891098992 |
| 12 | (1/6,1/5) | 0 | −131 | |
| | | 1 | −217800 | 7920 |
| | | 2 | −40837500 | 3630000 |
| | | 3 | −1905750000 | 326700000 |
| | | 4 | −25523437500 | 8167500000 |
| | | 5 | −85078125000 | 56718750000 |
| | | 6 | −32226562500 | 77343750000 |

TABLE 3:   PIECEWISE POLYNOMIALS FOR P(n;N,p)

n = 3

| N | p range | F | coefficient of $p^{2F}$ | coefficient of $p^{2F+1}$ |
|---|---|---|---|---|
| 13 | (0,1/8) | 0 | 0 | 0 |
| | | 1 | 858 | −12012 |
| | | 2 | −80080 | 3294720 |
| | | 3 | −30086628 | 62925720 |
| | | 4 | 909369747 | −8607341600 |
| | | 5 | 34250637564 | −68277827280 |
| | | 6 | 54680029443 | 3365355508 |
| 13 | (1/8,1/7) | 0 | 1 | −104 |
| | | 1 | 5850 | −158444 |
| | | 2 | 2848560 | −38877696 |
| | | 3 | 419752476 | −3535787112 |
| | | 4 | 22501646739 | −104573017120 |
| | | 5 | 341340799228 | −738292725456 |
| | | 6 | 948033227011 | −546390458380 |
| 13 | (1/7,1/6) | 0 | 144 | −11440 |
| | | 1 | 417846 | −9241804 |
| | | 2 | 138278855 | −1479488868 |
| | | 3 | 11654720220 | −68485295736 |
| | | 4 | 300346819344 | −971335553760 |
| | | 5 | 2252457979200 | −3547733904000 |
| | | 6 | 3402622080000 | −1501156800000 |
| 14 | (0,1/8) | 0 | 0 | 0 |
| | | 1 | 1092 | −16744 |
| | | 2 | −156156 | 6498492 |
| | | 3 | −63159096 | 56566224 |
| | | 4 | 3999887892 | −38992765812 |
| | | 5 | 174352258080 | −360697473864 |
| | | 6 | 4817094828 | 1349321055012 |
| | | 7 | −1753807624920 | |
| 14 | (1/8,1/7) | 0 | −64 | 6552 |
| | | 1 | −309036 | 8975512 |
| | | 2 | −178510332 | 2564958396 |
| | | 3 | −27418999608 | 221377405392 |
| | | 4 | −1356313562604 | 6275548683404 |
| | | 5 | −21590662949856 | 53575501829304 |
| | | 6 | −90670532458324 | 93708297788196 |
| | | 7 | −44634761108184 | |
| 14 | (1/7,1/6) | 0 | −428 | 36036 |
| | | 1 | −1405404 | 33729696 |
| | | 2 | −556539984 | 6678479808 |
| | | 3 | −60106318272 | 412157611008 |
| | | 4 | −2163827457792 | 8655309831168 |
| | | 5 | −25965929493504 | 56652937076736 |
| | | 6 | −84979405615104 | 78442528260096 |
| | | 7 | −33618226397184 | |
| 15 | (0,1/9) | 0 | 0 | 0 |
| | | 1 | 1365 | −22750 |
| | | 2 | −281190 | 12048036 |
| | | 3 | −122727605 | −112857030 |
| | | 4 | 14073724665 | −144064260340 |
| | | 5 | 680385878172 | −1025143195440 |
| | | 6 | −5203298446255 | 30893050642020 |
| | | 7 | −64426034589465 | 49991961279274 |

TABLE 3:   PIECEWISE POLYNOMIALS FOR P(n;N,p)

n = 3

| N | p range | F | coefficient of $p^{2F}$ | coefficient of $p^{2F+1}$ |
|---|---------|---|------------------------|---------------------------|
| 15 | (1/9,1/8) | 0 | 13 | -1680 |
| | | 1 | 102480 | -3781960 |
| | | 2 | 96242055 | -1800598800 |
| | | 3 | 25589273710 | -280538329500 |
| | | 4 | 2384010949320 | -15656380639900 |
| | | 5 | 78629775685461 | -296111707052070 |
| | | 6 | 808664482513160 | -1511172218544240 |
| | | 7 | 1727589374382480 | -910833321829088 |
| 15 | (1/8,1/7) | 0 | 573 | -60060 |
| | | 1 | 2927925 | -87957870 |
| | | 2 | 1821799980 | -27567900360 |
| | | 3 | 314953228590 | -2767033861020 |
| | | 4 | 18852497543880 | -99631266835100 |
| | | 5 | 405144364726101 | -1245107820746790 |
| | | 6 | 2799752684568840 | -4349065819578480 |
| | | 7 | 4173546405908880 | -1865389803365088 |
| 16 | (0,1/10) | 0 | 0 | 0 |
| | | 1 | 1680 | -30240 |
| | | 2 | -476840 | 21219744 |
| | | 3 | -223503280 | -776890400 |
| | | 4 | 42454861020 | -455446053920 |
| | | 5 | 2127697620048 | 379715175840 |
| | | 6 | -61267866277800 | 351389469344480 |
| | | 7 | -966875556265680 | 1311400166155168 |
| | | 8 | -643197414535750 | |
| 16 | (1/10,1/9) | 0 | -1 | 160 |
| | | 1 | -10320 | 529760 |
| | | 2 | -18676840 | 458019744 |
| | | 3 | -8231503280 | 113623109600 |
| | | 4 | -1244545138980 | 10984553946080 |
| | | 5 | -77952302379952 | 437179715175840 |
| | | 6 | -1881267866277800 | 5951389469344480 |
| | | 7 | -12966875556265680 | 17311400166155168 |
| | | 8 | -10643197414535750 | |
| 16 | (1/9,1/8) | 0 | -337 | 43680 |
| | | 1 | -2642640 | 99197280 |
| | | 2 | -2583129640 | 49447690656 |
| | | 3 | -719512954160 | 8116590241760 |
| | | 4 | -71732413492740 | 498332106752480 |
| | | 5 | -2712518153865520 | 11448103923503520 |
| | | 6 | -36730391718249320 | 86617170286847200 |
| | | 7 | -141604115105478480 | 143406020741608352 |
| | | 8 | -67784341235371670 | |
| 16 | (1/8,1/7) | 0 | -1429 | 160160 |
| | | 1 | -8408400 | 274674400 |
| | | 2 | -6248842600 | 104980555680 |
| | | 3 | -1347250464560 | 13472504645600 |
| | | 4 | -106095974084100 | 660152727634400 |
| | | 5 | -3234748365408560 | 12350857395196320 |
| | | 6 | -36023334069322600 | 77588719533925600 |
| | | 7 | -116383079300888400 | 108624207347495840 |
| | | 8 | -47523090714529430 | |

TABLE 3:  PIECEWISE POLYNOMIALS FOR $P(n;N,p)$

$n = 4$

| N | p range | F | coefficient of $p^{2F}$ | coefficient of $p^{2F+1}$ |
|---|---------|---|-------------------------|---------------------------|
| 4 | (0,1) | 0 | 0 | 0 |
|   |       | 1 | 0 | 4 |
|   |       | 2 | -3 | |
| 5 | (0,1/2) | 0 | 0 | 0 |
|   |         | 1 | 0 | 20 |
|   |         | 2 | -40 | 22 |
| 5 | (1/2,1) | 0 | 1 | -10 |
|   |         | 1 | 40 | -60 |
|   |         | 2 | 40 | -10 |
| 6 | (0,1/2) | 0 | 0 | 0 |
|   |         | 1 | 0 | 60 |
|   |         | 2 | -195 | 222 |
|   |         | 3 | -85 | |
| 6 | (1/2,1) | 0 | -4 | 30 |
|   |         | 1 | -75 | 100 |
|   |         | 2 | -75 | 30 |
|   |         | 3 | -5 | |
| 7 | (0,1/2) | 0 | 0 | 0 |
|   |         | 1 | 0 | 140 |
|   |         | 2 | -630 | 1092 |
|   |         | 3 | -840 | 240 |
| 8 | (0,1/3) | 0 | 0 | 0 |
|   |         | 1 | 0 | 280 |
|   |         | 2 | -1610 | 3752 |
|   |         | 3 | -4900 | 4952 |
|   |         | 4 | -3311 | |
| 8 | (1/3,1/2) | 0 | 15 | -280 |
|   |           | 1 | 2212 | -9296 |
|   |           | 2 | 22960 | -34048 |
|   |           | 3 | 29120 | -12544 |
|   |           | 4 | 1792 | |
| 9 | (0,1/4) | 0 | 0 | 0 |
|   |         | 1 | 0 | 504 |
|   |         | 2 | -3528 | 10332 |
|   |         | 3 | -21000 | 48528 |
|   |         | 4 | -87354 | 64948 |
| 9 | (1/4,1/3) | 0 | 1 | -36 |
|   |           | 1 | 576 | -4872 |
|   |           | 2 | 28728 | -118692 |
|   |           | 3 | 323064 | -541296 |
|   |           | 4 | 502470 | -197196 |
| 9 | (1/3,1/2) | 0 | -41 | 756 |
|   |           | 1 | -6048 | 28224 |
|   |           | 2 | -84672 | 169344 |
|   |           | 3 | -225792 | 193536 |
|   |           | 4 | -96768 | 21504 |

TABLE 3:  PIECEWISE POLYNOMIALS FOR $P(n;N,p)$

$n = 4$

| N | p range | F | coefficient of $p^{2F}$ | coefficient of $p^{2F+1}$ |
|---|---------|---|--------------------------|----------------------------|
| 10 | (0,1/4) | 0 | 0 | 0 |
| | | 1 | 0 | 840 |
| | | 2 | −6930 | 24444 |
| | | 3 | −72450 | 291240 |
| | | 4 | −878895 | 1348480 |
| | | 5 | −796302 | |
| 10 | (1/4,1/3) | 0 | −26 | 900 |
| | | 1 | −13860 | 125640 |
| | | 2 | −732690 | 2862972 |
| | | 3 | −7598850 | 13562280 |
| | | 4 | −15624495 | 10523520 |
| | | 5 | −3155598 | |
| 11 | (0,1/4) | 0 | 0 | 0 |
| | | 1 | 0 | 1320 |
| | | 2 | −12540 | 51744 |
| | | 3 | −212520 | 1272480 |
| | | 4 | −5409690 | 12624040 |
| | | 5 | −14943192 | 7068120 |
| 11 | (1/4,1/3) | 0 | 199 | −6600 |
| | | 1 | 97680 | −848760 |
| | | 2 | 4813380 | −18694368 |
| | | 3 | 50644440 | −95372640 |
| | | 4 | 121648230 | −99114840 |
| | | 5 | 45612072 | −8660520 |
| 12 | (0,1/5) | 0 | 0 | 0 |
| | | 1 | 0 | 1980 |
| | | 2 | −21285 | 100584 |
| | | 3 | −549780 | 4462920 |
| | | 4 | −24488145 | 77146960 |
| | | 5 | −140527332 | 143840988 |
| | | 6 | −68801337 | |
| 12 | (1/5,1/4) | 0 | 45 | −2376 |
| | | 1 | 57024 | −819720 |
| | | 2 | 7886340 | −53359416 |
| | | 3 | 259325220 | −908812080 |
| | | 4 | 2272621230 | −3944728040 |
| | | 5 | 4500097668 | −3020221512 |
| | | 6 | 897995538 | |
| 12 | (1/4,1/3) | 0 | −461 | 16632 |
| | | 1 | −274428 | 2744280 |
| | | 2 | −18523890 | 88914672 |
| | | 3 | −311201352 | 800232048 |
| | | 4 | −1500435090 | 2000580120 |
| | | 5 | −1800522108 | 982102968 |
| | | 6 | −245525742 | |

TABLE 3: PIECEWISE POLYNOMIALS FOR P(n;N,p)

n = 4

| N | p range | F | coefficient of $p^{2F}$ | coefficient of $p^{2F+1}$ |
|---|---------|---|---|---|
| 13 | (0,1/6) | 0 | 0 | 0 |
|    |         | 1 | 0 | 2860 |
|    |         | 2 | -34320 | 182754 |
|    |         | 3 | -1287000 | 13323024 |
|    |         | 4 | -89557182 | 357462820 |
|    |         | 5 | -907046712 | 1575736344 |
|    |         | 6 | -1891122740 | 1202871156 |
| 13 | (1/6,1/5) | 0 | 1 | -78 |
|    |         | 1 | 2808 | -58916 |
|    |         | 2 | 892320 | -9824958 |
|    |         | 3 | 78774696 | -467047152 |
|    |         | 4 | 2072108610 | -6848089820 |
|    |         | 5 | 16386279624 | -26722434024 |
|    |         | 6 | 26407047628 | -11857822860 |
| 13 | (1/5,1/4) | 0 | -571 | 30030 |
|    |         | 1 | -720720 | 10441860 |
|    |         | 2 | -101698740 | 702014742 |
|    |         | 3 | -3527399304 | 13033582848 |
|    |         | 4 | -35299153890 | 68861472680 |
|    |         | 5 | -93276845376 | 81868190976 |
|    |         | 6 | -40522639872 | 7946864640 |
| 14 | (0,1/6) | 0 | 0 | 0 |
|    |         | 1 | 0 | 4004 |
|    |         | 2 | -53053 | 314314 |
|    |         | 3 | -2777775 | 35153976 |
|    |         | 4 | -279672393 | 1359454096 |
|    |         | 5 | -4552890342 | 12010240788 |
|    |         | 6 | -25655271733 | 37088876090 |
|    |         | 7 | -24738404015 | |
| 14 | (1/6,1/5) | 0 | -64 | 4914 |
|    |         | 1 | -174447 | 3797612 |
|    |         | 2 | -56485429 | 607448842 |
|    |         | 3 | -487152963 | 29577919800 |
|    |         | 4 | -136464617289 | 475484817808 |
|    |         | 5 | -1230217394406 | 2290012955412 |
|    |         | 6 | -2897919564085 | 2231285470778 |
|    |         | 7 | -738789003951 | |
| 14 | (1/5,1/4) | 0 | 2575 | -138138 |
|    |         | 1 | 3408405 | -51243192 |
|    |         | 2 | 524257734 | -3858686832 |
|    |         | 3 | 21051654624 | -86367790080 |
|    |         | 4 | 267263732736 | -619122696192 |
|    |         | 5 | 1052757049344 | -1266634457088 |
|    |         | 6 | 1009607639040 | -469182185472 |
|    |         | 7 | 92666855424 | |
| 15 | (0,1/6) | 0 | 0 | 0 |
|    |         | 1 | 0 | 5460 |
|    |         | 2 | -79170 | 516516 |
|    |         | 3 | -5605600 | 84118320 |
|    |         | 4 | -772882110 | 4452688240 |
|    |         | 5 | -18987224256 | 70500961440 |
|    |         | 6 | -226095067380 | 524975268720 |
|    |         | 7 | -719993726400 | 426455993600 |

TABLE 3:   PIECEWISE POLYNOMIALS FOR P(n;N,p)

n = 4

| N | p range | F | coefficient of $p^{2F}$ | coefficient of $p^{2F+1}$ |
|---|---------|---|---|---|
| 15 | (1/6,1/5) | 0 | 1301 | −98280 |
| | | 1 | 3439800 | −73950240 |
| | | 2 | 1091639640 | −11713093392 |
| | | 3 | 94304570360 | −579737695680 |
| | | 4 | 2741840591490 | −9969833553680 |
| | | 5 | 27626556589632 | −57250980390240 |
| | | 6 | 858180029207820 | −87757153345680 |
| | | 7 | 54673674768960 | −15618606605056 |
| 15 | (1/5,1/4) | 0 | −6005 | 360360 |
| | | 1 | −10090080 | 174894720 |
| | | 2 | −2098736640 | 18468882432 |
| | | 3 | −123125882880 | 633218826240 |
| | | 4 | −2532875304960 | 7880056504320 |
| | | 5 | −18912135610368 | 34385701109760 |
| | | 6 | −45847601479680 | 42320862904320 |
| | | 7 | −24183350231040 | 6448893394944 |
| 16 | (0,1/7) | 0 | 0 | 0 |
| | | 1 | 0 | 7280 |
| | | 2 | −114660 | 816816 |
| | | 3 | −10690680 | 185911440 |
| | | 4 | −1937076570 | 12966416320 |
| | | 5 | −68530237776 | 338281791120 |
| | | 6 | −1473371443180 | 4759678410480 |
| | | 7 | −10010870240280 | 12130972436176 |
| | | 8 | −6493806802845 | |
| 16 | (1/7,1/6) | 0 | 91 | −9360 |
| | | 1 | 449640 | −13379520 |
| | | 2 | 276254160 | −4192712160 |
| | | 3 | 48345737440 | −431849452320 |
| | | 4 | 3018777470280 | −16559156430960 |
| | | 5 | 70978675904136 | −234974899635840 |
| | | 6 | 588205009854280 | −1074860723267520 |
| | | 7 | 1350310835874000 | −1040471236625472 |
| | | 8 | 369919998628350 | |
| 16 | (1/6,1/5) | 0 | −9437 | 720720 |
| | | 1 | −25585560 | 560079520 |
| | | 2 | −8455246800 | 93259534368 |
| | | 3 | −776583968160 | 4973393499360 |
| | | 4 | −24711688081080 | 95352624166800 |
| | | 5 | −283811064873336 | 641589458590080 |
| | | 6 | −1071724905251000 | 1262786364840000 |
| | | 7 | −966994893150000 | 408451843800000 |
| | | 8 | −61179649781250 | |

TABLE 3:   PIECEWISE POLYNOMIALS FOR $P(n;N,p)$

<u>n = 5</u>

| N | p range | F | coefficient of $p^{2F}$ | coefficient of $p^{2F+1}$ |
|---|---------|---|------------------------|--------------------------|
| 5 | (0,1) | 0 | 0 | 0 |
|   |       | 1 | 0 | 0 |
|   |       | 2 | 5 | −4 |
| 6 | (0,1/2) | 0 | 0 | 0 |
|   |        | 1 | 0 | 0 |
|   |        | 2 | 30 | −60 |
|   |        | 3 | 32 | |
| 6 | (1/2,1) | 0 | −1 | 12 |
|   |        | 1 | −60 | 160 |
|   |        | 2 | −210 | 132 |
|   |        | 3 | −32 | |
| 7 | (0,1/2) | 0 | 0 | 0 |
|   |        | 1 | 0 | 0 |
|   |        | 2 | 105 | −336 |
|   |        | 3 | 371 | −138 |
| 7 | (1/2,1) | 0 | 8 | −84 |
|   |        | 1 | 357 | −770 |
|   |        | 2 | 945 | −672 |
|   |        | 3 | 259 | −42 |
| 8 | (0,1/2) | 0 | 0 | 0 |
|   |        | 1 | 0 | 0 |
|   |        | 2 | 280 | −1232 |
|   |        | 3 | 2072 | −1552 |
|   |        | 4 | 434 | |
| 8 | (1/2,1) | 0 | −13 | 112 |
|   |        | 1 | −392 | 784 |
|   |        | 2 | −980 | 784 |
|   |        | 3 | −392 | 112 |
|   |        | 4 | −14 | |
| 9 | (0,1/2) | 0 | 0 | 0 |
|   |        | 1 | 0 | 0 |
|   |        | 2 | 630 | −3528 |
|   |        | 3 | 7980 | −9000 |
|   |        | 4 | 5040 | −1120 |
| 10 | (0,1/3) | 0 | 0 | 0 |
|    |        | 1 | 0 | 0 |
|    |        | 2 | 1260 | −8568 |
|    |        | 3 | 24360 | −36720 |
|    |        | 4 | 27720 | 180 |
|    |        | 5 | −12924 | |
| 10 | (1/3,1/2) | 0 | −56 | 1260 |
|    |          | 1 | −12420 | 70320 |
|    |          | 2 | −250740 | 585648 |
|    |          | 3 | −905520 | 915840 |
|    |          | 4 | −584640 | 218880 |
|    |          | 5 | −39168 | |
| 11 | (0,1/4) | 0 | 0 | 0 |
|    |        | 1 | 0 | 0 |
|    |        | 2 | 2310 | −18480 |
|    |        | 3 | 63294 | −119460 |
|    |        | 4 | 99330 | 146740 |
|    |        | 5 | −507188 | 409864 |

TABLE 3: PIECEWISE POLYNOMIALS FOR P(n;N,p)

n = 5

| N | p range | F | coefficient of $p^{2F}$ | coefficient of $p^{2F+1}$ |
|---|---------|---|-----------------------|-------------------------|
| 11 | (1/4,1/3) | 0 | 1 | -44 |
| | | 1 | 880 | -10560 |
| | | 2 | 86790 | -491568 |
| | | 3 | 1955646 | -5526180 |
| | | 4 | 10912770 | -14271180 |
| | | 5 | 11027148 | -3784440 |
| 11 | (1/3,1/2) | 0 | 265 | -5808 |
| | | 1 | 55770 | -306900 |
| | | 2 | 1061280 | -2350656 |
| | | 3 | 3148992 | -1837440 |
| | | 4 | -1267200 | 3210240 |
| | | 5 | -2348544 | 642048 |
| 12 | (0,1/4) | 0 | 0 | 0 |
| | | 1 | 0 | 0 |
| | | 2 | 3960 | -36432 |
| | | 3 | 145992 | -331056 |
| | | 4 | 256410 | 1502600 |
| | | 5 | -6414144 | 10191504 |
| | | 6 | -5941892 | |
| 12 | (1/4,1/3) | 0 | -43 | 1848 |
| | | 1 | -36168 | 425920 |
| | | 2 | -3354120 | 18616752 |
| | | 3 | -74601912 | 217019088 |
| | | 4 | -453908070 | 664726920 |
| | | 5 | -646569792 | 375095952 |
| | | 6 | -98216580 | |
| 12 | (1/3,1/2) | 0 | -461 | 11088 |
| | | 1 | -121968 | 813120 |
| | | 2 | -3659040 | 11708928 |
| | | 3 | -27320832 | 46835712 |
| | | 4 | -58544640 | 52039680 |
| | | 5 | -31223808 | 11354112 |
| | | 6 | -1892352 | |
| 13 | (0,1/4) | 0 | 0 | 0 |
| | | 1 | 0 | 0 |
| | | 2 | 6435 | -66924 |
| | | 3 | 307164 | -813384 |
| | | 4 | 468468 | 8940360 |
| | | 5 | -47519472 | 112484112 |
| | | 6 | -130898911 | 60573348 |
| 13 | (1/4,1/3) | 0 | 573 | -24024 |
| | | 1 | 459888 | -5317312 |
| | | 2 | 41396355 | -228885228 |
| | | 3 | 923707356 | -2752563528 |
| | | 4 | 6052210164 | -9690715320 |
| | | 5 | 10973538576 | -8311775472 |
| | | 6 | 3767706657 | -769898844 |

TABLE 3:   PIECEWISE POLYNOMIALS FOR $P(n;N,p)$

$n = 5$

| N | p range | F | coefficient of $p^{2F}$ | coefficient of $p^{2F+1}$ |
|---|---------|---|------------------------|---------------------------|
| 14 | (0,1/4) | 0 | 0 | 0 |
|    |         | 1 | 0 | 0 |
|    |         | 2 | 10010 | −116116 |
|    |         | 3 | 600600 | −1818960 |
|    |         | 4 | 420420 | 39419380 |
|    |         | 5 | −253265012 | 796541200 |
|    |         | 6 | −1388737350 | 1283049460 |
|    |         | 7 | −491262200 | |
| 14 | (1/4,1/3) | 0 | −3431 | 140140 |
|    |         | 1 | −2622620 | 29789760 |
|    |         | 2 | −229339110 | 1265596332 |
|    |         | 3 | −5162637480 | 15817230000 |
|    |         | 4 | −36592936380 | 63672909300 |
|    |         | 5 | −82156782708 | 76417220880 |
|    |         | 6 | −48621843270 | 19045926900 |
|    |         | 7 | −3490189560 | |
| 15 | (0,1/5) | 0 | 0 | 0 |
|    |         | 1 | 0 | 0 |
|    |         | 2 | 15015 | −192192 |
|    |         | 3 | 1106105 | −3770910 |
|    |         | 4 | −990990 | 142081940 |
|    |         | 5 | −1074365292 | 4214956200 |
|    |         | 6 | −9774469705 | 13380614940 |
|    |         | 7 | −9740526270 | 2623856564 |
| 15 | (1/5,1/4) | 0 | 286 | −18480 |
|    |         | 1 | 553245 | −10172890 |
|    |         | 2 | 128406915 | −1177368192 |
|    |         | 3 | 8093565480 | −42434552160 |
|    |         | 4 | 170888477760 | −527885418060 |
|    |         | 5 | 1238836181583 | −2165921762550 |
|    |         | 6 | 2727334905295 | −2332713135060 |
|    |         | 7 | 1211206739355 | −287903487186 |
| 15 | (1/4,1/3) | 0 | 9010 | −360360 |
|    |         | 1 | 6531525 | −70480410 |
|    |         | 2 | 497702205 | −2351024676 |
|    |         | 3 | 6987965985 | −8428369950 |
|    |         | 4 | −29980915965 | 196501425120 |
|    |         | 5 | −574116479937 | 1069264606410 |
|    |         | 6 | −1337319608625 | 1095863228460 |
|    |         | 7 | −534632303205 | 118097881902 |
| 16 | (0,1/6) | 0 | 0 | 0 |
|    |         | 1 | 0 | 0 |
|    |         | 2 | 21840 | −305760 |
|    |         | 3 | 1937936 | −7344480 |
|    |         | 4 | −6666660 | 441561120 |
|    |         | 5 | −3847107264 | 18087206592 |
|    |         | 6 | −52232180000 | 93136871520 |
|    |         | 7 | −91102848000 | 24509067904 |
|    |         | 8 | 22677254400 | |

TABLE 3: PIECEWISE POLYNOMIALS FOR P(n;N,p)

n = 5

| N | p range | F | coefficient of $p^{2F}$ | coefficient of $p^{2F+1}$ |
|---|---------|---|---|---|
| 16 | (1/6,1/5) | 0 | -1 | 96 |
| | | 1 | -4320 | 120960 |
| | | 2 | -2336880 | 33659808 |
| | | 3 | -371683312 | 3195123360 |
| | | 4 | -21623324580 | 115730403360 |
| | | 5 | -488060244672 | 1602784747200 |
| | | 6 | -4013976031520 | 7407125520480 |
| | | 7 | -9494802539520 | 7547468821120 |
| | | 8 | -2798432653056 | |
| 16 | (1/5,1/4) | 0 | -5433 | 349440 |
| | | 1 | -10439520 | 192162880 |
| | | 2 | -2437496880 | 22574348160 |
| | | 3 | -157768282672 | 848023107360 |
| | | 4 | -3539933834580 | 11505065503360 |
| | | 5 | -28999743444672 | 56084575747200 |
| | | 6 | -81632426031520 | 86620438020480 |
| | | 7 | -63477927539520 | 28952468821120 |
| | | 8 | -6282417028056 | |
| 16 | (1/4,1/3) | 0 | -24023 | 1153152 |
| | | 1 | -25945920 | 363242880 |
| | | 2 | -3541618080 | 25459650176 |
| | | 3 | -140248075968 | 601063182720 |
| | | 4 | -2028588241680 | 5409568644480 |
| | | 5 | -11360094153408 | 18589244978304 |
| | | 6 | -23236556222880 | 21449128821120 |
| | | 7 | -13738725670720 | 5515490268288 |
| | | 8 | -1034154425304 | |
| 17 | (0,1/6) | 0 | 0 | 0 |
| | | 1 | 0 | 0 |
| | | 2 | 30940 | -470288 |
| | | 3 | 3254888 | -13574704 |
| | | 4 | -23143120 | 1223279200 |
| | | 5 | -12076118912 | 66176848192 |
| | | 6 | -228189597636 | 491872248560 |
| | | 7 | -530564356000 | -215539011936 |
| | | 8 | 1360956862204 | -1186387008320 |
| 17 | (1/6,1/5) | 0 | 103 | -9792 |
| | | 1 | 436968 | -12154320 |
| | | 2 | 235993660 | -3392781392 |
| | | 3 | 37390853864 | -322582147888 |
| | | 4 | 2204875964720 | -12003227419040 |
| | | 5 | 52023542826112 | -178410383662784 |
| | | 6 | 477954293280828 | -978668358130960 |
| | | 7 | 1478201212135520 | -1550885618143584 |
| | | 8 | 1008497193823996 | -305866257013568 |
| 17 | (1/5,1/4) | 0 | 40223 | -2581280 |
| | | 1 | 77100712 | -1422267600 |
| | | 2 | 18128241040 | -169214226272 |
| | | 3 | 1195955935720 | -6525151224592 |
| | | 4 | 27759000260420 | -92328990651040 |
| | | 5 | 239168875486112 | -477278407662784 |
| | | 6 | 719631293618328 | -794790559380960 |
| | | 7 | 612643874635520 | -305432680643584 |
| | | 8 | 87092545386496 | -11373132013568 |

TABLE 3:   PIECEWISE POLYNOMIALS FOR P(n;N,p)

n = 6

| N | p range | F | coefficient of $p^{2F}$ | coefficient of $p^{2F+1}$ |
|---|---------|---|-------------------------|---------------------------|
| 6 | (0,1) | 0 | 0 | 0 |
|   |       | 1 | 0 | 0 |
|   |       | 2 | 0 | 6 |
|   |       | 3 | -5 |  |
| 7 | (0,1/2) | 0 | 0 | 0 |
|   |       | 1 | 0 | 0 |
|   |       | 2 | 0 | 42 |
|   |       | 3 | -84 | 44 |
| 7 | (1/2,1) | 0 | 1 | -14 |
|   |       | 1 | 84 | -280 |
|   |       | 2 | 560 | -630 |
|   |       | 3 | 364 | -84 |
| 8 | (0,1/2) | 0 | 0 | 0 |
|   |       | 1 | 0 | 0 |
|   |       | 2 | 0 | 168 |
|   |       | 3 | -532 | 576 |
|   |       | 4 | -210 |  |
| 8 | (1/2,1) | 0 | -13 | 168 |
|   |       | 1 | -924 | 2800 |
|   |       | 2 | -5040 | 5544 |
|   |       | 3 | -3668 | 1344 |
|   |       | 4 | -210 |  |
| 9 | (0,1/2) | 0 | 0 | 0 |
|   |       | 1 | 0 | 0 |
|   |       | 2 | 0 | 504 |
|   |       | 3 | -2184 | 3600 |
|   |       | 4 | -2646 | 728 |
| 9 | (1/2,1) | 0 | 43 | -504 |
|   |       | 1 | 2520 | -7056 |
|   |       | 2 | 12348 | -14112 |
|   |       | 3 | 10584 | -5040 |
|   |       | 4 | 1386 | -168 |
| 10 | (0,1/2) | 0 | 0 | 0 |
|   |       | 1 | 0 | 0 |
|   |       | 2 | 0 | 1260 |
|   |       | 3 | -6930 | 15360 |
|   |       | 4 | -17010 | 9380 |
|   |       | 5 | -2058 |  |
| 10 | (1/2,1) | 0 | -41 | 420 |
|   |       | 1 | -1890 | 5040 |
|   |       | 2 | -8820 | 10584 |
|   |       | 3 | -8820 | 5040 |
|   |       | 4 | -1890 | 420 |
|   |       | 5 | -42 |  |
| 11 | (0,1/2) | 0 | 0 | 0 |
|   |       | 1 | 0 | 0 |
|   |       | 2 | 0 | 2772 |
|   |       | 3 | -18480 | 51480 |
|   |       | 4 | -76230 | 63140 |
|   |       | 5 | -27720 | 5040 |

TABLE 3: PIECEWISE POLYNOMIALS FOR P(n;N,p)

n = 6

| N | p range | F | coefficient of $p^{2F}$ | coefficient of $p^{2F+1}$ |
|---|---------|---|------------------------|--------------------------|
| 12 | (0,1/3) | 0 | 0 | 0 |
| | | 1 | 0 | 0 |
| | | 2 | 0 | 5544 |
| | | 3 | -43428 | 145728 |
| | | 4 | -270270 | 298760 |
| | | 5 | -213444 | 144840 |
| | | 6 | -92290 | |
| 12 | (1/3,1/2) | 0 | 210 | -5544 |
| | | 1 | 65604 | -458920 |
| | | 2 | 2107710 | -6671016 |
| | | 3 | 14875476 | -23504976 |
| | | 4 | 26112240 | -19909120 |
| | | 5 | 9890496 | -2846976 |
| | | 6 | 340736 | |
| 13 | (0,1/4) | 0 | 0 | 0 |
| | | 1 | 0 | 0 |
| | | 2 | 0 | 10296 |
| | | 3 | -92664 | 363792 |
| | | 4 | -810810 | 1121120 |
| | | 5 | -1201200 | 1994304 |
| | | 6 | -3456648 | 2512592 |
| 13 | (1/4,1/3) | 0 | 1 | -52 |
| | | 1 | 1248 | -18304 |
| | | 2 | 183040 | -1307592 |
| | | 3 | 6936072 | -27751152 |
| | | 4 | 83534022 | -186311840 |
| | | 5 | 298691536 | -325161408 |
| | | 6 | 214647160 | -64596272 |
| 13 | (1/3,1/2) | 0 | -1429 | 37180 |
| | | 1 | -435864 | 3045328 |
| | | 2 | -14154140 | 46383480 |
| | | 3 | -111334080 | 202240896 |
| | | 4 | -287793792 | 330753280 |
| | | 5 | -306775040 | 215255040 |
| | | 6 | -98548736 | 21379072 |
| 14 | (0,1/4) | 0 | 0 | 0 |
| | | 1 | 0 | 0 |
| | | 2 | 0 | 18018 |
| | | 3 | -183183 | 823680 |
| | | 4 | -2144142 | 3559556 |
| | | 5 | -5423418 | 16574376 |
| | | 6 | -48150830 | 71463630 |
| | | 7 | -40554033 | |
| 14 | (1/4,1/3) | 0 | -64 | 3276 |
| | | 1 | -77532 | 1124032 |
| | | 2 | -11147136 | 79969890 |
| | | 3 | -427618191 | 1729892736 |
| | | 4 | -5315868558 | 12336648324 |
| | | 5 | -21260321082 | 26352609192 |
| | | 6 | -22185687342 | 11345752782 |
| | | 7 | -2657799729 | |

TABLE 3:    PIECEWISE POLYNOMIALS FOR P(n;N,p)

n = 6

| N | p range | F | coefficient $p^{2F}$ | coefficient of $p^{2F+1}$ |
|---|---------|---|----------------------|---------------------------|
| 14 | (1/3,1/2) | 0 | 3004 | -72072 |
|    |           | 1 | 738738 | -3987984 |
|    |           | 2 | 9513504 | 19027008 |
|    |           | 3 | -250522272 | 1065512448 |
|    |           | 4 | -2815997184 | 5141520384 |
|    |           | 5 | -6646768128 | 6014840832 |
|    |           | 6 | -3640885248 | 1328431104 |
|    |           | 7 | -221405184 | |
| 15 | (0,1/4) | 0 | 0 | 0 |
|    |         | 1 | 0 | 0 |
|    |         | 2 | 0 | 30030 |
|    |         | 3 | -340340 | 1724580 |
|    |         | 4 | -5135130 | 9949940 |
|    |         | 5 | -20684664 | 98236320 |
|    |         | 6 | -401306360 | 901465530 |
|    |         | 7 | -1023905220 | 465101364 |
| 15 | (1/4,1/3) | 0 | 1301 | -65520 |
|    |           | 1 | 1528800 | -21912800 |
|    |           | 2 | 215648160 | -1542502962 |
|    |           | 3 | 8279291020 | -33933891420 |
|    |           | 4 | 107007370470 | -259444124940 |
|    |           | 5 | 479395340424 | -662596552800 |
|    |           | 6 | 662770584840 | -452712512070 |
|    |           | 7 | 138726407740 | -36176338380 |
| 15 | (1/3,1/2) | 0 | -6005 | 180180 |
|    |           | 1 | -2522520 | 21861840 |
|    |           | 2 | -131171040 | 577152576 |
|    |           | 3 | -1923841920 | 4947022080 |
|    |           | 4 | -9894044160 | 15390735360 |
|    |           | 5 | -18468882432 | 16789893120 |
|    |           | 6 | -11193262080 | 5166120960 |
|    |           | 7 | -1476034560 | 196804608 |
| 16 | (0,1/4) | 0 | 0 | 0 |
|    |         | 1 | 0 | 0 |
|    |         | 2 | 0 | 48048 |
|    |         | 3 | -600600 | 3386240 |
|    |         | 4 | -11351340 | 25145120 |
|    |         | 5 | -68948880 | 458072160 |
|    |         | 6 | -2410557240 | 7254864400 |
|    |         | 7 | -12363566280 | 11217891776 |
|    |         | 8 | -4226754350 | |
| 16 | (1/4,1/3) | 0 | -12349 | 611520 |
|    |           | 1 | -14064960 | 199355520 |
|    |           | 2 | -1948157120 | 13912372656 |
|    |           | 3 | -75075920920 | 312187624320 |
|    |           | 4 | -1010726016300 | 2555860987680 |
|    |           | 5 | -5031181947792 | 7629320331360 |
|    |           | 6 | -8739103846200 | 7313229115920 |
|    |           | 7 | -4219586022600 | 1501303108032 |
|    |           | 8 | -248503019310 | |

TABLE 3:   PIECEWISE POLYNOMIALS FOR $P(n;N,p)$

$n = 6$

| N | p range | F | coefficient of $p^{2F}$ | coefficient of $p^{2F+1}$ |
|---|---------|---|------------------------|--------------------------|
| 17 | (0,1/4) | 0 | 0 | 0 |
|    |         | 1 | 0 | 0 |
|    |         | 2 | 0 | 74256 |
|    |         | 3 | -1014832 | 6301152 |
|    |         | 4 | -23483460 | 58538480 |
|    |         | 5 | -205837632 | 1785114240 |
|    |         | 6 | -11488640800 | 43348682160 |
|    |         | 7 | -98506259280 | 133945695520 |
|    |         | 8 | -100792077750 | 32368692720 |
| 17 | (1/4,1/3) | 0 | 64091 | -3118752 |
|    |         | 1 | 70691712 | -951070080 |
|    |         | 2 | 9623577600 | -68680191216 |
|    |         | 3 | 373021525968 | -1575001647648 |
|    |         | 4 | 5234656206780 | -13776120518160 |
|    |         | 5 | 28715751320256 | -47155246879104 |
|    |         | 6 | 60274796216544 | -58724544125520 |
|    |         | 7 | 42114395260080 | -20940989063904 |
|    |         | 8 | 6439704916554 | -920308740624 |
| 18 | (0,1/5) | 0 | 0 | 0 |
|    |         | 1 | 0 | 0 |
|    |         | 2 | 0 | 111384 |
|    |         | 3 | -1652196 | 11202048 |
|    |         | 4 | -45945900 | 127287160 |
|    |         | 5 | -560539980 | 6049042272 |
|    |         | 6 | -46008015144 | 208727078616 |
|    |         | 7 | -592898344740 | 1074825303552 |
|    |         | 8 | -1217819108142 | 800636730714 |
|    |         | 9 | -245482512311 | |
| 18 | (1/5,1/4) | 0 | 1820 | -139230 |
|    |         | 1 | 4998357 | -111840144 |
|    |         | 2 | 1746987660 | -20219511816 |
|    |         | 3 | 179600406804 | -1251109577952 |
|    |         | 4 | 6924384066600 | -30654934650340 |
|    |         | 5 | 108734650866270 | -308052757207728 |
|    |         | 6 | 691354218547356 | -1211527366671384 |
|    |         | 7 | 1620131320405260 | -1594268924696448 |
|    |         | 8 | 1086516555891858 | -457439109363036 |
|    |         | 9 | 89506714753314 | |
| 18 | (1/4,1/3) | 0 | -189617 | 9189180 |
|    |         | 1 | -208594386 | 2953256592 |
|    |         | 2 | -29320047900 | 217823998392 |
|    |         | 3 | -1263325403340 | 5896473115104 |
|    |         | 4 | -22647153136608 | 72632748988800 |
|    |         | 5 | -195662083318416 | 441044896715280 |
|    |         | 6 | -820861927137252 | 1233947669618664 |
|    |         | 7 | -1452520351725300 | 1282203953686656 |
|    |         | 8 | -794655112641390 | 3072888864803492 |
|    |         | 9 | -55648725100254 | |

TABLE 3:   PIECEWISE POLYNOMIALS FOR P(n;N,p)

n = 7

| N | p range | F | coefficient of $p^{2F}$ | coefficient of $p^{2F+1}$ |
|---|---------|---|-------------------------|---------------------------|
| 7 | (0,1) | 0 | 0 | 0 |
|   |       | 1 | 0 | 0 |
|   |       | 2 | 0 | 0 |
|   |       | 3 | 7 | −6 |
| 8 | (0,1/2) | 0 | 0 | 0 |
|   |         | 1 | 0 | 0 |
|   |         | 2 | 0 | 0 |
|   |         | 3 | 56 | −112 |
|   |         | 4 | 58 | |
| 8 | (1/2,1) | 0 | −1 | 16 |
|   |         | 1 | −112 | 448 |
|   |         | 2 | −1120 | 1792 |
|   |         | 3 | −1736 | 912 |
|   |         | 4 | −198 | |
| 9 | (0,1/2) | 0 | 0 | 0 |
|   |         | 1 | 0 | 0 |
|   |         | 2 | 0 | 0 |
|   |         | 3 | 252 | −792 |
|   |         | 4 | 846 | −304 |
| 9 | (1/2,1) | 0 | 19 | −288 |
|   |         | 1 | 1908 | −7224 |
|   |         | 2 | 17136 | −26208 |
|   |         | 3 | 25788 | −15768 |
|   |         | 4 | 5454 | −816 |
| 10 | (0,1/2) | 0 | 0 | 0 |
|    |         | 1 | 0 | 0 |
|    |         | 2 | 0 | 0 |
|    |         | 3 | 840 | −3600 |
|    |         | 4 | 5850 | −4240 |
|    |         | 5 | 1152 | |
| 10 | (1/2,1) | 0 | −101 | 1440 |
|    |         | 1 | −9000 | 32400 |
|    |         | 2 | −74340 | 113904 |
|    |         | 3 | −118440 | 82800 |
|    |         | 4 | −37350 | 9840 |
|    |         | 5 | −1152 | |
| 11 | (0,1/2) | 0 | 0 | 0 |
|    |         | 1 | 0 | 0 |
|    |         | 2 | 0 | 0 |
|    |         | 3 | 2310 | −12540 |
|    |         | 4 | 27390 | −29920 |
|    |         | 5 | 16302 | −3540 |
| 11 | (1/2,1) | 0 | 199 | −2640 |
|    |         | 1 | 15510 | −53460 |
|    |         | 2 | 120780 | −188496 |
|    |         | 3 | 207900 | −162360 |
|    |         | 4 | 88110 | −31680 |
|    |         | 5 | 6798 | −660 |

TABLE 3:   PIECEWISE POLYNOMIALS FOR P(n;N,p)

n = 7

| N | p range | F | coefficient of $p^{2F}$ | coefficient of $p^{2F+1}$ |
|---|---------|---|------------------------|---------------------------|
| 12 | (0,1/2) | 0 | 0 | 0 |
|    |         | 1 | 0 | 0 |
|    |         | 2 | 0 | 0 |
|    |         | 3 | 5544 | −36432 |
|    |         | 4 | 99990 | −146080 |
|    |         | 5 | 119592 | −51984 |
|    |         | 6 | 9372 | |
| 12 | (1/2,1) | 0 | −131 | 1584 |
|    |         | 1 | −8712 | 29040 |
|    |         | 2 | −65340 | 104544 |
|    |         | 3 | −121968 | 104544 |
|    |         | 4 | −65340 | 29040 |
|    |         | 5 | −8712 | 1584 |
|    |         | 6 | −132 | |
| 13 | (0,1/2) | 0 | 0 | 0 |
|    |         | 1 | 0 | 0 |
|    |         | 2 | 0 | 0 |
|    |         | 3 | 12012 | −92664 |
|    |         | 4 | 306306 | −560560 |
|    |         | 5 | 612612 | −399672 |
|    |         | 6 | 144144 | −22176 |
| 14 | (0,1/3) | 0 | 0 | 0 |
|    |         | 1 | 0 | 0 |
|    |         | 2 | 0 | 0 |
|    |         | 3 | 24024 | −212784 |
|    |         | 4 | 822822 | −1809808 |
|    |         | 5 | 2474472 | −2153424 |
|    |         | 6 | 1081080 | 11144 |
|    |         | 7 | −360464 | |
| 14 | (1/3,1/2) | 0 | −792 | 24024 |
|    |         | 1 | −331968 | 2764944 |
|    |         | 2 | −15479464 | 61509448 |
|    |         | 3 | −178570392 | 384157488 |
|    |         | 4 | −614744130 | 727919192 |
|    |         | 5 | −628011384 | 383143488 |
|    |         | 6 | −156540384 | 38589824 |
|    |         | 7 | −4454528 | |
| 15 | (0,1/4) | 0 | 0 | 0 |
|    |         | 1 | 0 | 0 |
|    |         | 2 | 0 | 0 |
|    |         | 3 | 45045 | −450450 |
|    |         | 4 | 1994850 | −5125120 |
|    |         | 5 | 8414406 | −9156420 |
|    |         | 6 | 5345340 | 5378520 |
|    |         | 7 | −18028785 | 14007074 |
| 15 | (1/4,1/3) | 0 | 1 | −60 |
|    |         | 1 | 1680 | −29120 |
|    |         | 2 | 349440 | −3075072 |
|    |         | 3 | 20545525 | −105881490 |
|    |         | 4 | 423719010 | −1317155840 |
|    |         | 5 | 3157238134 | −5734381380 |
|    |         | 6 | 7638978620 | −7041052200 |
|    |         | 7 | 4008503055 | −1059734750 |

## TABLE 3:  PIECEWISE POLYNOMIALS FOR P(n;N,p)

$n = 7$

| N | p range | F | coefficient of $p^{2F}$ | coefficient of $p^{2F+1}$ |
|---|---------|---|-------------------------|---------------------------|
| 15 | (1/3,1/2) | 0 | 6995 | −209820 |
|  |  | 1 | 2877420 | −23896600 |
|  |  | 2 | 134190420 | −538642104 |
|  |  | 3 | 1591820230 | −3506225580 |
|  |  | 4 | 5725605600 | −6687961280 |
|  |  | 5 | 4927610688 | −938071680 |
|  |  | 6 | −2594475520 | 3347635200 |
|  |  | 7 | −1871301120 | 431961088 |
| 16 | (0,1/4) | 0 | 0 | 0 |
|  |  | 1 | 0 | 0 |
|  |  | 2 | 0 | 0 |
|  |  | 3 | 80080 | −892320 |
|  |  | 4 | 4453020 | −13087360 |
|  |  | 5 | 25081056 | −32760000 |
|  |  | 6 | 19459440 | 72300480 |
|  |  | 7 | −297508560 | 455175584 |
|  |  | 8 | −256734870 |  |
| 16 | (1/4,1/3) | 0 | −89 | 5280 |
|  |  | 1 | −146400 | 2517760 |
|  |  | 2 | −30051840 | 263897088 |
|  |  | 3 | −1762961200 | 9136464480 |
|  |  | 4 | −37107273060 | 118444543360 |
|  |  | 5 | −295969049376 | 572489736000 |
|  |  | 6 | −839680201360 | 902015432640 |
|  |  | 7 | −668701794000 | 305397853600 |
|  |  | 8 | −64681244310 |  |
| 16 | (1/3,1/2) | 0 | −24595 | 755040 |
|  |  | 1 | −10845120 | 97697600 |
|  |  | 2 | −626305680 | 3078531456 |
|  |  | 3 | −12170238080 | 39709063680 |
|  |  | 4 | −107547897600 | 239454392320 |
|  |  | 5 | −430128771072 | 608618250240 |
|  |  | 6 | −658700922880 | 523992268800 |
|  |  | 7 | −287967313920 | 97521369088 |
|  |  | 8 | −15322644480 |  |
| 17 | (0,1/4) | 0 | 0 | 0 |
|  |  | 1 | 0 | 0 |
|  |  | 2 | 0 | 0 |
|  |  | 3 | 136136 | −1672528 |
|  |  | 4 | 9286420 | −30727840 |
|  |  | 5 | 67328976 | −102621792 |
|  |  | 6 | 53909856 | 557986240 |
|  |  | 7 | −2850846280 | 6500478224 |
|  |  | 8 | −7318536686 | 3284515808 |
| 17 | (1/4,1/3) | 0 | 2551 | −149600 |
|  |  | 1 | 4107200 | −70088960 |
|  |  | 2 | 832352640 | −7297285632 |
|  |  | 3 | 48886326216 | −255507940688 |
|  |  | 4 | 1054296257300 | −3452452134560 |
|  |  | 5 | 8967759813584 | −18362807476832 |
|  |  | 6 | 29263587088736 | −35516976057920 |
|  |  | 7 | 31689981266360 | −19579624133360 |
|  |  | 8 | 7476661976594 | −1328155345952 |

TABLE 3:   PIECEWISE POLYNOMIALS FOR P(n;N,p)

n = 7

| N | p range | F | coefficient of $p^{2F}$ | coefficient of $p^{2F+1}$ |
|---|---------|---|------------------------|--------------------------|
| 17 | (1/3,1/2) | 0 | 9725 | 77792 |
| | | 1 | -8596016 | 167641760 |
| | | 2 | -1802440640 | 12940543616 |
| | | 3 | -67222867712 | 263600370176 |
| | | 4 | -799166551040 | 1897353103360 |
| | | 5 | -3544098967552 | 5195699052544 |
| | | 6 | -5918783782912 | 5138404311040 |
| | | 7 | -3287527260160 | 1462220750848 |
| | | 8 | -404030488576 | 52256309248 |
| 18 | (0,1/4) | 0 | 0 | 0 |
| | | 1 | 0 | 0 |
| | | 2 | 0 | 0 |
| | | 3 | 222768 | -2991456 |
| | | 4 | 18290844 | -67290080 |
| | | 5 | 165930336 | -289089216 |
| | | 6 | 107819712 | 3146546592 |
| | | 7 | -19404892080 | 58796516064 |
| | | 8 | -99174454038 | 88883270496 |
| | | 9 | -33090054464 | |
| 18 | (1/4,1/3) | 0 | -34849 | 2019600 |
| | | 1 | -54896400 | 929652480 |
| | | 2 | -10987701120 | 96217406208 |
| | | 3 | -646733894352 | 3410674379424 |
| | | 4 | -14301860431716 | 48027941813280 |
| | | 5 | -129438871394976 | 279243059943744 |
| | | 6 | -478565721361728 | 642642913033632 |
| | | 7 | -661294967367600 | 503187640132320 |
| | | 8 | -266647529198358 | 878350651277776 |
| | | 9 | -13540762200384 | |
| 18 | (1/3,1/2) | 0 | -87515 | 3150576 |
| | | 1 | -53559792 | 571304448 |
| | | 2 | -4284783360 | 23994786816 |
| | | 3 | -103977409536 | 356493975552 |
| | | 4 | -980358432768 | 2178574295040 |
| | | 5 | -3921433731072 | 5703903608832 |
| | | 6 | -6654554210304 | 6142665424896 |
| | | 7 | -4387618160640 | 2340063019008 |
| | | 8 | -877523632128 | 206476148736 |
| | | 9 | -22941794304 | |
| 19 | (0,1/4) | 0 | 0 | 0 |
| | | 1 | 0 | 0 |
| | | 2 | 0 | 0 |
| | | 3 | 352716 | -5139576 |
| | | 4 | 34314228 | -138936512 |
| | | 5 | 380782116 | -746447832 |
| | | 6 | 99768240 | 14312021472 |
| | | 7 | -103795935012 | 392325750984 |
| | | 8 | -881620942446 | 1183957925760 |
| | | 9 | -880315617404 | 275574086024 |

TABLE 3:    PIECEWISE POLYNOMIALS FOR $P(n;N,p)$

$n = 8$

| N | p range | F | coefficient of $p^{2F}$ | coefficient of $p^{2F+1}$ |
|---|---------|---|------------------------|---------------------------|
| 8 | (0,1) | 0 | 0 | 0 |
|   |       | 1 | 0 | 0 |
|   |       | 2 | 0 | 0 |
|   |       | 3 | 0 | 8 |
|   |       | 4 | $-7$ | |
| 9 | (0,1/2) | 0 | 0 | 0 |
|   |         | 1 | 0 | 0 |
|   |         | 2 | 0 | 0 |
|   |         | 3 | 0 | 72 |
|   |         | 4 | $-144$ | 74 |
| 9 | (1/2,1) | 0 | 1 | $-18$ |
|   |         | 1 | 144 | $-672$ |
|   |         | 2 | 2016 | $-4032$ |
|   |         | 3 | 5376 | $-4536$ |
|   |         | 4 | 2160 | $-438$ |
| 10 | (0,1/2) | 0 | 0 | 0 |
|    |         | 1 | 0 | 0 |
|    |         | 2 | 0 | 0 |
|    |         | 3 | 0 | 360 |
|    |         | 4 | $-1125$ | 1190 |
|    |         | 5 | $-423$ | |
| 10 | (1/2,1) | 0 | $-26$ | 450 |
|    |         | 1 | $-3465$ | 15600 |
|    |         | 2 | $-45360$ | 88704 |
|    |         | 3 | $-117600$ | 104040 |
|    |         | 4 | $-58725$ | 19110 |
|    |         | 5 | $-2727$ | |
| 11 | (0,1/2) | 0 | 0 | 0 |
|    |         | 1 | 0 | 0 |
|    |         | 2 | 0 | 0 |
|    |         | 3 | 0 | 1320 |
|    |         | 4 | $-5610$ | 9020 |
|    |         | 5 | $-6468$ | 1740 |
| 11 | (1/2,1) | 0 | 199 | $-3300$ |
|    |         | 1 | 24420 | $-106260$ |
|    |         | 2 | 301620 | $-585816$ |
|    |         | 3 | 794640 | $-753720$ |
|    |         | 4 | 490710 | $-209220$ |
|    |         | 5 | 52668 | $-5940$ |
| 12 | (0,1/2) | 0 | 0 | 0 |
|    |         | 1 | 0 | 0 |
|    |         | 2 | 0 | 0 |
|    |         | 3 | 0 | 3960 |
|    |         | 4 | $-21285$ | 45980 |
|    |         | 5 | $-49698$ | 26820 |
|    |         | 6 | $-5775$ | |

TABLE 3:  PIECEWISE POLYNOMIALS FOR P(n;N,p)

n = 8

| N | p range | F | coefficient of $p^{2F}$ | coefficient of $p^{2F+1}$ |
|---|---------|---|------------------------|---------------------------|
| 12 | (1/2,1) | 0 | −626 | 9900 |
| | | 1 | −70290 | 296340 |
| | | 2 | −827145 | 1613304 |
| | | 3 | −2259180 | 2292840 |
| | | 4 | −1676565 | 862620 |
| | | 5 | −296802 | 61380 |
| | | 6 | −5775 | |
| 13 | (0,1/2) | 0 | 0 | 0 |
| | | 1 | 0 | 0 |
| | | 2 | 0 | 0 |
| | | 3 | 0 | 10296 |
| | | 4 | −66924 | 181610 |
| | | 5 | −262548 | 212940 |
| | | 6 | −91806 | 16434 |
| 13 | (1/2,1) | 0 | 859 | −12870 |
| | | 1 | 87516 | −358644 |
| | | 2 | 990990 | −1953666 |
| | | 3 | 2831400 | −3057912 |
| | | 4 | 2463318 | −1462890 |
| | | 5 | 622908 | −180180 |
| | | 6 | 31746 | −2574 |
| 14 | (0,1/2) | 0 | 0 | 0 |
| | | 1 | 0 | 0 |
| | | 2 | 0 | 0 |
| | | 3 | 0 | 24024 |
| | | 4 | −183183 | 598598 |
| | | 5 | −1084083 | 1173900 |
| | | 6 | −759759 | 272118 |
| | | 7 | −41613 | |
| 14 | (1/2,1) | 0 | −428 | 6006 |
| | | 1 | −39039 | 156156 |
| | | 2 | −429429 | 858858 |
| | | 3 | −1288287 | 1472328 |
| | | 4 | −1288287 | 858858 |
| | | 5 | −429429 | 156156 |
| | | 6 | −39039 | 6006 |
| | | 7 | −429 | |
| 15 | (0,1/2) | 0 | 0 | 0 |
| | | 1 | 0 | 0 |
| | | 2 | 0 | 0 |
| | | 3 | 0 | 51480 |
| | | 4 | −450450 | 1721720 |
| | | 5 | −3747744 | 5077800 |
| | | 6 | −4384380 | 2356200 |
| | | 7 | −720720 | 96096 |

TABLE 3:   PIECEWISE POLYNOMIALS FOR $P(n;N,p)$

$n = 8$

| N | p range | F | coefficient of $p^{2F}$ | coefficient of $p^{2F+1}$ |
|---|---------|---|------------------------|--------------------------|
| 16 | (0,1/3) | 0 | 0 | 0 |
|    |         | 1 | 0 | 0 |
|    |         | 2 | 0 | 0 |
|    |         | 3 | 0 | 102960 |
|    |         | 4 | -1016730 | 4450160 |
|    |         | 5 | -11315304 | 18411120 |
|    |         | 6 | -19879860 | 14248080 |
|    |         | 7 | -6949800 | 3550000 |
|    |         | 8 | -2196795 | |
| 16 | (1/3,1/2) | 0 | 3003 | -102960 |
|    |         | 1 | 1627560 | -15726480 |
|    |         | 2 | 103832820 | -496078128 |
|    |         | 3 | 1771970200 | -4821754080 |
|    |         | 4 | 10092293640 | -16292367520 |
|    |         | 5 | 20212880688 | -19071299520 |
|    |         | 6 | 13421728320 | -6816015360 |
|    |         | 7 | 2357372160 | -495610880 |
|    |         | 8 | 47109120 | |
| 17 | (0,1/4) | 0 | 0 | 0 |
|    |         | 1 | 0 | 0 |
|    |         | 2 | 0 | 0 |
|    |         | 3 | 0 | 194480 |
|    |         | 4 | -2139280 | 10550540 |
|    |         | 5 | -30688944 | 58290960 |
|    |         | 6 | -75555480 | 67701480 |
|    |         | 7 | -48425520 | 63033824 |
|    |         | 8 | -107019182 | 76910044 |
| 17 | (1/4,1/3) | 0 | 1 | -68 |
|    |         | 1 | 2176 | -43520 |
|    |         | 2 | 609280 | -6336512 |
|    |         | 3 | 50692096 | -318441552 |
|    |         | 4 | 1591040880 | -6362170100 |
|    |         | 5 | 20362017104 | -51850415344 |
|    |         | 6 | 103741857128 | -159651354840 |
|    |         | 7 | 182487684560 | -145965854240 |
|    |         | 8 | 72907424850 | -17102959140 |
| 17 | (1/3,1/2) | 0 | -32605 | 1108604 |
|    |         | 1 | -17433568 | 168253760 |
|    |         | 2 | -1115277520 | 5384995616 |
|    |         | 3 | -19607692832 | 55023993024 |
|    |         | 4 | -120739699080 | 209527112080 |
|    |         | 5 | -291327617152 | 331643137280 |
|    |         | 6 | -320360185600 | 272608017920 |
|    |         | 7 | -203064494080 | 121069785088 |
|    |         | 8 | -48614590464 | 9434370048 |

TABLE 3:   PIECEWISE POLYNOMIALS FOR P(n;N,p)

n = 8

| N | p range | F | coefficient of $p^{2F}$ | coefficient of $p^{2F+1}$ |
|---|---------|---|---|---|
| 18 | (0,1/4) | 0 | 0 | 0 |
|  |  | 1 | 0 | 0 |
|  |  | 2 | 0 | 0 |
|  |  | 3 | 0 | 350064 |
|  |  | 4 | -4244526 | 23288980 |
|  |  | 5 | -76270194 | 165643920 |
|  |  | 6 | -250558308 | 269455032 |
|  |  | 7 | -269111700 | 669873168 |
|  |  | 8 | -1907262351 | 2780989272 |
|  |  | 9 | -1544183474 |  |
| 18 | (1/4,1/3) | 0 | -118 | 7956 |
|  |  | 1 | -252756 | 5026560 |
|  |  | 2 | -70110720 | 728211456 |
|  |  | 3 | -5835927552 | 36889712496 |
|  |  | 4 | -186406323246 | 758377045140 |
|  |  | 5 | -2489260952178 | 6574032520848 |
|  |  | 6 | -13859875141476 | 22959819325112 |
|  |  | 7 | -29365765820820 | 27819173049360 |
|  |  | 8 | -18401547158415 | 7579103299416 |
|  |  | 9 | -1461833064114 |  |
| 18 | (1/3,1/2) | 0 | 135253 | -4471272 |
|  |  | 1 | 67586220 | -612495312 |
|  |  | 2 | 3626066340 | -13848149952 |
|  |  | 3 | 26184113592 | 53653354560 |
|  |  | 4 | -648455753088 | 2889951018240 |
|  |  | 5 | -8661683961216 | 19352425936896 |
|  |  | 6 | -33200606100480 | 43852832464896 |
|  |  | 7 | -43963333539840 | 32411226292224 |
|  |  | 8 | -16583397801984 | 5261622312960 |
|  |  | 9 | -779738578944 |  |
| 19 | (0,1/4) | 0 | 0 | 0 |
|  |  | 1 | 0 | 0 |
|  |  | 2 | 0 | 0 |
|  |  | 3 | 0 | 604656 |
|  |  | 4 | -8011692 | 48406072 |
|  |  | 5 | -176257224 | 430817400 |
|  |  | 6 | -744936192 | 934751664 |
|  |  | 7 | -1257079824 | 5016288192 |
|  |  | 8 | -20066398902 | 44151807384 |
|  |  | 9 | -49011877880 | 21760302600 |
| 19 | (1/4,1/3) | 0 | 4523 | -302328 |
|  |  | 1 | 9534960 | -188559648 |
|  |  | 2 | 2620672128 | -27191287296 |
|  |  | 3 | 218361808896 | -1388483061264 |
|  |  | 4 | 7091939328084 | -29344059688904 |
|  |  | 5 | 98729835530040 | -270016145670024 |
|  |  | 6 | 597570428153088 | -1060366124649040 |
|  |  | 7 | 1486159835462640 | -1606979009217600 |
|  |  | 8 | 1292226071301450 | -727050861732456 |
|  |  | 9 | 255209484457992 | -42034559459832 |

TABLE 3:   PIECEWISE POLYNOMIALS FOR P(n;N,p)

$\underline{n = 8}$

| N | p range | F | coefficient of $p^{2F}$ | coefficient of $p^{2F+1}$ |
|---|---------|---|------------------------|---------------------------|
| 19 | (1/3,1/2) | 0 | -503879 | 20255976 |
| | | 1 | -397863648 | 5083342992 |
| | | 2 | -47258098992 | 337875726240 |
| | | 3 | -1913588703936 | 8726255111808 |
| | | 4 | -32326138420992 | 97700368077312 |
| | | 5 | -241135845350400 | 484966276882432 |
| | | 6 | -790564365766656 | 1034547879665664 |
| | | 7 | -1070634523705344 | 856166581370880 |
| | | 8 | -510201651068928 | 213276044427264 |
| | | 9 | -55797745975296 | 6875238629376 |
| 20 | (0,1/4) | 0 | 0 | 0 |
| | | 1 | 0 | 0 |
| | | 2 | 0 | 0 |
| | | 3 | 0 | 1007760 |
| | | 4 | -14486550 | 95569240 |
| | | 5 | -382999188 | 1040512200 |
| | | 6 | -2024463870 | 2903511600 |
| | | 7 | -5104808280 | 29198543664 |
| | | 8 | -150306527055 | 442325636160 |
| | | 9 | -736227976340 | 652835506200 |
| | | 10 | -240685752990 | |
| 20 | (1/4,1/3) | 0 | -83656 | 5542680 |
| | | 1 | -173528520 | 3412585440 |
| | | 2 | -47267587440 | 490034059776 |
| | | 3 | -3944487989760 | 25237377478800 |
| | | 4 | -130318829431830 | 548313034802520 |
| | | 5 | -1889583108594324 | 5341618956989640 |
| | | 6 | -12361958467231230 | 23288818317660720 |
| | | 7 | -35359733829677400 | 42596030660635440 |
| | | 8 | -39753366334052175 | 27698600518194240 |
| | | 9 | -135544074736365 00 | 41537144120463 60 |
| | | 10 | -599543242367646 | |
| 20 | (1/3,1/2) | 0 | -831401 | 44341440 |
| | | 1 | -1069737240 | 15763381920 |
| | | 2 | -160560354240 | 1208499342336 |
| | | 3 | -7001336010240 | 32057796925440 |
| | | 4 | -118063872875520 | 353721244631040 |
| | | 5 | -867871380000768 | 1748536932679680 |
| | | 6 | -2889977828474880 | 3899617518551040 |
| | | 7 | -4256233372385280 | 3701394680315904 |
| | | 8 | -2505573726289920 | 1272810747985920 |
| | | 9 | -456599061135360 | 103161601720320 |
| | | 10 | -11042650324992 | |

TABLE 3:  PIECEWISE POLYNOMIALS FOR P(n;N,p)

n = 9

| N | p range | F | coefficient of $p^{2F}$ | coefficient of $p^{2F+1}$ |
|---|---------|---|-------------------------|---------------------------|
| 9 | (0, 1) | 0 | 0 | 0 |
|   |        | 1 | 0 | 0 |
|   |        | 2 | 0 | 0 |
|   |        | 3 | 0 | 0 |
|   |        | 4 | 9 | −8 |
| 10 | (0,1/2) | 0 | 0 | 0 |
|   |        | 1 | 0 | 0 |
|   |        | 2 | 0 | 0 |
|   |        | 3 | 0 | 0 |
|   |        | 4 | 90 | −180 |
|   |        | 5 | 92 | |
| 10 | (1/2,1) | 0 | −1 | 20 |
|   |        | 1 | −180 | 960 |
|   |        | 2 | −3360 | 8064 |
|   |        | 3 | −13440 | 15360 |
|   |        | 4 | −11430 | 4940 |
|   |        | 5 | −932 | |
| 11 | (0,1/2) | 0 | 0 | 0 |
|   |        | 1 | 0 | 0 |
|   |        | 2 | 0 | 0 |
|   |        | 3 | 0 | 0 |
|   |        | 4 | 495 | −1540 |
|   |        | 5 | 1617 | −570 |
| 11 | (1/2,1) | 0 | 34 | −660 |
|   |        | 1 | 5775 | −30030 |
|   |        | 2 | 102960 | −243936 |
|   |        | 3 | 406560 | −475200 |
|   |        | 4 | 380655 | −198660 |
|   |        | 5 | 60753 | −8250 |
| 12 | (0,1/2) | 0 | 0 | 0 |
|   |        | 1 | 0 | 0 |
|   |        | 2 | 0 | 0 |
|   |        | 3 | 0 | 0 |
|   |        | 4 | 1980 | −8360 |
|   |        | 5 | 13332 | −9480 |
|   |        | 6 | 2530 | |
| 12 | (1/2,1) | 0 | −351 | 6600 |
|   |        | 1 | −56100 | 284680 |
|   |        | 2 | −959310 | 2258784 |
|   |        | 3 | −3806880 | 4625280 |
|   |        | 4 | −4021380 | 2441560 |
|   |        | 5 | −983532 | 236280 |
|   |        | 6 | −25630 | |
| 13 | (0,1/2) | 0 | 0 | 0 |
|   |        | 1 | 0 | 0 |
|   |        | 2 | 0 | 0 |
|   |        | 3 | 0 | 0 |
|   |        | 4 | 6435 | −34320 |
|   |        | 5 | 73502 | −78780 |
|   |        | 6 | 42185 | −9020 |

TABLE 3:   PIECEWISE POLYNOMIALS FOR P(n;N,p)

n = 9

| N | p range | F | coefficient of $p^{2F}$ | coefficient of $p^{2F+1}$ |
|---|---------|---|------------------------|---------------------------|
| 13 | (1/2,1) | 0 | 1574 | −28600 |
|    |         | 1 | 235950 | −1169740 |
|    |         | 2 | 3887455 | −9147996 |
|    |         | 3 | 15692820 | −19888440 |
|    |         | 4 | 18642195 | −12778480 |
|    |         | 5 | 6232798 | −2050620 |
|    |         | 6 | 408265 | −37180 |
| 14 | (0,1/2) | 0 | 0 | 0 |
|    |         | 1 | 0 | 0 |
|    |         | 2 | 0 | 0 |
|    |         | 3 | 0 | 0 |
|    |         | 4 | 18018 | −116116 |
|    |         | 5 | 312312 | −447720 |
|    |         | 6 | 360360 | −154308 |
|    |         | 7 | 27456 | |
| 14 | (1/2,1) | 0 | −3431 | 60060 |
|    |         | 1 | −480480 | 2330328 |
|    |         | 2 | −7663656 | 18102084 |
|    |         | 3 | −31711680 | 41904720 |
|    |         | 4 | −42017976 | 31843812 |
|    |         | 5 | −17969952 | 7327320 |
|    |         | 6 | −2042040 | 348348 |
|    |         | 7 | −27456 | |
| 15 | (0,1/2) | 0 | 0 | 0 |
|    |         | 1 | 0 | 0 |
|    |         | 2 | 0 | 0 |
|    |         | 3 | 0 | 0 |
|    |         | 4 | 45045 | −340340 |
|    |         | 5 | 1102101 | −1979250 |
|    |         | 6 | 2127125 | −1367520 |
|    |         | 7 | 486915 | −74074 |
| 15 | (1/2,1) | 0 | 3576 | −60060 |
|    |         | 1 | 465465 | −2212210 |
|    |         | 2 | 7222215 | −17177160 |
|    |         | 3 | 30775745 | −42329430 |
|    |         | 4 | 45090045 | −37217180 |
|    |         | 5 | 23618595 | −11321310 |
|    |         | 6 | 3968965 | −960960 |
|    |         | 7 | 143715 | −10010 |
| 16 | (0,1/2) | 0 | 0 | 0 |
|    |         | 1 | 0 | 0 |
|    |         | 2 | 0 | 0 |
|    |         | 3 | 0 | 0 |
|    |         | 4 | 102960 | −892320 |
|    |         | 5 | 3379376 | −7294560 |
|    |         | 6 | 9809800 | −8414560 |
|    |         | 7 | 4495920 | −1368224 |
|    |         | 8 | 181610 | |

TABLE 3:   PIECEWISE POLYNOMIALS FOR $P(n;N,p)$

$n = 9$

| N | p range | F | coefficient of $p^{2F}$ | coefficient of $p^{2F+1}$ |
|---|---------|---|------------------------|----------------------------|
| 16 | (1/2,1) | 0 | $-1429$ | 22880 |
|    |         | 1 | $-171600$ | 800800 |
|    |         | 2 | $-2602600$ | 6246240 |
|    |         | 3 | $-11451440$ | 16359200 |
|    |         | 4 | $-18404100$ | 16359200 |
|    |         | 5 | $-11451440$ | 6246240 |
|    |         | 6 | $-2602600$ | 800800 |
|    |         | 7 | $-171600$ | 22880 |
|    |         | 8 | $-1430$ | |
| 17 | (0,1/2) | 0 | 0 | 0 |
|    |         | 1 | 0 | 0 |
|    |         | 2 | 0 | 0 |
|    |         | 3 | 0 | 0 |
|    |         | 4 | 218790 | $-2139280$ |
|    |         | 5 | 9276696 | $-23390640$ |
|    |         | 6 | 37777740 | $-40526640$ |
|    |         | 7 | 28880280 | $-13185744$ |
|    |         | 8 | 3500640 | $-411840$ |
| 18 | (0,1/3) | 0 | 0 | 0 |
|    |         | 1 | 0 | 0 |
|    |         | 2 | 0 | 0 |
|    |         | 3 | 0 | 0 |
|    |         | 4 | 437580 | $-4764760$ |
|    |         | 5 | 23279256 | $-67148640$ |
|    |         | 6 | 126606480 | $-163049040$ |
|    |         | 7 | 145276560 | $-88449504$ |
|    |         | 8 | 33256080 | 330228 |
|    |         | 9 | $-8597444$ | |
| 18 | (1/3,1/2) | 0 | $-11440$ | 437580 |
|    |         | 1 | $-7788924$ | 85648992 |
|    |         | 2 | $-651388320$ | 3635145360 |
|    |         | 3 | $-15412946640$ | 50725737504 |
|    |         | 4 | $-131269361652$ | 268952031920 |
|    |         | 5 | $-437142770064$ | 562216150080 |
|    |         | 6 | $-567941149776$ | 444644722368 |
|    |         | 7 | $-263983262400$ | 114750045696 |
|    |         | 8 | $-34418292480$ | 6366993408 |
|    |         | 9 | $-550667264$ | |
| 19 | (0,1/4) | 0 | 0 | 0 |
|    |         | 1 | 0 | 0 |
|    |         | 2 | 0 | 0 |
|    |         | 3 | 0 | 0 |
|    |         | 4 | 831402 | $-9976824$ |
|    |         | 5 | 54225886 | $-17610660$ |
|    |         | 6 | 379673580 | $-570640224$ |
|    |         | 7 | 610249068 | $-464476584$ |
|    |         | 8 | 209236170 | 159839172 |
|    |         | 9 | $-526046920$ | 399361840 |

TABLE 3: PIECEWISE POLYNOMIALS FOR P(n;N,p)

$$n = 9$$

| N | p range | F | coefficient of $p^{2F}$ | coefficient of $p^{2F+1}$ |
|---|---------|---|---|---|
| 19 | (1/4,1/3) | 0 | 1 | −76 |
| | | 1 | 2736 | −62016 |
| | | 2 | 992256 | −11907072 |
| | | 3 | 111132672 | −825556992 |
| | | 4 | 4954173354 | −24226315256 |
| | | 5 | 96919579614 | −317189990988 |
| | | 6 | 845750033388 | −1821368338272 |
| | | 7 | 3121977731436 | −4162287786408 |
| | | 8 | 4162032545994 | −2937597791292 |
| | | 9 | 1305144011064 | −274478545104 |
| 19 | (1/3,1/2) | 0 | 147289 | −5596944 |
| | | 1 | 99227538 | −1090183140 |
| | | 2 | 8315671176 | −46758932208 |
| | | 3 | 200870513928 | −674183579472 |
| | | 4 | 1792555921728 | −3803839298688 |
| | | 5 | 6449938009944 | −8677498675248 |
| | | 6 | 9053843220864 | −6871682135808 |
| | | 7 | 2983107771648 | 619204953600 |
| | | 8 | −2287223506944 | 1949699690496 |
| | | 9 | −863129757696 | 167957188608 |
| 20 | (0,1/4) | 0 | 0 | 0 |
| | | 1 | 0 | 0 |
| | | 2 | 0 | 0 |
| | | 3 | 0 | 0 |
| | | 4 | 1511640 | −19819280 |
| | | 5 | 118613352 | −428298000 |
| | | 6 | 1039252500 | −1785595680 |
| | | 7 | 2228157360 | −2035272096 |
| | | 8 | 976897350 | 2664542520 |
| | | 9 | −10725746240 | 16049461200 |
| | | 10 | −8879600220 | |
| 20 | (1/4,1/3) | 0 | −151 | 11400 |
| | | 1 | −408120 | 9211200 |
| | | 2 | −146977920 | 1762246656 |
| | | 3 | −16471449600 | 122880983040 |
| | | 4 | −742999781160 | 3676460651760 |
| | | 5 | −14965578537624 | 50193436815600 |
| | | 6 | −138428357166060 | 312134962641120 |
| | | 7 | −569647337374800 | 828200803382880 |
| | | 8 | −936409267813050 | 793197224767800 |
| | | 9 | −473316121765440 | 1773122994440080 |
| | | 10 | −31344960991836 | |
| 20 | (1/3,1/2) | 0 | −792641 | 30310320 |
| | | 1 | −546360960 | 6199894560 |
| | | 2 | −50037725880 | 308791214496 |
| | | 3 | −1535776782240 | 6419203911360 |
| | | 4 | −23257867805640 | 74214554331840 |
| | | 5 | −208398303018048 | 508239118283520 |
| | | 6 | −1056658149761920 | 1835698793379840 |
| | | 7 | −2612793962649600 | 2979882358185984 |
| | | 8 | −2648569593968640 | 17637843352780 80 |
| | | 9 | −826687766937600 | 242948404346880 |
| | | 10 | −33652274921472 | |

TABLE 4:  MEAN AND VARIANCE OF SHORTEST INTERVAL

| n | N | Expectation of $\tilde{p}_n$ | Variance of $\tilde{p}_n$ |
|---|---|---|---|
| **3** | 3 | 0.500000000000000 | 0.0500000000000000 |
| | 4 | 0.300000000000000 | 0.0266666666666667 |
| | 5 | 0.1875000000000000 | 0.0094866071428571 |
| | 6 | 0.1349206349206349 | 0.0052752582514487 |
| | 7 | 0.1022376543209876 | 0.0030576929488645 |
| | 8 | 0.0810185185185185 | 0.0019622342249657 |
| | 9 | 0.0661601562500000 | 0.0013245737817938 |
| | 10 | 0.0553212121212121 | 0.0009362638261402 |
| | 11 | 0.0471220000000000 | 0.0006848607057436 |
| | 12 | 0.0407472010646614 | 0.0005154833648441 |
| | 13 | 0.0356763286707881 | 0.0003972380109327 |
| | 14 | 0.0315658052058808 | 0.0003122855943299 |
| | 15 | 0.0281799710716697 | 0.0002497288306823 |
| | 16 | 0.0253525480772281 | 0.0002026817412087 |
| | 17 | 0.0229631817083678 | 0.0001666431413987 |
| | 18 | 0.0209228610010472 | 0.0001385887878027 |
| | 19 | 0.0191644960122135 | 0.0001164360642017 |
| **4** | 4 | 0.5999999999999999 | 0.0400000000000000 |
| | 5 | 0.4166666666666667 | 0.0287698412698413 |
| | 6 | 0.3035714285714286 | 0.0172193877551021 |
| | 7 | 0.2265625000000000 | 0.0085652669270833 |
| | 8 | 0.1815843621399177 | 0.0057616873274738 |
| | 9 | 0.1496913580246913 | 0.0039813862216126 |
| | 10 | 0.1261665003948954 | 0.0028323677942849 |
| | 11 | 0.1084426440329218 | 0.0021121823769069 |
| | 12 | 0.0945827448918269 | 0.0016161778827786 |
| | 13 | 0.0834986212448846 | 0.0012645803368797 |
| | 14 | 0.0744702216666666 | 0.0010094693731120 |
| | 15 | 0.0669926566666667 | 0.0008190673122192 |
| | 16 | 0.0607141820836753 | 0.0006740480373527 |
| | 17 | 0.0553798168034931 | 0.0005616229569546 |
| | 18 | 0.0508003057988069 | 0.0004730594238187 |
| | 19 | 0.0468328964499156 | 0.0004023206735709 |
| **5** | 5 | 0.6666666666666666 | 0.0317460317460318 |
| | 6 | 0.4999999999999999 | 0.0267857142857143 |
| | 7 | 0.3906250000000000 | 0.0192871093750000 |
| | 8 | 0.3125000000000000 | 0.0127604166666667 |
| | 9 | 0.2539062500000000 | 0.0076196843927557 |
| | 10 | 0.2151374859708193 | 0.0056452641752983 |
| | 11 | 0.1855852766346593 | 0.0042735477470648 |
| | 12 | 0.1623469979951461 | 0.0032869887656446 |
| | 13 | 0.1437234406413076 | 0.0025765486636072 |
| | 14 | 0.1286073721612559 | 0.0020730601610179 |
| | 15 | 0.1160716446911847 | 0.0016943459371204 |
| | 16 | 0.1055270952914369 | 0.0014034172756388 |
| | 17 | 0.0965534426986652 | 0.0011766005664124 |
| | 18 | 0.0888380582398300 | 0.0009972974190270 |
| | 19 | 0.0821416639580000 | 0.0008533033724787 |

TABLE 4:   MEAN AND VARIANCE OF SHORTEST INTERVAL

| n | N | Expectation of $\tilde{p}_n$ | Variance of $\tilde{p}_n$ |
|---|---|---|---|
| 6 | 6 | 0.7142857142857142 | 0.0255102040816327 |
|   | 7 | 0.5625000000000000 | 0.0238715277777778 |
|   | 8 | 0.4583333333333333 | 0.0190972222222222 |
|   | 9 | 0.3812500000000000 | 0.0143075284090909 |
|   | 10 | 0.3217329545454545 | 0.0101834552018767 |
|   | 11 | 0.2744140625000000 | 0.0068032924945538 |
|   | 12 | 0.2405156893004115 | 0.0053439036895021 |
|   | 13 | 0.2134406231628454 | 0.0042748986724446 |
|   | 14 | 0.1912748310724991 | 0.0034588262422554 |
|   | 15 | 0.1728396715545564 | 0.0028292059039393 |
|   | 16 | 0.1573321079226934 | 0.0023440773489541 |
|   | 17 | 0.1441761699411364 | 0.0019739255754510 |
| 7 | 7 | 0.7500000000000000 | 0.0208333333333333 |
|   | 8 | 0.6111111111111110 | 0.0209876543209877 |
|   | 9 | 0.5124999999999999 | 0.0180255681818182 |
|   | 10 | 0.4375000000000000 | 0.0145596550509091 |
|   | 11 | 0.3782552083333333 | 0.0113200326251169 |
|   | 12 | 0.3302283653846154 | 0.0085015962008995 |
|   | 13 | 0.2905273437500000 | 0.0061212062835693 |
|   | 14 | 0.2604813243026977 | 0.0050004840262183 |
|   | 15 | 0.2356802090477823 | 0.0041502676961809 |
|   | 16 | 0.2147943806390107 | 0.0034762512902042 |
|   | 17 | 0.1969741133854307 | 0.0029340342574702 |
|   | 18 | 0.1816207410884795 | 0.0024960294254902 |
|   | 19 | 0.1682917562077311 | 0.0021428038920591 |
|   | 20 | 0.1566499649630870 | 0.0018598075886828 |
| 8 | 8 | 0.7777777777777777 | 0.0172839506172840 |
|   | 9 | 0.6499999999999999 | 0.0184090909090909 |
|   | 10 | 0.5568181818181818 | 0.0166580578512397 |
|   | 11 | 0.4843749999999999 | 0.0141789362980769 |
|   | 12 | 0.4260817307692308 | 0.0116634933200814 |
|   | 13 | 0.3780691964285714 | 0.0093516291404257 |
|   | 14 | 0.3378255208333333 | 0.0073135333591038 |
|   | 15 | 0.3036193847656250 | 0.0055523010189919 |
|   | 16 | 0.2766730347188601 | 0.0046652722706790 |
|   | 17 | 0.2538780689554438 | 0.0039752088256099 |
|   | 18 | 0.2342726374215527 | 0.0034137451726650 |
|   | 19 | 0.2172243749776551 | 0.0029491960805685 |
|   | 20 | 0.2022759654225256 | 0.0025622223491800 |
|   | 21 | 0.1890812506226816 | 0.0022392946877980 |
|   | 22 | 0.1773706687491786 | 0.0019702422357506 |
|   | 23 | 0.1669300273858039 | 0.0017471138090148 |